T3-BUM-893

THE MOLECULAR BASIS OF SKELETOGENESIS

The Novartis Foundation is an international scientific and educational charity (UK Registered Charity No. 313574). Known until September 1997 as the Ciba Foundation, it was established in 1947 by the CIBA company of Basle, which merged with Sandoz in 1996, to form Novartis. The Foundation operates independently in London under English trust law. It was formally opened on 22 June 1949.

The Foundation promotes the study and general knowledge of science and in particular encourages international co-operation in scientific research. To this end, it organizes internationally acclaimed meetings (typically eight symposia and allied open meetings and 15–20 discussion meetings each year) and publishes eight books per year featuring the presented papers and discussions from the symposia. Although primarily an operational rather than a grant-making foundation, it awards bursaries to young scientists to attend the symposia and afterwards work with one of the other participants.

The Foundation's headquarters at 41 Portland Place, London W1N 4BN, provide library facilities, open to graduates in science and allied disciplines. Media relations are fostered by regular press conferences and by articles prepared by the Foundation's Science Writer in Residence. The Foundation offers accommodation and meeting facilities to visiting scientists and their societies.

Information on all Foundation activities can be found at http://www.novartisfound.org.uk

Novartis Foundation Symposium 232

THE MOLECULAR BASIS OF SKELETOGENESIS

2001

JOHN WILEY & SONS, LTD

Chichester · New York · Weinheim · Brisbane · Singapore · Toronto

Published in 2001 by John Wiley & Sons Ltd,
Baffins Lane, Chichester,
West Sussex PO19 1UD, England

National 01243 779777
International (+44) 1243 779777
e-mail (for orders and customer service enquiries): cs-books@wiley.co.uk
Visit our Home Page on http://www.wiley.co.uk
or http://www.wiley.com

Other Wiley Editorial Offices

John Wiley & Sons, Inc., 605 Third Avenue,
New York, NY 10158-0012, USA

WILEY-VCH Verlag GmbH, Pappelallee 3,
D-69469 Weinheim, Germany

Jacaranda Wiley Ltd, 33 Park Road, Milton,
Queensland 4064, Australia

John Wiley & Sons (Asia) Pte Ltd, 2 Clementi Loop #02-01,
Jin Xing Distripark, Singapore 129809

John Wiley & Sons (Canada) Ltd, 22 Worcester Road,
Rexdale, Ontario M9W 1L1, Canada

Novartis Foundation Symposium 232
ix+293 pages, 46 figures, 7 tables

Library of Congress Cataloging-in-Publication Data

Cardew, Gail.
The molecular basis of skeletogenesis / [Cardew, Gail, Jamie Goode].
p. cm. – (Novartis Foundation symposium ; 232)
Includes bibliographical references and indexes.
ISBN 0-471-49433-X (alk. paper)
1. Bones–Growth–Molecular aspects–Congresses. 2. Human skeleton–Molecular aspects–Congresses. I. Goode, Jamie II. Title III. Series.
QP88.2.C367 2001
612.7′5–dc21 00-047754

British Library Cataloguing in Publication Data

A catalogue record for this book is available from the British Library

ISBN 0 471 49433 X

Typeset in 10½ on 12½ pt Garamond by Dobbie Typesetting Limited, Tavistock, Devon.
Printed and bound in Great Britain by Biddles Ltd, Guildford and King's Lynn.
This book is printed on acid-free paper responsibly manufactured from sustainable forestry, in which at least two trees are planted for each one used for paper production.

Contents

Symposium on The molecular basis of skeletogenesis, held at the Novartis Foundation, London, 9–11 November 1999

Editors: Gail Cardew (Organizer) and Jamie A. Goode

This symposium is based on a proposal made by Gillian Morriss-Kay and Adam Wilkins

Brian Hall Introduction 1

Gerard Karsenty Genetic control of skeletal development 6
Discussion 17

Sandrine Pizette and **Lee Niswander** Early steps in limb patterning and chondrogenesis 23
Discussion 36

General discussion I 44

Martin J. Cohn Developmental mechanisms of vertebrate limb evolution 47
Discussion 57

David M. Ornitz Regulation of chondrocyte growth and differentiation by fibroblast growth factor receptor 3 63
Discussion 76

Stefan Mundlos Defects of human skeletogenesis — models and mechanisms 81
Discussion 91

G. M. Morriss-Kay, S. Iseki and **D. Johnson** Genetic control of the cell proliferation–differentiation balance in the developing skull vault: roles of fibroblast growth factor receptor signalling pathways 102
Discussion 116

Andrew O. M. Wilkie, Michael Oldridge, Zequn Tang and **Robert E. Maxson Jr** Craniosynostosis and related limb anomalies 122
Discussion 133

Henry M. Kronenberg and **Ung-il Chung** The parathyroid hormone-related protein and Indian hedgehog feedback loop in the growth plate 144
Discussion 152

William Wu, Fackson Mwale, Elena Tchetina, Tadashi Yasuda and **A. Robin Poole** Cartilage matrix resorption in skeletogenesis 158
Discussion 166

T. Michael Underhill, Arthur V. Sampaio and **Andrea D. Weston** Retinoid signalling and skeletal development 171
Discussion 185

General discussion II 189

Daniel H. Cohn Defects in extracellular matrix structural proteins in the osteochondrodysplasias 195
Discussion 210

David M. Kingsley Genetic control of bone and joint formation 213
Discussion 222

Tatsuo Suda, Kanichiro Kobayashi, Eijiro Jimi, Nobuyuki Udagawa and **Naoyuki Takahashi** The molecular basis of osteoclast differentiation and activation 235
Discussion 247

Graham Russell, Gabrielle Mueller, Claire Shipman and **Peter Croucher** Clinical disorders of bone resorption 251
Discussion 267

Final discussion 272

Index of contributors 283

Subject index 285

Participants

J. Bard Department of Biomedical Sciences, University of Edinburgh, Hugh Robson Building, George Square, Edinburgh EH8 9XD, UK

J. Beresford Department of Bone Research, School of Pharmacy and Pharmacology, University of Bath, Claverton Down, Bath BA2 7AY, UK

H. Blair Center for Metabolic Bone Disease, The University of Alabama at Birmingham, 509 LHRB 701 South 19th Street, Birmingham, AL 35294-0007, USA

E. H. Burger Department of Oral Cell Biology, ACTA-Vrije Universiteit Amsterdam, Van der Boechorststraat 7, 1081 BT Amsterdam, The Netherlands

Q. Chen (*Novartis Foundation Bursar*) Department of Orthopaedics & Rehabilitation, and Cellular & Molecular Physiology, Penn State University School of Medicine, 500 University Drive, Hershey, PA 17033, USA

D. H. Cohn Medical Genetics, Steven Spielberg Pediatric Research Center, Ahmanson Department of Pediatrics, Cedars-Sinai Medical Center, and Departments of Human Genetics and Pediatrics, UCLA School of Medicine, Los Angeles, CA 90048, USA

M. J. Cohn Division of Zoology, School of Animal and Microbial Sciences, University of Reading, Whiteknights, Reading RG6 6AJ, UK

B. K. Hall (*Chair*) Department of Biology, Dalhousie University, 1355 Oxford Street, Halifax, Nova Scotia, Canada B3H 4J1

G. Karsenty Department of Molecular and Human Genetics, Baylor College of Medicine, One Baylor Plaza, Houston, TX 77030, USA

D. M. Kingsley Howard Hughes Medical Institute and Department of Developmental Biology, Beckman Center B300, Stanford University School of Medicine, Stanford CA 94305-5329, USA

H. M. Kronenberg Endocrine Unit, Massachusetts General Hospital and Harvard Medical School, Boston, MA 02114, USA

M. C. Meikle Department of Orthodontics and Paediatric Dentistry, Guy's, King's and St Thomas' Dental Institute, 22nd Floor, Guy's Tower, Guy's Hospital, London SE1 9RT, UK

G. M. Morriss-Kay Department of Human Anatomy and Genetics, University of Oxford, South Parks Road, Oxford OX1 3QX, UK

S. Mundlos Universitätsklinikum der Humboldt-Universität zu Berlin, Campus Charité Mitte, Institut für Medizinische Genetik, Schumannstrasse 20/21, 10117 Berlin, Germany

S. A. Newman Department of Cell Biology and Anatomy, New York Medical College, Valhalla, NY 10595, USA

D. M. Ornitz Department of Molecular Biology and Pharmacology, Washington University School of Medicine, Campus Box 8103, St Louis, MO 63110, USA

F. Perrin-Schmitt Faculté de Médecine, U184 de l'INSERM-LGME du CNRS, 11 rue Humann, Strasbourg, 67085-Cedex, France

S. Pizette Molecular Biology Program and Howard Hughes Medical Institute, Memorial Sloan–Kettering Cancer Center, 1275 York Avenue, New York, NY 10021, USA

R. Poole Joint Diseases Laboratory, Shriners Hospitals for Children and Division of Surgical Research, Department of Surgery, McGill University, 1529 Cedar Avenue, Montreal, Quebec, Canada H3G 1A6

A. H. Reddi Center for Tissue Regeneration and Repair, University of California-Davis, Medical Center, Sacramento, CA 95817, USA

G. Russell Department of Human Metabolism and Clinical Biochemistry, University of Sheffield Medical School, Beech Hill Road, Sheffield S10 2RX, UK

T. Suda Department of Biochemistry, School of Dentistry, Showa University, Tokyo, 142-8555, Japan

C. Tickle Department of Anatomy and Physiology, University of Dundee, MSI/WTB Complex, Dow Street, Dundee DD1 5EH, UK

T. M. Underhill School of Dentistry and Department of Physiology, Faculty of Medicine and Dentistry, The University of Western Ontario, London, Ontario, Canada N6A 5C1

A. O. M. Wilkie Institute of Molecular Medicine, John Radcliffe Hospital, Headington, Oxford OX3 9DS, UK

A. Wilkins BioEssays Editorial Office, 10/11 Tredgold Lane, Napier Street, Cambridge CB1 1HN, UK

Introduction

Brian Hall

Department of Biology, Dalhousie University, 1355 Oxford Street, Halifax, Nova Scotia, Canada B3H 4J1

The topic of this symposium is 'The molecular basis of skeletogenesis'. As we will be hearing a lot about molecules, I thought I should begin by giving a brief overview of the skeleton as a whole. I'll begin by quoting from the proposal that initiated this symposium (interspersing the quotes with my own commentary), put together by Adam Wilkins, in which he stated that, 'Our aim is to bring about a cross-fertilization of ideas between human molecular genetics, developmental biology, tissue biology and the biochemistry of cell signalling pathways, in order to create new insights into the mechanisms of normal and abnormal skeletogenesis.' This is a broad and ambitious aim. 'These areas are only just beginning to be investigated in ways that integrate tissue biology with molecular genetics. The main focus of the meeting is on the molecular control of differentiation. Skeletal patterning is altered by mutations affecting differentiation. [I suspect the first paper will tell us that control of patterning is quite distinct from differentiation.] We have left this aspect to emerge in an integrated manner within the differentiation-based studies.' There may be some discussion here about whether we can actually leave patterning and morphogenesis to emerge from studies on differentiation. 'Remodelling is secondary to differentiation, but there are a number of osteoclast-related developmental disorders, and it is also appropriate to include this topic.'

Skeletogenic tissues and rates of bone formation

Though the skeleton exists as an organ system, we are primarily going to be talking about the tissue level of organization. In this context we can include four different tissues: bone, cartilage, chondroid and chondroid bone, although it may be that neither of these two latter tissues will emerge in our discussion. There are two different types of bone, membrane and endrochondral, and at least three different types of cartilage: the primary cartilage that makes up the primary skeleton, secondary cartilage (that emerges from periostea on membrane bones) and all the varieties of ectopic cartilages, some of which will come up in the following papers.

There are both direct and indirect modes of bone formation. We normally think of direct bone formation as being solely intramembranous, but there are several other types of direct bone formation. Intramembranous bone formation takes place in the skull, for example, but there is also sub-periosteal bone formation, which most of us who work on mammals tend not to think about. Most of the long bones in bird embryos, however, develop in this way, with only a very small amount developing endochondrally. And, of course, ectopic bone can often develop directly.

There are also various modes of indirect ossification, only one of which is referred to in any of the abstracts, and that is endochondral ossification in which cartilage is replaced by bone. However, bone can also develop by the replacement of ligaments, tendons and a whole variety of other tissues during ectopic osteogenesis, although we have very little idea of the identity of the cells that are replaced during this process.

Origins of skeletogenic cells

When we look at the origins of skeletal cells, we can take evolutionary perspective (which at least one of the papers will do) and see that there really are two separate skeletal systems within vertebrates, the endoskeleton and the exoskeleton. The endoskeleton which comprises the skeleton on the inside, is based on cartilage and often replaced by bone. But then there is the exoskeleton, which organisms such as sharks have in great prominence, causing the rough surface of their skin. All humans really have left of this exoskeleton is our teeth, elements of the craniofacial skeleton, and the clavicle. This second skeletal system is based not in cartilage, but in dentine and associated bone and enamel: cartilage is very much a secondary feature of the exoskeleton.

When we look at the origins of skeletal cells, we can also take an embryological perspective. There are two major embryological origins of skeletal tissues: the mesoderm which produces most of the internal skeleton in the trunk, and the neural crest which produces most of the skeleton in the head (the craniofacial skeleton). I've left out the cellular origin of skeletal tissues in the adult, which is a much more daunting problem. Which cells present in the adult are capable of forming skeletal tissues? Obviously, there are the cells that are associated with the skeleton itself, in the periostea and the perichondria, but what about those cells that make ectopic cartilage—that put down bone and cartilage where they are not wanted? These cells are much more demanding to identify, and I think issues relating to that will certainly come up during the course of this meeting.

Major stages of skeletogenesis

The first two papers identify major stages in skeletogenesis which are in fact different to the ones that I will outline here, so there is obviously a diversity of approaches to the stages of skeletogenesis. I don't think it's a terribly important issue, but it may turn out to emerge as such during the course of the meeting. In my version of events, the first major stage is cell migration: in most cases the cells that make skeletal tissues have to move to a new position to make those tissues. This may be a short migration in the embryo, or it may be a long migration as in the neural crest cells that come out of the developing brain and have to get all the way down into the developing jaws and branchial arches. Then there is the phase of aggregation, or condensation, which is under completely separate molecular control from cell migration and differentiation. Thirdly, there is cell differentiation, which of course is the main aim and thrust of our meeting, and which is clearly multistep, involving a whole variety of genetic cascades and signals at different steps. Finally, there is morphogenesis, which I have divided rather simplistically into first-order and second-order. The first order is the establishment of the fundamental shape of the skeletal element; the second order is later adaptations of form to function.

Coupling

I have identified another topic which I feel will turn out to be important in this meeting—the notion of coupling. In all of the stages and processes there is coupling between a variety of events. There is coupling between different tissues (usually epithelia and mesenchyme) in terms of the initial induction of the skeleton and the embryo. There is cell–cell coupling in all sorts of situations—in condensation, differentiation and morphogenesis. There is the cellular coupling that occurs in replacement of cartilage by bone, which involves hypertrophic chondrocytes coupling to the invading vascular system. There is the hormonal coupling that occurs between bone formation and bone resorption: the paradoxical fact that the best way to stimulate bone resorption is to stimulate the cells that lay down bone and have them interact with osteoclasts. Then of course there is extensive coupling in growth, including coupling with other organ systems (musculoskeletal systems), with hormones and with molecules, including a whole variety of growth factors and signalling molecules that are involved in controlling these processes.

Changing views

Rather than look ahead and anticipate what you are all going to say over the next couple of days, let me look back at the last Novartis Foundation Symposium (or

Ciba Foundation, as it was then known) that was held on this topic in 1988, entitled the 'Cell and molecular biology of vertebrate hard tissues'. It had a very heated discussion section. It is interesting to see what the aim of this meeting was, as a way of illustrating the difficulty of trying to predict how a field is going to change. In his introduction, Gideon Rodan says, 'In last few years we have seen considerable advances in identifying molecules that are part of the extracellular matrix. Our discussion will focus on their biochemical characterization, their structure, the interaction of these molecules with cells and with other similar molecules, and the regulation of their production and degradation' (Rodan 1988, p 1). The emphasis was on the extracellular matrix as the controlling element of skeletogenesis. We are not talking about the extracellular matrix in this context at this meeting at all — we're talking about cells and what's happening at the cell surface. So there has been a major shift in emphasis away from the extracellular matrix as containing the control, to the cells. This was identified by Melvyn Glimcher in the final general discussion to the 1988 meeting, where he said, 'To sum up, we need a greater understanding of the complex system of cell–cell and cell–matrix systems of communication and other factors which regulate cell proliferation and differentiation to specific cell types, so that we can tailor the repair and healing process for different types of bony tissue, in order to achieve the regulation of bone cell populations and formation to fit the specific task of repair at hand' (Ciba Foundation 1988, p 280). This could stand as a summary of the proposal that Adam Wilkins and Gillian Morriss-Kay have put together for this meeting, with the emphasis on cell–cell interactions.

Under the heading of 'Future perspectives', H. Fleisch said, 'We should take advantage of the knowledge developed in the haemopoietic field, on the development from stem cells to mature blood cells, and we should investigate whether a cascade of factors is necessary in bone differentiation from stem cells to bone-forming cells' (Ciba Foundation 1988, p 288–289). Clearly, a lot of the present meeting is going to be about those cascades of factors which are involved in going from stem cell to bone-forming cell, with major advances in identifying the cascades of molecules that are involved at each of the steps I have outlined.

We have made much less progress on the topic that was raised by J. D. Termine in 1988: 'Over the next few years the definition of the lineage of osteoblast phenotype should be achieved at the molecular level' (Ciba Foundation 1988, p 288). I don't think we're quite there yet. Continuing Termine's conclusion: 'The big task for this group, and the field as a whole, is to develop *in vitro* systems of bone resorption and bone formation that more accurately reflect what's going on in the adult. . . . Presently our techniques are basically bone modelling ones, based on embryos or neonates' (Ciba Foundation 1988, p 289–290). I think this is still a major issue: how well can we model what's happening in the adult, with studies based on systems that are derived from embryos and neonates?

With this brief outline of skeletogenesis, I would now like to introduce the first paper.

References

Ciba Foundation 1988 Final general discussion. In: Cell and molecular biology of vertebrate hard tissues. Wiley, Chichester (Ciba Found Symp 136) p 275–296

Rodan GA 1988 Introduction. In: Cell and molecular biology of vertebrate hard tissues. Wiley, Chichester (Ciba Found Symp 136) p 1

Genetic control of skeletal development

Gerard Karsenty

Department of Molecular and Human Genetics, Baylor College of Medicine, Houston, TX 77030, USA

Abstract. There are three major topics of skeleton biology. The first is skeleton patterning, which addresses how the shape and the location of each specific skeletal element is achieved. The second topic is cell differentiation in the skeleton. There are three specific cell types in the skeleton: the chondrocyte in cartilage, and the osteoblast and osteoclast in bone. The first two cell types are from mesenchymal origin while the third is from monocytic origin. The genes controlling skeleton patterning and cell differentiation are for the most part different. The third aspect of skeleton biology addresses the molecular control of the major function of the skeleton such as skeleton growth, bone mineralization and bone remodelling. Our current knowledge in each of these areas of skeleton biology will be presented in broad terms to set the course for other presentations during the symposium.

2001 The molecular basis of skeletogenesis. Wiley, Chichester (Novartis Foundation Symposium 232) p 6–22

Skeleton is composed of multiple elements of various shapes and origins spread throughout the body. These skeletal elements are formed by two different tissues, cartilage and bone, and each of these two tissues has its own specific cell types: the chondrocyte in cartilage, the osteoblast and the osteoclast in bone. Skeleton biology can now be viewed as encompassing three independently evolving fields of investigation, each relying on its own concepts and making use of its specific sets of genes. These areas of research are skeleton patterning, cell differentiation and cell function. This arbitrary distinction largely reflects the current state of knowledge and, from a conceptual point of view, helps to define the issues that are at stake for the future.

Skeleton patterning

How is the three-dimensional structure of the skeleton, position, number and shape of the skeletal elements achieved? This process, called pattern formation, has been best analysed in the developing limb and we are beginning to understand how growth is coupled to the establishment of three axes: the proximal–distal (P–D), dorsal–ventral (D–V) and anterior–posterior axis (A–P).

The definition of an area where the limbs are destined to form is the first step in this complex process, followed by the establishment of specific fields (Johnson & Tabin 1997) that will ensure the linkage of growth and pattern formation. The fields are interpreted within the limb by specific genes which define the formation of the different structures (Johnson & Tabin 1997).

Limb outgrowth is thought to be initiated by fibroblast growth factor 8 (FGF8) expressed in the intermediate mesoderm (Crossley et al 1996, Vogel et al 1996). *Fgf8* induces *Fgf10* which in turn causes the formation of the apical ectodermal ridge (AER) (Ohuchi et al 1997), a thickened epithelial structure that forms along the most distal part of the limb bud at its A–P axis. A main function of the AER is to mediate bud outgrowth by maintaining the mesenchyme at the limb bud tip in an undifferentiated state called the progress zone. This ensures that the proximal bones of the limb form first followed by the more distal structures. *Ffg2*, *Ffg4* and *Ffg8* are expressed in the AER and each of them has the ability to induce limb outgrowth (Niswander et al 1993, Fallon et al 1994). Shortly after the formation of the AER, cartilage blastemas condense serving as a blueprint for the later bones.

Establishment of the D–V axis requires the transcription factor Engrailed 1 (En1), which is present in ventral ectoderm, and acts to restrict the expression of *Wnt7a* and *Radical fringe* (*Rfng*) to the dorsal ectoderm (Loomis et al 1996, Laufer et al 1997, Rodriguez-Estaban et al 1997). Misexpression of *En1* and *Rfng* in the chick revealed that *En1* represses *Rfng*. The knockout of *En1* resulted in dorsal transformation of ventral paw structures accompanied by ventrally extended *Wnt7a* expression domains. *Wnt7a* induces the expression of the transcription factor LMX1 in the underlying mesoderm which in turn controls the dorsalization of the limb.

The A–P axis is determined by signals present in the transient cell population located at the distal posterior region of the limb bud, called the zone of polarizing activity (ZPA) (Saunders & Gasseling 1968). One critical molecule present in the ZPA is the secreted protein Sonic hedgehog (SHH) (Echelard et al 1993, Krauss et al 1993, Riddle et al 1993). SHH acts in the mesoderm indirectly through FGF4 produced by the AER. FGF4 induces the competence of the mesoderm to respond to SHH which then causes the expression of *Bmp2* that determines specific skeletal elements. *Shh* expression is both AER and *Wnt7a* dependent since absence of *Wnt7a* expression in the dorsal ectoderm of the limb bud reduces the *Shh* expression domain causing a lack of posterior skeletal elements (Yang & Niswander 1995).

The three-dimensional structure superimposed by the AER and ZPA is interpreted, in part, by Hox genes. Hox genes are organized into four paralog clusters. Genes of a cluster that are more 5′ in the genome seem to have a dominant role if several of them are expressed in the same cell (Duboule &

Morata 1994). This leads to a distinct pattern formation within different regions of the limb since each gene has a different effect on proliferation and differentiation. It has also been suggested that the level of Hox gene expression can determine cell fate.

During skeleton development the cartilage forms as a template for bone. The size and shape of this template is determined by the number of mesenchymal precursors recruited to become chondrocytes and their subsequent proliferation rate and deposition of extracellular matrix. The formation of the mesenchymal condensation is determined by Hox genes and members of the TGFβ superfamily (Goff & Tabin 1997, Kanzler et al 1998). For example, overexpression of *Gdf5*, *Bmp2*, *Bmp4* or *Bmp7* causes an increased recruitment of mesenchymal cells resulting in the formation of longer and/or wider cartilage anlagen in chick (Macias et al 1997, Francis-West et al 1999). On the other hand, mutations in *Gdf5* cause shorter distal bones in the mouse mutant brachypod (Storm et al 1994). Similarly, mutations in the human homologue of GDF5, cartilage-derived morphogenetic protein 1, result in three allelic chondrodysplasia (Thomas et al 1996, 1997). BMPs influence mesenchymal condensations through the BMP receptor type 1A (BMPR1A), and the effect of BMP on growth plate size, is in turn controlled by its endogenous antagonist noggin.

Cell differentiation

Chondrocytes

During mouse embryonic development the mesenchymal condensations prefiguring the future skeletal elements appear between 10.5 and 12.5 days post coitum (dpc). The cells present in the anlagen initially express both type I collagen, a marker of mesenchymal cells, and type IIa collagen, a transcript of α1(II) collagen that is not chondrocyte-specific (Lui et al 1995). Thereafter, the mesenchymal cells present in the anlagen differentiate along the chondrocytic pathway. They start expressing type IIb collagen, the chondrocyte-specific transcript of the α1(II) collagen gene (Ryan & Sandell 1990), type IX and XI collagen, and several other matrix genes, while the expression of type I collagen is turned off. These chondrocytes then undergo a program of differentiation including hypertrophy and expression of type X collagen associated with a decreased expression of type II collagen. At the time of vascular invasion from the perichondrium, the hypertrophic chondrocytes die through apoptosis and the osteoblasts start depositing bone tissue. This process of chondrogenesis, cartilage hypertrophy, degradation and replacement by bone is called endochondral ossification. Much of the cartilage anlagen is consumed by this process. Chondrocytes only subsist at either end of the growing bones, getting organized in a structure called the growth plate cartilage which controls longitudinal skeletal growth. Another mechanism of skeletogenesis, occurring in some cranial bones and the clavicle is called intramembranous ossification. In this type of ossification mesenchymal cells condense, the tissue vascularizes and cells differentiate directly into osteoblasts (Thorogood 1993).

Several growth factors and transcription factors have been implicated in the control of chondrocyte differentiation. The first factor to be shown to control chondrocyte differentiation is the parathyroid hormone related peptide (PTHrP). In PTHrP-deficient mice the main phenotype is a premature differentiation of the chondrocyte into hypertrophic chondrocytes leading to dwarfism (Karaplis et al 1994). In contrast, bone explants exposed to high concentrations of PTHrP have a delayed differentiation of the hypertrophic chondrocytes (Lanske et al 1996). These two experiments indicate that PTHrP controls the rate of chondrocyte differentiation into hypertrophic chondrocytes. PTHrP binds to a seven-transmembrane-domain receptor common for PTH and PTHrP. Patients with an activating mutation in the PTH/PTHrP receptor exhibit growth abnormalities similar to those seen in bone explants exposed to excess PTHrP (Schipani et al 1995).

PTHrp expression and function is under the control of Indian hedgehog (*Ihh*), a member of the hedgehog family of growth factors expressed in prehypertrophic chondrocytes (Bitgood & McMahon 1995). Overexpression of *Ihh* in developing cartilage in chick leads to the formation of cartilage elements that are broader, shorter and lacking hypertrophic chondrocytes (Vortkamp et al 1996). The phenotypic abnormalities are exactly opposite to the phenotype of the PTHrP-deficient mice suggesting that the two genes could be part of the same pathway. Indeed, in the case of *Ihh* overexpression there is an increased expression of *PTHrp* in the perichondrium indicating that *PTHrp* is downstream of *Ihh*. *Ihh*-deficient mice have been generated and are currently being analysed. Analysis of *Ihh*-deficient mice confirmed that *PTHrp* is downstream of *Ihh* (St Jacques et al 1999).

FGF and their receptors also play a role during chondrocyte differentiation. Activating mutations in FGFR3 were shown to cause achondroplasia (Chiang et al 1996). Milder activating mutations in FGFR3 cause hypochondroplasia and thanatophoric dysplasia, two conditions similar to achondroplasia (Naski et al 1996). FGFR3-deficient mice have a phenotype characterized by a prolonged endochondral bone growth with enlargement of the growth plate cartilage (Colvin et al 1996). This complementary genetic evidence indicates that FGFR3 or rather FGFR3 ligand, negatively controls bone growth by limiting chondrocyte proliferation.

The transcriptional control of chondrocyte differentiation has also recently begun to be elucidated. SOX9, a homologue of SRY that shares the same HMG-type DNA binding domain, is expressed in mesenchymal condensations before and during chondrogenesis (Ng et al 1997). Heterozygous mutations in SOX9 in humans lead to campomelic dysplasia, a skeletal dysplasia characterized by abnormalities in all skeletal elements formed through the process of endochondral ossification (Wagner et al 1994). Using chimeric mice it was shown that SOX9 is required for cells of the mesenchymal condensation to acquire chondrocyte phenotype (Li et al 1995). It will be important to determine whether SOX9 controls PTHrP expression.

Osteoblasts

We know much less about the molecular control of osteoblast differentiation and function compared to what we know about osteoclast and chondrocyte differentiation. To date only two genes have been shown to control osteoblast differentiation directly or indirectly, *Cbfa1* and *Ihh*.

CBFA1, one of the mammalian homologues of the Runt *Drosophila* transcription factors is the earliest and most specific marker of osteoblast differentiation known. Three different lines of evidence reinforcing each other demonstrate that *Cbfa1* plays a role apparently not redundant with the function of any other gene during osteoblast differentiation. CBFA1 was identified as the factor binding to one OSE2, an osteoblast-specific *cis*-acting element present in the promoter most genes expressed in osteoblasts. Its expression during development and after birth is highly osteoblast-specific, is regulated by the bone morphogenetic proteins and other growth factors, and forced expression of *Cbfa1* in non-osteoblastic cells leads to the expression of osteoblast-specific genes such as *Osteocalcin* in these cells (Ducy et al 1997).

Deletion of *Cbfa1* from the mouse genome leads to mutant animals in which the skeleton was made only of chondrocytes producing a typical cartilaginous matrix without any evidence of bone (Komori et al 1997, Otto et al 1997). One of the two groups who performed *Cbfa1* deletion in mice explored *Cbfa1*$^{+/-}$ mice and realized that they had hypoplastic clavicles and delay in the suture of the fontanelles, two features observed in a classical mouse mutation called cleidocranial dysplasia (CCD) (Otto et al 1997). *Cbfa1* maps at the same location as does CCD and the two mutations are allelic. The third line of evidence demonstrating the role of *Cbfa1* in osteoblast differentiation came from human genetics. Two groups, one looking for a gene causing human CCD and one looking for a disease in which *CBFA1* was mutated, identified at the same time deletions, stop codon insertions, and missense mutations in CBFA1 in patients affected with CCD (Mundlos et al 1997, Lee et al 1997).

IHH also affects osteoblast differentiation. IHH-deficient mice have a disorganized growth plate as expected (see above) but they also have no osteoblasts in bone formed through endochondral ossification. In contrast, osteoblasts appear to be present in bones of the skull, the mandibles and the clavicles. At this point it is not known whether this is a direct consequence of the absence of IHH signalling or whether it is an indirect effect. The same group that generated IHH-deficient mice also generated transgenic mice overexpressing hedgehog interacting protein (HIP) in chondrocytes. HIP is a cell surface protein that binds IHH with the same affinity as Patched, a receptor for IHH, and antagonizes IHH signalling. Transgenic mice that overexpress HIP in chondrocytes show a chondrocyte phenotype similar to the IHH-deficient mice and also have osteoblasts (St Jacques et al 1999). Thus, it seems clear that the failure of osteoblast differentiation in bones forming through endochondral ossification is not a consequence of the defect in chondrocyte differentiation.

Another family of growth factors that has received a tremendous amount of attention are the bone morphogenetic proteins (BMPs). These are members of the TGFβ superfamily of growth factors that can, when applied locally, induce *de novo* bone formation by recapitulating all the events that occur during skeletal development (Urist 1965). The analysis of their functions *in vivo* has relied mostly on gene deletion experiments. Although several of these mutant animals have skeletal patterning and joint formation defects, it has not been possible yet using this approach to show that they affect osteoblast differentiation. It may possibly be due to functional redundancy (Hogan 1996).

TGFβ itself plays a complex role during bone remodeling. *In vitro* TGFβ induces extracellular matrix synthesis by osteoblasts and affects osteoblast differentiation. *Osteocalcin*-promoter TGFβ2 transgenic mice developed a complex low bone mass phenotype characterized by an overall increase in bone resorption, a large increase of osteocyte numbers (osteoblasts embedded in matrix) and the presence of an hypomineralized matrix (Erlebacher & Dernyck 1996). An increase in osteocyte number was also reported by the same group in a mouse model overexpressing a dominant negative forms of the type II TGFβ receptor in osteoblasts, suggesting that TGFβ could control the steady-state rate of osteoblast differentiation.

Osteoclasts

Multiple transcription factors and secreted molecules control osteoclast differentiation. PU1 is an ETS-domain containing transcription factor that is expressed specifically in the monocytic and B lymphoid lineages (Klemsz et al 1990). Deletion of PU1 results in a multi-lineage defect in the generation of progenitors for B and T lymphocytes, monocytes and granulocytes (Scott et al 1994, McKercher et al 1996). Also there are no osteoclasts and no macrophages in the bone marrow of PU1-deficient mice (Tondravi et al 1997). This is a cell autonomous defect that could be corrected by bone marrow transplantation. To date PU1 is the earliest known marker of osteoclast differentiation as it controls both macrophage and osteoclast differentiation.

Another transcription factor that plays a critical role during osteoclast differentiation is c-Fos. c-*fos* is the cellular homologue of the v-*fos* oncogene and is a major component of the A–P1 transcription factor. The first indication that c-*fos* might play a role in bone cell differentiation came from the observation that v-*fos*-containing constructs injected into rodents led to the appearance of osteosarcomas (Ward & Young 1976). Likewise, transgenic mice expressing high levels of c-*fos* in multiple tissues and cell types eventually developed only one type of tumour: chondroblastic osteosarcoma (Grigoriadis et al 1993). Given these findings it came as a surprise that the deletion of c-*fos* in mice led to an early arrest of osteoclast differentiation without any overt consequences on osteoblast differentiation (Johnson et al 1992, Wang et al 1992). The presence of a large number of macrophages in c-*fos*-deficient mice places c-*fos* downstream of *Pu1* in the genetic pathway controlling osteoclast differentiation. c-*fos* may not be the only member of the *fos* gene family to contribute to osteoclast differentiation as the Fos-related protein Fra1 favours osteoclast differentiation in osteoclast macrophage precursor cell lines.

Two other transcription factors are involved in osteoclast differentiation they are NFκB and mi. NFκB is a dimer composed of various combinations of proteins: p50, p52, p65, c-Rel and RelB (Verma et al 1995). These proteins are all related by the Rel homology domain that contains the DNA-binding motif. Mice deficient in both p50 and p52 harbour an osteopetrotic phenotype due to an arrest of osteoclast differentiation (Franzoso et al 1997, Iotsova et al 1997). Those findings are important also because NFκB seems to be one of the targets of recently identified growth factors that regulate osteoclast differentiation. The last factor found to play a role in osteoclast differentiation was identified by searching for the gene mutated in the mouse mutant microphthalmia (mi). Mi mice present among other defects a failure of secondary bone resorption (Hertwig 1942). The gene mutated in these mice encodes a bHLH transcription factor called mi (Hodgkinson et al 1993). In mi mice osteoclasts differentiate normally but they fail to resorb bones, thus placing mi downstream of the other transcription factors mentioned above.

The requirement for secreted molecules in osteoclast differentiation was first established with the genetic elucidation of a mouse mutation called op/op. Mice homozygous for this recessive mutation lack osteoclasts and macrophages. The osteopetrotic phenotype of the op/op mice is not cured by bone marrow transplantation indicating that it is a non-cell autonomous defect (Marks et al 1984). The gene mutated in op/op mice encodes the growth factor macrophage colony-stimulating factor (M-CSF) (Yoshida et al 1990) whose receptor synthesis is thought to be regulated by PU1.

In the last three years a group of secreted molecules regulating positively or negatively osteoclast differentiation has been identified. Two groups identified a novel soluble receptor related to the TNF super-family that it is a called either osteoprotegerin (OPG) or osteoclastogenesis inhibitory factor (OCIF) (Simonet et al 1997, Yasuda et al 1998). Overexpression in transgenic mice of this molecule resulted in osteopetrosis due to an arrest of osteoclast differentiation. OPG/OCIF-deficient mice develop an osteoporosis due to an increase in osteoclasts (Bucay et al 1998, Mizuno et al 1998).

The identification of a soluble receptor with such a powerful inhibitory effect on osteoclast differentiation suggested that it might be titrating out an osteoclast differentiation activity. This factor was cloned by the same two groups nearly at the same time. The secreted protein binding to OPG had previously been cloned and is called TRANCE or RANK ligand or osteoclast differentiation factor (ODF). RANKL/ODF is present on the membrane of the osteoblast progenitor but can also be found as a soluble molecule in the bone microenvironment. Systemic administration of RANK/ODF leads to increased bone resorption (Burgess et al 1999). *In vitro* RANKL/ODF has all the attributes of a real osteoclast differentiation factor: it favours osteoclast differentiation in conjunction with m-CSF, it bypasses the need for stromal cells and 1,25(OH)2 vitamin D3 to induce osteoclast differentiation, and it activates mature osteoclasts to resorb mineralized bone (Horwood et al 1998, Rifas et al 1998). Consistent with these cell and molecular biology data RANKL-deficient mice lack osteoclasts and develop a severe osteopetrosis besides immunological defects (Kong et al 1999).

RANKL/ODF binds to a receptor present on T cells and bone marrow stromal cells called RANK. RANK-deficient mice and transgenic mice expressing a soluble form of RANK develop an osteopetrosis and polyclonal antibodies against RANK extracellular domain promote osteoclastogenesis in bone marrow culture, suggesting that RANK activation mediates the effect of OPGL/ODF (Dougall et al 1999). The signal transduction pathway initiated following binding of RANKL/ODF to RANK has also been partly elucidated. RANK's intracellular domain contains two binding sites for TNF receptor-associated factors (TRAFs) (Wong et al 1998). TRAFs have been implicated in mediating signals induced by a subset of TNF receptor family members. RANK contains a binding site for TRAF6 in its intracellular region. Importantly TRAF6-deficient mice exhibit an osteopetrotic phenotype due to defective osteoclast function (Lomaga et al 1999). This observation is even more important as TRAFs seem to control the activation of NFκB, a transcription factor required for osteoclast differentiation.

Other cytokines such as IL-1, IL-6, oncostatin-M and IL-11 that all transduce their signals through a composite receptor containing a gp130 subunit, can induce osteoclast differentiation *in vitro* (Manolagas et al 1996, Mundy et al 1996). Their role *in vivo* has been difficult to establish possibly because of functional redundancy.

References

Bitgood MJ, McMahon AP 1995 Hedgehog and Bmp genes are coexpressed at many diverse sites of cell–cell interaction in the mouse embryo. Dev Biol 172:126–138

Bucay N, Sarosi I, Dunstan CR et al 1998 Osteoprotegerin-deficient mice develop early onset osteoporosis and arterial calcification. Genes Dev 12:1260–1268

Burgess TL, Qian Y, Kaufman S et al 1999 The ligand for osteoprotegerin (OPGL) directly activates mature osteoclasts. J Cell Biol 145:527–538

Chiang C, Litingtung Y, Lee E et al 1996 Cyclopia and defective axial patterning in mice lacking Sonic hedgehog gene function. Nature 383:407–413

Colvin JS, Bohne BA, Harding GW, McEwen DG, Ornitz DM 1996 Skeletal overgrowth and deafness in mice lacking fibroblast growth factor receptor 3. Nat Genet 12:390–397

Crossley PH, Minowada G, MacArthur CA, Martin GR 1996 Roles for FGF8 in the induction, initiation, and maintenance of chick limb development. Cell 84:127–136

Dougall WC, Gladdum M, Charrier K et al 1999 RANK is essential for osteoclast and lymph node development. Genes Dev 13:2412–2424

Duboule D, Morata G 1994 Colinearity and functional hierarchy among genes of the homeotic complexes. Trends Genet 10:358–364

Ducy P, Zhang R, Geoffroy V, Ridall AL, Karsenty G 1997 Osf2/Cbfa1: a transcriptional activator of osteoblast differentiation. Cell 89:747–754

Echelard Y, Epstein DJ, St-Jacques B et al 1993 Sonic hedgehog, a member of a family of putative signaling molecules, is implicated in the regulation of CNS polarity. Cell 75: 1417–1430

Erlebacher A, Derynck R 1996 Increased expression of TGF-β2 in osteoblasts results in an osteoporosis-like phenotype. J Cell Biol 132:195–210

Fallon JF, López A, Ros MA, Savage MP, Olwin BB, Simandl BK 1994 FGF-2: apical ectodermal ridge growth signal for chick limb development. Science 264:104–107

Francis-West PH, Abdelfattah A, Chen P et al 1999 Mechanisms of GDF-5 action during skeletal development. Development 126:1305–1315

Franzoso G, Carlson L, Xing L et al 1997 Requirement for NF-κB in osteoclast and B-cell development. Genes Dev 11:3482–3496

Goff DJ, Tabin CJ 1997 Analysis of Hoxd13 and Hoxd11 misexpression in chick limb buds reveals that Hox genes affect both bone condensations and growth. Development 124: 627–636

Grigoriadis AE, Schellander K, Wang ZQ, Wagner EF 1993 Osteoblasts are target cells for transformation in c-fos transgenic mice. J Cell Biol 122:685–701

Hertwig P 1942 Neue Mutationen und Koppplungsgruppen bei der Hausmaus. Z Indukt Abstammungs-Vererbungsl 80:220–246

Hodgkinson CA, Moore KJ, Nakayama A et al 1993 Mutations at the mouse microphthalmia locus are associated with defects in a gene encoding a novel basic-helix-loop-helix-zipper protein. Cell 74:395–404

Hogan BL 1996 Bone morphogenetic proteins: multifunctional regulators of vertebrate development. Genes Dev 10:1580–1594

Horwood NJ, Kartsogiannis V, Lam MHX et al 1998 Activated T cells are capable of inducing osteoclast formation: a mechanism for rheumatoid arthritis. In: Drezner MC (ed) Second joint meeting of the American Society for Bone and Mineral Research and the International Bone and Mineral Society. Elsevier Science, San Francisco, p S215

Iostova V, Caamaño J, Loy J, Yang Y, Lewin A, Bravo R 1997 Osteopetrosis in mice lacking NF-κB1 and NF-κB2. Nat Med 3:1285–1289

Johnson RL, Tabin CJ 1997 Molecular models for vertebrate limb development. Cell 90: 979–990

Johnson RS, Spiegelman BM, Papaioannou V 1992 Pleiotropic effects of a null mutation in the c-fos proto-oncogene. Cell 71:577–586

Kanzler B, Kuschert SJ, Liu YH, Mallo M 1998 Hoxa2 restricts the chondrogenic domain and inhibits bone formation during development of the branchial area. Development 125: 2587–2597

Karaplis AC, Luz A, Glowaski J et al 1994 Lethal skeletal dysplasia from targeted disruption of the parathyroid hormone-related peptide gene. Genes Dev 8:277–289

Klemsz MJ, McKercher SR, Celada A, Van Beveren V, Make RA 1990 The macrophage and B cell-specific transcription factor PU.1 is related to the ets oncogene. Cell 61:113–124

Komori T, Yagi H, Nomura S et al 1997 Targeted disruption of Cbfa1 results in a complete lack of bone formation owing to maturational arrest of osteoblasts. Cell 89:755–764

Kong YY, Yoshida H, Sarosi I et al 1999 OPGL is a key regulator of osteoclastogenesis, lymphocyte development and lymph-node organogenesis. Nature 397:315–323

Krauss S, Concordet JP, Ingham PW 1993 A functionally conserved homolog of the Drosophila segment polarity gene hh is expressed in tissues with polarizing activity in zebrafish embryos. Cell 75:1431–1444

Lanske B, Karaplis AC, Lee K et al 1996 PTh/PThrP receptor in early development and Indian hedgehog-regulated bone growth. Science 273:663–666

Laufer E, Dahn R, Orozco OE et al 1997 Expression of radical fringe in limb-bud ectoderm regulates apical ectodermal ridge formation. Nature 386:366–373 (erratum: 1997 Nature 388:400)

Lee B, Thirunavukkarasu K, Zhou L et al 1997 Missense mutations abolishing DNA binding of the osteoblast-specific transcription factor OSF2/CBFA1 in cleidocranial dysplasia. Nat Genet 16:307–310

Li Y, Lacerda DA, Warman ML et al 1995 A fibrillar collagen gene, Col11a1, is essential for skeletal morphogenesis. Cell 80:423–430

Lomaga MA, Yeh WC, Sarosi I et al 1999 TRAF6 deficiency results in osteopetrosis and defective interleukin-1, CD40, and LPS signaling. Genes Dev 13:1015–1024

Loomis CA, Harris E, Michaud J, Wurst W, Hanks M, Hoyner AL 1996 The mouse Engrailed-1 gene and ventral limb patterning. Nature 382:360–363

Lui VC, Ng LJ, Nicholls J, Tam PP, Cheah KS 1995 Tissue-specific and differential expression of alternatively spliced a1 (II) collagen mRNAs in early human embryos. Dev Dyn 203: 198–211

Macias D, Gañan Y, Sampath TK, Piedra ME, Ros MA, Hurle JM 1997 Role of BMP-2 and OP-1 (BMP-7) in programmed cell death and skeletogenesis during chick limb development. Development 124:1109–1117

Manolagas SC, Jilka RL, Bellido T, O'Brien CA, Parfitt AM 1996 Interleukin-6-type cytokines and their receptors. In: Bilezikian JP, Raisz LG, Rodan GA (eds) Principles of bone biology. Academic Press, San Diego, CA, p 701–713

Marks SC Jr, Seifert MF, McGuire JL 1984 Congenitally osteopetrotic (oplop) mice are not cured by transplants of spleen or bone marrow cells from normal littermates. Metab Bone Dis Relat Res 5:183–186

Mizuno A, Amizuka N, Irie K et al 1998 Severe osteoporosis in mice lacking osteoclastogenesis inhibitory factor/osteoprotegerin. Biochem Biophys Res Comm 247:610–615

McKercher SR, Torbett BE, Anderson KL et al 1996 Targeted disruption of the PU.1 gene results in mulitple hematopoietic abnormalities. EMBO J 15:5647–5658

Mundlos S, Otto F, Mundlos C et al 1997 Mutations involving the transcription factor CBFA1 cause cleidocranial dysplasia. Cell 89:773–779

Mundy GR, Boyce BF, Yoneda T, Bonewald LF, Roodman GD 1996 Cytokines and bone remodeling. In: Marcus R, Feldman D, Kelsey J (eds) Osteoporosis. Academic Press, San Diego, CA, p 302–313

Naski MC, Wang Q, Xu J, Ornitz DM 1996 Graded activation of fibroblast growth factor receptor 3 by mutations causing achondroplasia and thanatophoric dysplasia. Nat Genet 13:233–237

Ng LJ, Wheatley S, Muscat GE et al 1997 SOX9 binds DNA, activates transcription, and coexpresses with type II collagen during chondrogenesis in the mouse. Dev Biol 183:108–121

Niswander L, Tickle C, Vogel A, Booth I, Martin GR 1993 FGF-4 replaces the apical ectodermal ridge and directs outgrowth and patterning of the limb. Cell 75:579–587

Ohuchi H, Nakagawa T, Yamamoto A et al 1997 The mesenchymal factor, FGF10, initiates and maintains the outgrowth of the chick limb bud through interaction with FGF8, an apical ectodermal factor. Development 124:2235–2244

Otto F, Thornell AP, Crompton T et al 1997 Cbfa1, candidate gene for cleidocranial dysplasia syndrome, is essential for osteoblast differentiation and bone development. Cell 89:765–771

Riddle RD, Johnson RL, Laufer E, Tabin C 1993 Sonic hedgehog mediates the polarizing activity of the ZPA. Cell 75:1401–1416

Rifas L, Holliday LS, Gluck SL, Aviolo LV 1998 Activated T cells induce osteoclastogenesis. In: Drezner MC (ed) Second meeting of the American Society for Bone and Mineral Research and the International Bone and Mineral Society. Elsevier Science, San Francisco, p S195

Rodriguez-Esteban C, Schwabe JW, De La Peña J, Foys B, Eshelman B, Belmonte JC 1997 Radical fringe positions the apical ectodermal ridge at the dorsoventral boundary of the vertebrate limb. Nature 386:360–366 (erratum: Nature 388:906)

Ryan MC, Sandell LJ 1990 Differential expression of a cysteine-rich domain in the amino-terminal propeptide of type II (cartilage) procollagen by alternative splicing of mRNA. J Biol Chem 265:10334–10339

Saunders JW Jr, Gasseling MT 1968 Ectoderm–mesenchymal interaction in the origins of wing symmetry. In: Fleischmajer R, Billingham RE (eds) Epithelial–mesenchymal interactions. Williams & Wilkins, Baltimore, PA, p 78–97

Schipani E, Kruse K, Jüppner H 1995 A constitutively active mutant PTH-PTHrP receptor in Jansen-type metaphyseal chondrodysplasia. Science 268:98–100

Scott EW, Simon MC, Anastasi J, Singh H 1994 Requirement of transcription factor PU.1 in the development of multiple hematopoietic lineages. Science 265:1573–1577

Simonet WS, Lacey DL, Dunstan CR et al 1997 Osteoprotegerin: a novel secreted protein involved in the regulation of bone density. Cell 89:309–319

St-Jacques B, Hammerschmidt M, McMahon AP 1999 Indian hedgehog signaling regulates proliferation and differentiation of chondrocytes and is essential for bone formation. Genes Dev 13:2072–2086

Storm EE, Huynh TV, Copeland NG, Junkins NA, Kingsley DM, Lee SJ 1994 Limb alterations in *brachypodism* mice due to mutations in a new member of the TGFβ-superfamily. Nature 368:639–643

Thomas JT, Lin K, Nandedkar M, Camargo M, Cervenka J, Luyten FP 1996 A human chondrodysplasia due to a mutation in a TGFβ superfamily member. Nat Genet 12:315–317

Thomas JT, Kilpatrick MW, Lin K et al 1997 Disruption of human limb morphogenesis by a dominant negative mutation in CDMP1. Nat Genet 17:58–64

Thorogood P 1993 Differentiation and morphogenesis of cranial skeletal tissues. In: Hanken J, Hall BK (ed) The skull, vol 1: Development. University of Chicago Press, Chicago, p 112–152

Tondravi MM, McKercher SR, Anderson K et al 1997 Osteopetrosis in mice lacking haematopoietic transcription factor PU.1. Nature 386:81–84

Urist MR 1965 BoneL formation by autoinduction. Science 150:893–899

Verma IM, Stevenson JK, Schwarz EM, Van Antwerp D, Miyamoto S 1995 Rel/NF-κB/I κB family: intimate tales of association and dissociation. Genes Dev 9:2723–2735

Vogel A, Rodriguez C, Izpisúa-Belmonte JC 1996 Involvement of Fgf-9 in initiation, outgrowth and patterning of the vertebrate limb. Development 122:1737–1750

Vortkamp A, Lee K, Lanske B, Segre GV, Kronenberg HM, Tabin CJ 1996 Regulation of rate of cartilage differentiation by Indian hedgehog and PTH-related protein. Science 273:613–622
Wagner T, Wirth J, Meyer J et al 1994 Autosomal sex reversal and campomelic dysplasia are caused by mutations in and around the SRY-related gene SOX9. Cell 79:1111–1120
Wang ZQ, Ovitt C, Grigoriadis AE, Möhle-Steinlein U, Rüther U, Wagner EF 1992 Bone and haematopoietic defects in mice lacking c-fos. Nature 360:741–745
Ward JM, Young DM 1976 Histogenesis and morphology of periosteal sarcomas induced by FBJ virus in NIH Swiss mice. Cancer Res 36:3985–3992
Wong BR, Josien R, Lee SY, Vologodskaia M, Steinman RM, Choi Y 1998 The TRAF family of signal transducers mediates NF-κB activation by the TRANCE receptor. J Biol Chem 273:28355–28359
Yang Y, Niswander L 1995 Interaction between the signaling molecules WNT7a and SHH during vertebrate limb development: dorsal signals regulate anteroposterior patterning. Cell 80:939–947
Yasuda H, Shima N, Nakagawa N et al 1998 Identity of osteoclastogenesis inhibitory factor (OCIF) and osteoprotegerin (OPG): a mechanism by which OPG/OCIF inhibits osteoclastogenesis in vitro. Endocrinology 139:1329–1337
Yoshida H, Hayashi S, Kunisada T et al 1990 The murine mutation osteopetrosis is in the coding region of the macrophage colony stimulating factor gene. Nature 345:442–444

DISCUSSION

Newman: In your coverage of the genetic interplay that results in skeletal pattern formation, it seems to me that what was left out is perhaps the most central factor in both development and evolution of skeletal development, namely precartilage condensation, which involves extracellular matrix molecules and adhesive molecules. This might be the primary process that is modulated by the various Hox proteins, retinoid receptors and so forth.

Karsenty: You are probably right; the *Hox* genes do play a role.

Hall: The patterning that you discussed in terms of the limb, I think was patterning of the limb bud. How much do we know about patterning of the limb skeleton itself? We know a lot about what makes the P–D axis and A–P axis of the limb bud, but we don't know why this cartilage cell up here is part of the humerus and this one down there is part of the digit, as Lewis Wolpert likes to say.

Karsenty: I am sure that David Kingsley will discuss in his paper the role of certain members of the BMP family in the joint formation. This is a beginning of an understanding about skeletal patterning.

Kingsley: I think there has been a big gap between how much we know about the setting up of the basic axes and then getting the cells to lay in precise orientations to form skeletal elements. You could imagine that some mechanism exists to bridge the gap between the early and late steps.

Hall: It would be interesting to know whether the specification of the joints occurs early or late in that process.

Kingsley: We don't know this. One observation that has been intriguing researchers is that many of the joints appear to show up at boundaries of Hox gene expression. When I say that there is a link between early events and later morphological events, it is because you can see axes and patterns of gene expression in broad limb domains that in some way must be related to the differentiation of the specialized shapes that appear within those. But how you get from broad domains to gene expression, defining sub regions in the limb, is still a problem: we have a long way to go.

Reddi: In mouse genetics, it is commonly thought that genes involved in patterning are not involved in differentiation. Yet, in the earlier part of your paper, you said that BMPs may be involved in patterning *and* differentiation.

Karsenty: Life is full of exceptions! BMPs are certainly involved in patterning, but they are also involved in the early events of chondrogenesis *in vivo*.

Reddi: I thought from your work on the knockouts in Japan, that *CBFA1* is *the* critical gene in osteoblast differentiation. Since the first papers were published, do you now have more insight into what regulates *CBFA1* upstream? What is downstream of *CBFA1*? Is there anything after *CBFA1* prior to bone formation?

Karsenty: The work is just unfolding in many laboratories. The picture that is emerging is rather complex. *CBFA1* is not a cell-specific gene: it is a gene that is expressed in a very mobile way, first in the mesenchymal condensation, then in an osteochondral progenitor, and then in an oesteoblast.

Reddi: Is there direct evidence of the role of any BMP in regulating CBFA1?

Karsenty: Not direct. BMP will increase CBFA1 expression, but it takes at least 8 h.

Ornitz: I have a comment on genes involved in both patterning and differentiation. The *FGF*s are a good example of genes that early in development act as patterning molecules. The limb bud is one example where FGFs are involved in setting up early patterns. Brain development is another case where FGFs have been shown to have organizer activity. At later times in development, FGF clearly regulates growth and differentiation either in positive or negative ways, depending on the tissue.

Reddi: Cheryll Tickle, can you shed some more light on patterning in terms of skeletal elements in wing, other than post limb bud pattern?

Tickle: David Kingsley's point about the gap in understanding between the early events in setting up the different regions of the limb, and the later events in patterning is absolutely critical. There is a huge conceptual gap between the pattern of *Hox* gene expression (if *Hox* genes are really what determine what cells will do) and how you then get something that looks like a skeleton.

I favour the idea that BMPs are important patterning molecules in the limb. There is evidence that BMPs seem to specify tooth identity, for example (Tucker et al 1998). The BMPs could therefore well be multifunctional molecules that are involved in patterning, then later again in stimulating differentiation, and later on even downstream of IHH in controlling bone. BMPs could thus act at many different stages in skeletogenesis.

Newman: With regard to this gap between the setting up of the axes and the outgrowth of the limb bud, and the actual formation of the pattern, I would like to briefly summarize and update a hypothesis that my colleague Harry Frisch and I made 20 years ago (Newman & Frisch 1979). This concerns an interplay between molecules that activate precartilage condensation, and molecules that inhibit precartilage condensation. It is a well-known result first described by the mathematician Alan Turing, back in the early 1950s (Turing 1952), that such an interplay can give rise to periodic chemical patterns and structures. Indeed, with small changes in parameters you can have a harmonic progression of these peaks and valleys of chemical activation leading to patterns with increasing numbers of elements, as seen in the limbs. In the intervening time the TGFβs have been proposed to be components of such systems (Newman 1988, Leonard et al 1991). In particular, TGFβ2 is expressed at the right times in the right places to play the role of this activating molecule in the digital plate (Merino et al 1998, Miura & Shiota 2000). TGFβ2 is a molecule that induces a number of extracellular and adhesive proteins, such as fibronectin and N-cadherin. If centres of activation of these matrix and adhesive molecules are established by the local action of TGFβ, which also induces its own production, and (this is more hypothetical) if an inhibitor is released from those same centres (and we have no idea of what this inhibitor might be at the time), then you could have periodic structures just as you see in the limb. This might be an underlying process that gives rise to the particular pattern that we see in the limb bud.

Bard: You are making a key point here: the trouble with signals is that they are distributive. We almost have too many signals in too large an area to get the specificity of the pattern detail. There are two ways we can get localized events. One is by means of a Turing-type mechanism. This builds up concentration patterns over space but is actually quite tricky to do: the patterns are not particularly stable over time, and it's not clear that they easily exhibit the size invariance that we need in a limb. The alternative is to focus much more on the response to the signals. I find myself favouring this approach because the great thing about signals is that their presence can be identified. I wonder, however, whether we ought to think more in terms of the pattern of their receptors, which tells us where the signals are actually mediated. Perhaps not enough has been done on this. Also, once these receptors have been activated, the onus of patterning is placed on the cells that are responding and their patterns of transcription factors. Maybe it is at this that we should really be studying if we're thinking about the pattern of the skeleton.

Reddi: I don't think it just stops with the receptor. BMPs have very specific binding proteins, such as Noggin, which have the same affinity as the receptor. So the availability of the ligand of the signalling molecule is determined by another molecule, which specifically interacts with it. In thinking about skeletal patterning in cartilage and bone, we have increasingly paid attention to the extracellular matrix. With great respect for Turing and his legion of disciples, I have great difficulty in believing that these are all soluble molecules floating in a swimming pool-like environment. Instead, I suspect that they are really orientated in space and time by the extracellular matrix. Although the extracellular matrix is de-emphasized in this conference, I think it has an important role in putting molecules or constellations of molecules in the right place at the right time during morphogenesis.

Bard: Whoever, for example, would have predicted that extracellular matrix molecules such as heperan sulfate could act as low affinity co-receptors for signals?

Reddi: The availability of the ligand to the receptor is further modulated both by the extracellular matrix, and specific binding proteins such as Noggin and Corrin.

Hall: If you think of the condensation as a way of trapping these signals, then you have this pericellular matrix around these cells. You don't really have an extracellular matrix: you have the receptors in the pericellular matrix which is really a way of trapping these molecules at a location.

Ornitz: There are many factors that can actually modify the activity of a growth factor in the extracellular matrix. It has been shown for a number of molecules, such as FGF, that extracellular matrix heparan sulfate proteoglycans are essential for the ability of these factors to activate their receptors. In addition, there are tissue-specific modifications that can form unique sequences along the proteoglycan chain, which can then either activate or in some cases inhibit the activity of growth factors.

Tickle: Going back to Stuart Newman's hypothesis, I can see how you can make condensations in this way, but the question is, what determines then whether a particular condensation will be a radius or an ulna, for example?

Newman: This kind of reaction–diffusion or Turing mechanism can only go so far. It can only map out identical structures. It certainly can't determine by itself whether something will become a thumb or little finger, for instance. This of course has to depend on the inhomogeneities in terms of Hox proteins and retinoid receptors and so on that exist in the limb bud and create a kind of uneven response medium to whatever is activating these condensations.

Tickle: I think the question of scale is also important. The early limb bud is quite small. The cells that give rise to the digits comprise a tiny part of the limb bud, for instance. We have always thought of sweeping morphogen gradients within that small group of cells early on, but then later on, as the limb grows, you can imagine very much more local interactions actually building parts of the skeleton. I am thinking of the work of Le Douarin in the vertebrae, where local interactions make the arch and other bits (Watanbe et al 1998).

Newman: On the question of the scale, I think that the Turing-type of mechanism is relevant to the spatial scale of the early limb bud. This is illustrated by the observations made when these reaction–diffusion processes are made to occur in non-living systems. Chemists have looked at this in gelled media, and have measured the stripes and spots that these kinds of mechanisms can produce. The scales of these structures are precisely that of the skeletal primordia in the developing limb (Ouyang & Swinney 1991). The reaction–diffusion model does represent a good physical model for generic features of skeletogenesis.

Meikle: The discussants seem to be studiously avoiding using the word 'morphogen'. Are FGFs and BMPs, as secreted proteins, acting as endogenous morphogens?

Bard: You have to look at the expression pattern. When Turing said 'morphogen', he thought that all of the participating cells would make the morphogen themselves. If you look at the pattern of signals and receptors, they don't match this paradigm. So I deliberately avoided using the term 'morphogen'.

Newman: I think the term has become somewhat vague. Turing, who initiated the use of this term, used it in a very specific fashion, as part of a self-organizing system. But now since so many factors have been identified by which cells and tissues influence one another during development any factor that affects morphogenesis can be called a morphogen, whether or not through a self-organizing process.

Bard: The word isn't very useful now. It stems from a time when people thought that the instructions were simply linear—'A' told 'B' told 'C'—and now we know there is actually an awful lot of cross-talk going on.

Ornitz: If the term is used, it should be according to the original definition, which is that there is a graded response at different concentrations of a factor.

Bard: That cuts the list of potential Turing 'morphogens' down quite considerably.

Kronenberg: One thing that happens when you go from the very tiny condensations to much larger structures is a continuing definition of edges and ends. Once condensations from ends and borders are defined, this leads to unique areas where signals can be produced and act. These signals then diffuse in a way that is tightly regulated. Hedgehog regulation is one of the most vivid and least understood examples, where it seems that cellular processes have a lot to do with that molecule getting through space. Examples such as this one led to the idea that secreted paracrine factors use diffusion, but this diffusion is regulated by binding proteins, proteases and matrix interactions. Thus, functional diffusion is a carefully regulated biologic process, not a simple physical/mathematical one.

References

Leonard CM, Fuld HM, Frenz DA, Downie SA, Massagué J, Newman SA 1991 Role of transforming growth factor-β in chondrogenic pattern formation in the embryonic limb: stimulation of mesenchymal condensation and fibronectin gene expression by exogenous TGF-β and evidence for endogenous TGF-β-like activity. Dev Biol 145:99–109

Merino R, Gañan Y, Macias D, Economides AN, Sampath KT, Hurle JM 1998 Morphogenesis of digits in the avian limb is controlled by FGFs, TGFβs, and noggin through BMP signaling. Dev Biol 200:35–45

Miura T, Shiota K 2000 TGFβ2 acts as an 'activator' molecule in reactiondiffusion model and is involved in cell sorting phenomenon in mouse limb micromass culture. Dev Dyn, in press

Newman SA 1988 Lineage and pattern in the developing vertebrate limb. Trends Genet 4: 329–332

Newman SA, Frisch HL 1979 Dynamics of skeletal pattern formation in developing chick limb. Science 205:662–668

Ouyang Q, Swinney H 1991 Transition from a uniform state to hexagonal and striped Turing patterns. Nature 352:610–612

Tucker AS, Matthews KL, Sharpe PT 1998 Transformation of tooth type induced by inhibition of BMP signaling. Science 282:1136–1138

Turing A 1952 The chemical basis of morphogenesis. Phil Trans R Soc Lond B Biol Sci 237: 37–72

Watanbe Y, Duprez D, Monsoro-Burq AH, Vincent C, Le Douarin NM 1998 Two domains in vertebral development: antagonistic regulation by SHH and BMP4 proteins. Development 125:2631–2639

Early steps in limb patterning and chondrogenesis

Sandrine Pizette and Lee Niswander

Molecular Biology Program and Howard Hughes Medical Institute, Memorial Sloan-Kettering Cancer Center, 1275 York Avenue, New York, NY 10021, USA

Abstract. The interplay of a number of signalling molecules coordinates growth and patterning of the early embryonic vertebrate limb along the three axes. An unresolved question is how this information translates into proper positioning and patterning of the skeletal elements. More is known about how these elements develop. Cells first form precartilaginous condensations and then subsequently differentiate into chondrocytes. This provides a cartilage template that will ultimately be ossified to give rise to the bony skeleton. Several studies support a role for the bone morphogenetic proteins (BMPs) as extracellular signals that regulate early steps of limb chondrogenesis (cartilage formation). However, they have not clarified the step(s) at which BMPs act, and no genetic evidence is available to date. Here, we have used a retroviral vector to misexpress the BMP antagonist Noggin in the embryonic chick limb. We find that BMP signalling is necessary for the formation of precartilaginous condensations, their differentiation into chondrocytes, and for maintenance of chondrogenesis. These results also indicate that Noggin could be clinically useful to treat diseases involving ectopic cartilage formation.

2001 The molecular basis of skeletogenesis. Wiley, Chichester (Novartis Foundation Symposium 232) p 23–43

Vertebrate limb development requires coordination of growth and patterning along the three axes. Proximodistal elongation of the limb depends on a specialized epithelium located at its distal dorsoventral interface; anteroposterior patterning is dictated by the posterior mesenchymal zone of polarizing activity (ZPA); dorsoventral identity is imposed on the mesenchyme by the overlying ectoderm. The molecules governing these processes have been, for the most part, identified, and the interdependent expression of their genes has been shown to ensure proper limb morphogenesis (Johnson & Tabin 1997).

One major limb morphogenetic event is the formation of the bony skeleton, which occurs through ossification of a preexisting cartilage template (reviewed in Hall & Miyake 1995). The first sign of chondrogenesis is the aggregation of mesenchymal cells into prechondrogenic condensations and their subsequent expression of chondrogenic markers. The condensed chondroprogenitors then differentiate into chondrocytes, and this is characterized by changes in cell shape and gene expression. Concomitant with chondrocyte differentiation, a sheath of flattened fibroblastic-like cells — the perichondrium — forms around the cartilage rudiment. Chondrocyte maturation then proceeds, followed by ossification in the centre of the element.

The limb cartilage elements form in a temporal proximal to distal sequence, but are initially contiguous. The primary condensation of the stylopod (humerus/femur) bifurcates across the anteroposterior axis to form the zeugopodial (radius–ulna/tibia–fibula) and then autopodial elements (wrist/ankle and digits). The individual elements are generated by the progressive cleavage of this continuous condensation along the proximodistal axis at the presumptive joint regions (Fell & Canti 1934). In regions of the condensation where joints will form, condensed chondroprogenitors do not differentiate into chondrocytes and instead adopt a joint fate. This involves expression of new genes and a down-regulation of genes specific to the chondrogenic lineage.

The cascade of extracellular signals that translate the early patterning information into the formation of skeletal elements of the correct position and shape remains largely unknown. Grafting the ZPA to the anterior margin of the limb produces a mirror image duplication of the skeleton. Sonic hedgehog (SHH) is synthesized in the ZPA and mimics this activity (Johnson & Tabin 1997), but is not implicated in chondrogenesis. Members of the BMP family have been presented as downstream effectors of SHH in limb patterning and can also promote cartilage formation (Hogan 1996). They could thus provide a link between these two events.

In the developing chick limb, *Bmp2*, *Bmp4* and *Bmp7* are expressed in the mesenchyme in overlapping patterns, both prior to formation of precartilaginous condensations and during their aggregation, in cells adjacent to the condensations. Their two types of receptors are expressed ubiquitously in the mesenchyme throughout limb morphogenesis (*BmpRII* and *BmpRIA*), or more specifically in prechondrogenic aggregates (*BmpRIB*) (see references in Macias et al 1999). These expression profiles are compatible with BMP acting in early steps of chondrogenesis, but targeted mutations of *Bmps* and their receptors are not informative in this respect (Hogan 1996, Zhang & Bradley 1996, Dunn et al 1997, Katagiri et al 1998). Other studies indicate a role for BMPs in limb cartilage formation but either do not define a specific step or suggest a late function (reviewed in Macias et al 1999, and Brunet et al 1998).

Here we have used Noggin, a secreted BMP antagonist (Zimmerman et al 1996), to investigate whether BMPs play a role in early steps of chondrogenesis in the autopod and zeugopod. We have misexpressed Noggin prior to the onset of cartilage formation, and characterized the step of chondrogenesis that is inhibited and the mechanism that leads to the complete absence of skeletal elements. We also report the phenotype resulting from Noggin misexpression in cartilaginous rudiments containing immature chondrocytes.

Experimental procedures

Retroviral infection

Fertilized chicken eggs (SPAFAS) were incubated at 39 °C, and *Noggin*-RCAS(A) virus (Pizette & Niswander 1999) injected at Hamburger–Hamilton stage (Hamburger & Hamilton 1992) 14, 17 and 20 into the limb fields, and at stage 24 or 27 in the zeugopod or autopod, respectively. For molecular studies other than alcian blue staining, embryos were injected at stage 14.

RNA in situ hybridization and probes

RNA in situ hybridization with digoxygenin-labelled probes was performed as referenced in Zou et al (1997). Antisense probes were generated as described: *BmpRIB* and *colII* (Francis-West et al 1999), and *sox9* (Kent et al 1996).

Histology

BrdU incorporation and detection, wholemount nile blue staining, alcian blue staining on whole embryos or on sections, were executed as in (Zou et al 1997), TUNEL as in (Chen et al 1994) and PNA staining as in (Stringa & Tuan 1996).

Results

Noggin misexpression disrupts chondrogenesis

Widespread *Noggin* misexpression in the limbs before overt chondrogenesis (infection from stage 14 to 20) produced very short limbs completely lacking alcian blue-stained cartilage (Fig. 1A, $n=29/30$, assayed up to stage 31). Viral injections restricted to the anterior or posterior region of stage 17 limb buds caused loss of anterior or posterior skeletal elements, respectively. The remaining digits retained their proper identity, as scored by segmentation pattern ($n=17/18$, data not shown and Capdevila & Johnson 1998). Thus, the Noggin-induced phenotype likely is the result of impaired chondrogenesis but not of inhibition of earlier patterning events.

Noggin blocks formation of prechondrogenic condensations

We then addressed whether Noggin could affect the expression of early markers of chondrogenesis. At stage 27 in control autopods, *BmpRIB* and the *Sry*-related *Sox9* transcription factor (Ng et al 1997) were transcribed in, and peanut agglutinin

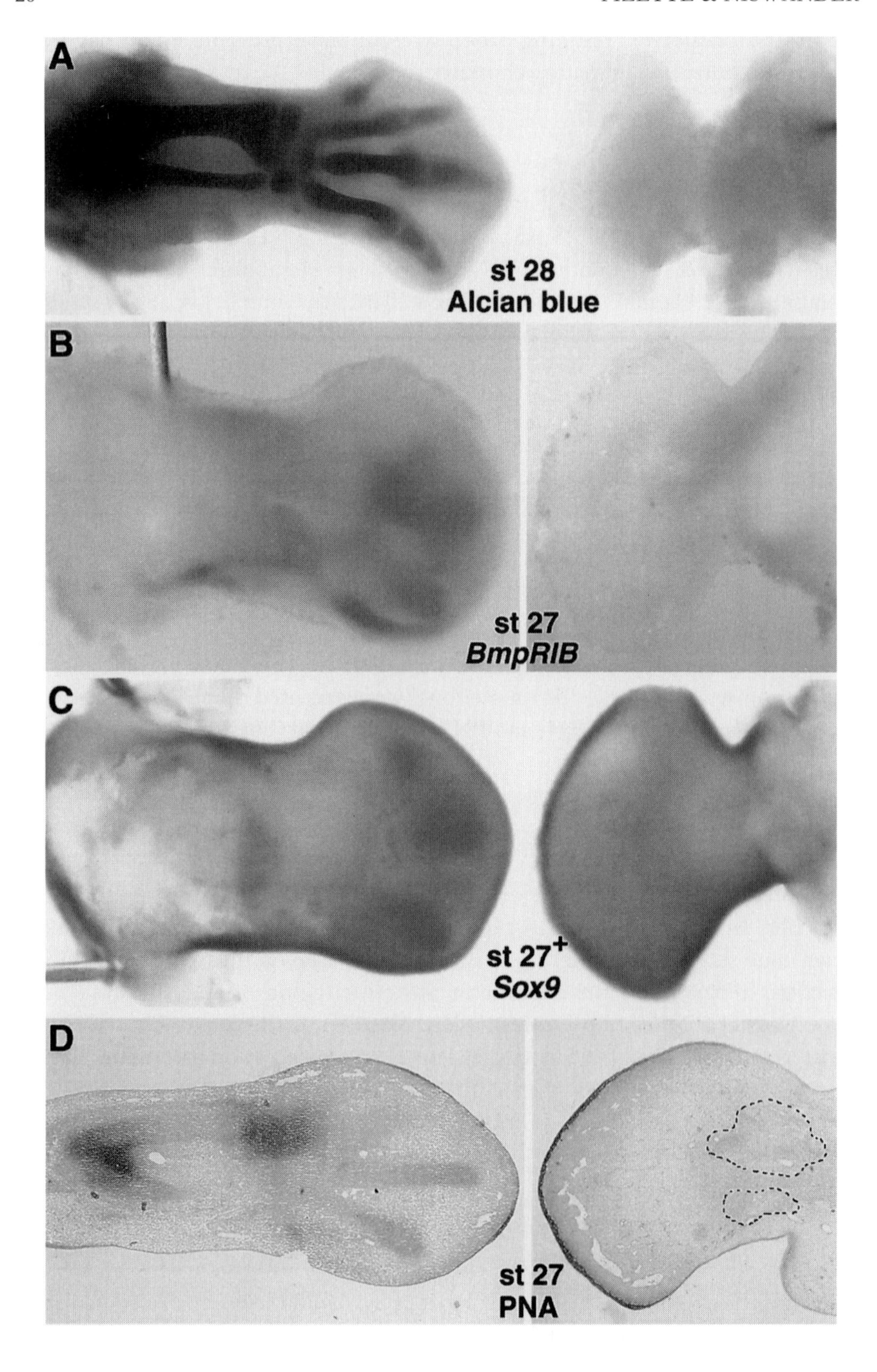
A
st 28
Alcian blue
B
st 27
BmpRIB
C
st 27+
Sox9
D
st 27
PNA

(PNA) (Aulthouse & Solursh 1987) bound to, the wrist/ankle prechondrogenic condensations and the digital rays. In stage 27 limbs infected at stage 14 with RCAS-*Noggin*, *BmpRIB* and *Sox9* expression were extremely reduced or below detection in chondrogenic regions (Fig. 1B,C, $n=10/10$ for both probes), and no PNA staining was present in the autopod ($n=10$, Fig. 1D). The lack of cellular aggregation in the autopod was confirmed by semi-thin plastic sections (not shown) and the faint methyl green counterstain which normally strongly highlights regions of high cell density (Fig. 1D and 2A). This indicates that *Noggin* misexpression compromises chondrogenesis at the condensation step or upstream of this event.

Fate of chondrogenic progenitors in the autopod of RCAS-Noggin infected limbs

To determine why prechondrogenic condensations do not form in infected autopods, we examined the patterns of cell death and proliferation. To assay cell death, we performed wholemount nile blue staining and TUNEL on sections from stage 24–27. No apoptosis was detected in stage 27 autopod of infected limbs ($n=7/7$, Fig. 2B and Pizette & Niswander 1999), or earlier in the distal mesenchyme underlying the AER (progress zone) where the progenitors of the autopodial elements should be present (Fig. 3A and data not shown). Interestingly, comparison of BrdU incorporation (which assesses proliferation) at stage 27 on sections of control and infected limbs demonstrated the existence of a zone within the proximal infected autopod where cells were largely BrdU-negative, not aggregated, and very sparsely distributed (Fig. 2A, boxed area). Control autopods also exhibited regions of low mitotic index, but these were restricted to the areas of condensations (Fig. 2A, $n=6/6$). Therefore, when *Noggin* is misexpressed, cells normally destined to form the prechondrogenic condensations of the autopod do not die and presumably have an altered proliferative capability.

Noggin prevents chondrocyte differentiation

Stage 14 viral infections lead to widespread *Noggin* misexpression in the limb prior to normal condensation of autopodial and zeugopodial skeletal elements, and

FIG. 1. Cartilage formation is dramatically inhibited following *Noggin* misexpression. Control uninfected and RCAS-*Noggin* infected hindlimbs on left and right respectively. Limb primordia were injected at stage 14 (B–D) or stage 17 (A) and embryos fixed at stages 27–28. (A–C) Alcian blue staining is absent from RCAS-*Noggin*-infected limbs and *BmpRIB* and *Sox9* expression are greatly reduced (superficial *Sox9* expression is unaffected). (D) In representative sections, the lack of PNA labelling and the faint methyl green counterstaining in the infected autopod indicates that chondrogenesis is blocked at or before the condensation step. A faint PNA signal is however present in the infected zeugopod (dotted outline). wholemount embryos shown in dorsal view, anterior to the top. Reprinted from Pizette & Niswander (2000).

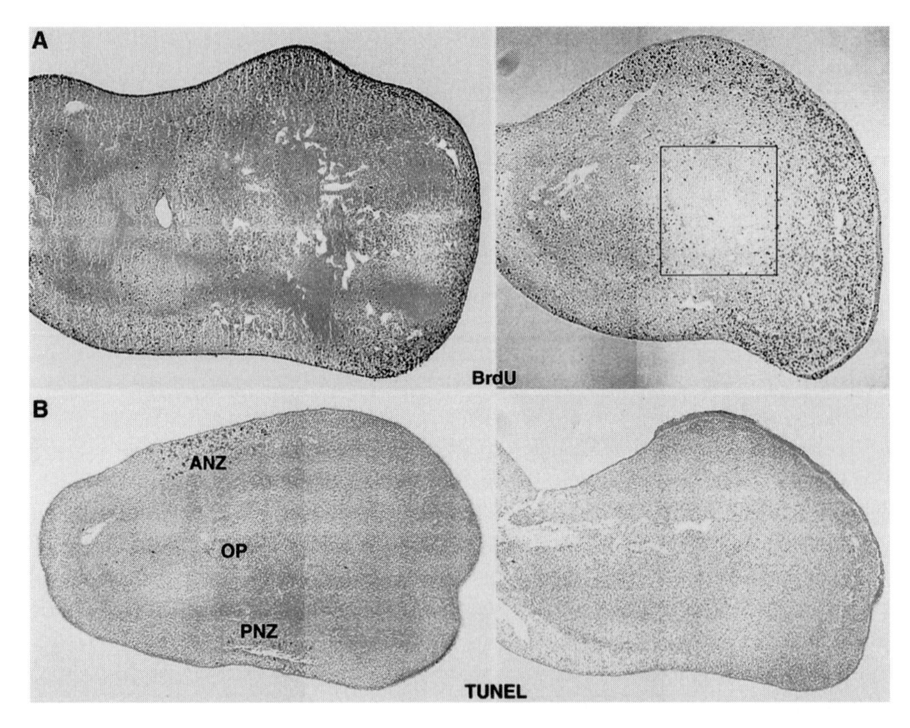

FIG. 2. Analysis of proliferation and cell death. RCAS-*Noggin* virus was injected into the limb primordia at stage 14, infected and uninfected limbs were fixed at stage 27, sectioned, and the tissue processed for BrdU incorporation (A, hindlimbs) or TUNEL (B, forelimbs). Representative sections of control and RCAS-*Noggin* infected limbs on left and right, respectively. Anterior is up and distal to right. (A) In control, brown BrdU-labelled cells are detected throughout except in condensed mesenchyme which stains intensely with methyl green. In infected limbs, BrdU is incorporated into cells in the distal but not proximal (square) part of the autopod, where light methyl green staining indicates a lack of mesenchyme condensation. (B) In control limbs, TUNEL-positive cells are present in the anterior and posterior necrotic zones (ANZ, PNZ), and opaque patch (OP). No apoptosis is observed in the autopod and zeugopod of RCAS-*Noggin*-infected limbs. Reprinted from Pizette & Niswander (2000).

results in total loss of alcian blue-stained cartilage (Fig. 1A). However, unlike in the autopod, we found evidence for prechondrogenic condensations in the zeugopod of infected limbs as a PNA signal was detected at stage 27 (Fig. 1D), and histology indicated mesenchyme condensation (concentrated methyl green staining) but no chondrocytes (Fig. 1D, 2A). This implies that in the zeugopod, Noggin blocks chondrocyte differentiation.

To explore whether condensed chondroprogenitors that would give rise to zeugopodial chondrocytes die, we performed wholemount nile blue staining. At stage 24, when condensations of the zeugopodial elements are normally laid down,

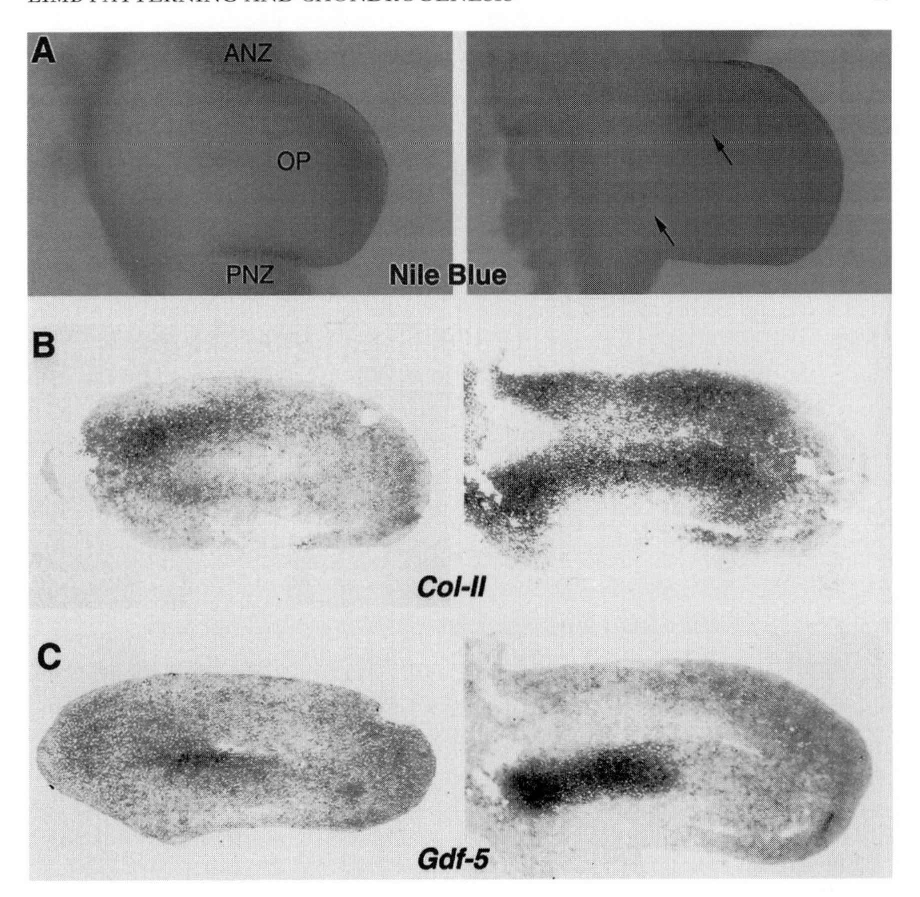

FIG. 3. Analysis of cell death and prechondrogenic condensation markers. Control uninfected (*left*) and infected (*right*) limbs, anterior at top and distal to right, dorsal views in (A). *Noggin* virus was injected at stage 14 and the embryos fixed at stage 24. Cell death was detected by wholemount staining with nile blue (A). In control forelimbs, the ANZ, PNZ and OP are visible, whereas they are not in infected forelimbs, and instead there is widespread ectopic cell death (arrows). (B,C) alternate sections of hindlimbs. Controls display *Gdf5* expression in between the zeugopod condensations which are marked by type II collagen (*ColII*). Infected hindlimbs show up-regulated *colII* expression distributed throughout the mesenchyme, abnormally overlapping with *Gdf5* expression which is also up-regulated and expanded (B,C and not shown). Reprinted from Pizette & Niswander (2000).

45% of the RCAS-*Noggin* infected limbs ($n = 18$) exhibited ectopic cell death in the zeugopod (Fig. 3A). This phenomenon was transient, observed at stage 25 in only 33% of the infected limbs ($n = 6/18$, not shown) and not at all at stage 27 (Pizette & Niswander 1999). Careful analysis of alternate sections at stage 24 of infected limbs, assayed for TUNEL and *ColII* expression which faintly marks precartilaginous condensations (Nah et al 1988), confirmed ectopic cell death which colocalized to a subset of *ColII* positive cells (not shown). However, *ColII* was up-regulated and misexpressed throughout the mesenchyme (Fig. 3B); the significance of this deregulation and its relationship to the state of mesenchyme condensation is unclear. It thus remains uncertain if dying cells are condensed chondroprogenitors.

We also addressed whether an aberrant conversion of zeugopodial chondroprogenitors to a joint fate could explain the absence of chondrocytes. Indeed, cells of the precartilaginous condensations normally give rise to both chondrocytes and joint cells, and BMPs may act as repressors of the joint fate (see Brunet et al 1998, Merino et al 1999a). Growth and differentiation factor 5 (*Gdf5*) expression, which marks the presumptive joint cells (Storm & Kingsley 1996), was detected in control limbs in a narrow region between the proximal ends of the *ColII* stained zeugopodial condensations (Fig. 3B,C; alternate sections). In infected limbs, *Gdf5* expression was increased and expanded in the centre of the zeugopod, overlapping that of *ColII*. This ectopic *Gdf5* expression suggests that cells of the precartilaginous condensations may be abnormally directed into a joint fate.

Noggin misexpression inhibits chondrogenesis during chondrocyte differentiation

To study whether Noggin could interfere with chondrogenesis when chondrocyte differentiation had already initiated at the time of misexpression, we injected RCAS-*Noggin* virus in the zeugopod of stage 24 limb buds. At stage 25–26 (a few hours before the onset of viral transcription), control zeugopods exhibited substantial alcian blue staining and contained immature chondrocytes (Fig. 4A and not shown). The injected embryos were allowed to develop until embryonic day 7 to 10 (E7–E10), and wholemount alcian blue-stained infected and contralateral uninfected limbs compared. As early as E7, the zeugopod that had been alcian blue-positive prior to Noggin protein misexpression frequently showed no (50%) or very small regions of staining (Fig. 4B,C; $n=25$). Section analysis of alcian blue staining and histology at E6 and E7 of these infected limbs ($n=9$), revealed that zeugopodial elements were clearly missing or much smaller than controls. Places where cartilage had disappeared were instead filled with loose mesenchyme (Fig. 4E,F). Therefore, Noggin inhibits chondrogenesis even after substantial amounts of cartilage matrix have been deposited and chondrocytes have formed, and the loss of alcian blue staining reflects the loss of chondrocytes, and not just impaired synthesis of cartilage proteoglycans.

Discussion

Bmps and *BmpRs* are expressed in the limb mesenchyme prior to aggregation of prechondrogenic condensations, and then, in or around condensations. BMPs

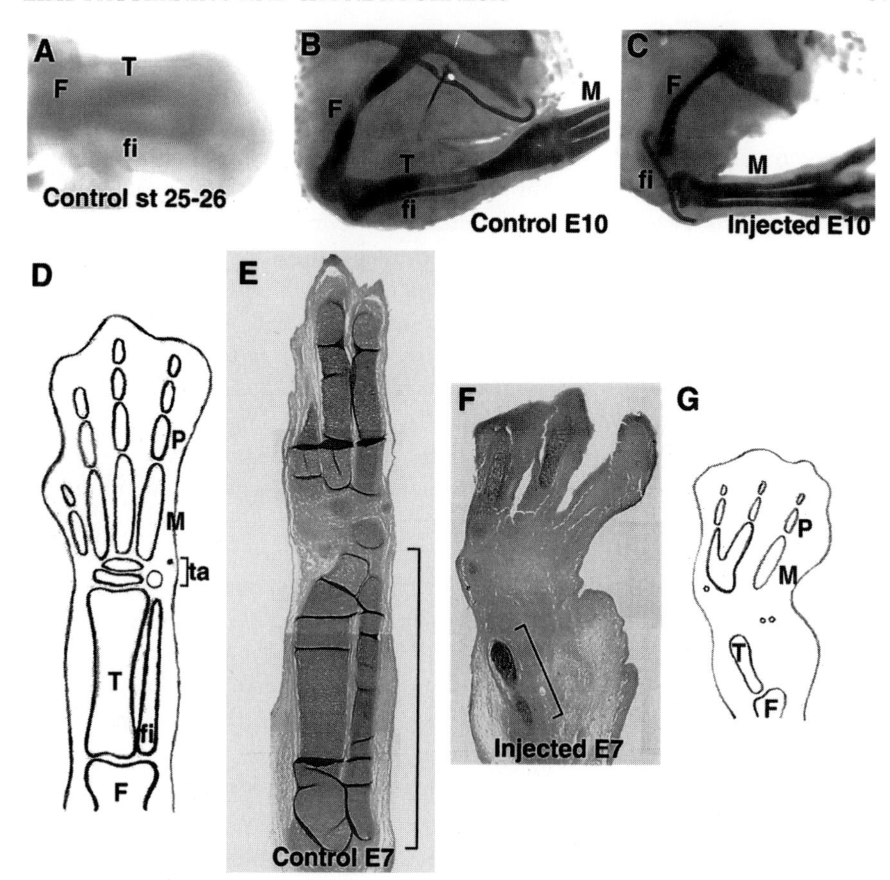

FIG. 4. *Noggin* misexpression results in disappearance of the cartilage after its formation. (A) Control limbs stained with alcian blue to indicate the extent of cartilage formation prior to *Noggin* misexpression. (B,C) E10 wholemount, and (E,F) E7 sections of limbs stained with alcian blue. (B,E) Non-injected control limbs. (C,F) RCAS-*Noggin* virus injection at stage 24 into the hindlimb zeugopod results in loss of the tibia (C) and loss of the fibula and truncation of the tibia (F). The length of the infected zeugopod is approximately one-third that of control (E, F; depicted by brackets). The overall extent of cartilage formation is represented schematically (not to scale) for control (D) and infected (G) limbs. (D–G) Anterior at left, distal at top. (E,F) Composites of 100×bright field photographs. F, femur; fi, fibula; M, metatarsals; P, phalanx; T, tibia; ta, tarsals.

are thus good candidates to direct early stages of cartilage formation. Here, we have examined their role(s) during limb chondrogenesis by misexpressing *Noggin*. This prevents the appearance of prechondrogenic condensations and also blocks their differentiation into chondrocytes. Moreover, once chondrocytes are present, *Noggin* misexpression leads to their disappearance.

Specificity of the Noggin-mediated antagonism

Activin and transforming growth factor (TGF)β2 have been shown to be potential regulators of early steps of chondrogenesis in the chick autopod, based on their expression pattern and ability to stimulate *de novo* chondrogenesis. Their effects are suggested to be mediated through induction of *BmpRIB* expression, only then allowing BMPs to perform later functions (Merino et al 1998, 1999b). GDF5 may also participate in the initial stages of cartilage formation and is expressed in both zeugopod and autopod (Francis-West et al 1999). Noggin binds BMP2, BMP4 and BMP7, preventing them from interacting with their receptors, but does not bind activin or TGFβ (Zimmerman et al 1996). However, Noggin also binds GDF5, blocking its activity (Merino et al 1999a). Hence, BMP signalling aside, GDF5 signalling could have been antagonized by *Noggin* misexpression. Could the loss of GDF5 function account for the described phenotypes?

In the autopod, *Gdf5* is expressed in presumptive joint regions and around the well-developed condensations, but not in or around early condensations (Merino et al 1999a, S. Pizette & L. Niswander, unpublished observations). Accordingly, digit condensations in the *Gdf5* loss-of-function mouse mutant, *Brachypodism*, are formed but are thinner than normal (Grüneberg & Lee 1973, Storm et al 1994). These data suggest a restricted role for GDF5 in recruitment of cells to *preexisting* cartilaginous condensations of the autopod. We thus conclude that the total absence of autopodial prechondrogenic condensations in the Noggin-induced phenotype arises from loss of BMP signalling.

In the zeugopod, *Gdf5* is expressed in the joint-forming area of the precartilaginous condensations (Fig. 3B). GDF5 has been implicated in joint formation as well as in proliferation of the adjacent immature chondrocytes, and chondrocyte differentiation is delayed in the autopod of the *Brachypodism* mutant (Storm & Kingsley 1996, 1999, Merino et al 1999a, Francis-West et al 1999). Therefore, the later effects of *Noggin* misexpression on chondrogenesis may occur through loss of GDF and/or BMP signalling and the specific contribution can not be inferred from other studies. Indeed, GDF5 and BMPs share the same receptors and may form heterodimers (Erlacher et al 1998, Chang & Hemmati-Brivanlou 1999), indicating that alterations of the activity of one of these factors could alter the activity of the others.

Effects of Noggin misexpression on early chondrogenesis

Autopod

In the adult, exogenous BMPs clearly induce *de novo* ectopic chondrogenesis, strongly suggesting *Bmps* are master genes able to trigger the entire series of events needed for normal chondrogenesis. However, the origin and state of differentiation of the cells recruited to form the adult ectopic cartilage have not been determined. In the RCAS-*Noggin*-infected embryonic autopod, condensed chondroprogenitors are absent and there is a defect in proliferation, but no ectopic cell death is detected at the time and location these cells would be condensing or specified. Thus, BMPs are not a survival signal for autopodial chondroprogenitors, but they may stimulate chondroprogenitor proliferation prior to condensation. Whether BMPs regulate chondrogenesis at the level of chondroprogenitor aggregation, or upstream of this, can not be distinguished due to a lack of pre-aggregation chondroprogenitor-specific markers.

Our results do not exclude a role for activin and TGFβ2 in autopodial chondrogenesis. However, in RCAS-*Noggin* infected autopods, the lack of *BmpRIB* expression that has been proposed to be controlled by these molecules, indicates a BMP requirement at this step. In addition, the fact that TGFβs do not promote ectopic cartilage formation in the adult (Roark & Greer 1994) even though BMPs do, also disagrees with TGFβ2 acting prior to BMPs. One discrepancy is that TGFβs, but not BMPs, can induce ectopic cartilage in the interdigits (Gañan et al 1996). However, since BMPs promote apoptosis in this region, it is possible that the apoptotic response supercedes a potential chondrogenic response.

Zeugopod

In the RCAS-*Noggin*-infected zeugopod, prechondrogenic condensations are observed, but not chondrocytes. BMP/GDF signalling is therefore required for chondrocyte differentiation. This signalling may maintain the chondrogenic potential of chondroprogenitors, for instance by repressing joint fate, as *Noggin* misexpression results in an expansion of *Gdf5* expression into areas that appear to abnormally overlap the cartilage forming regions. This could result from an increase in the number of cells legitimately expressing *Gdf5*, if BMP/GDF signalling normally limits proliferation of the joint cell population. Alternatively, because precartilaginous condensations give rise to both joint cells and chondrocytes, it is tempting to speculate that the chondroprogenitors have been converted to a joint fate. The ectopic cell death present in the infected zeugopods, in contrast to the lack of cell death in the autopod, may also indicate that BMP/GDF act as survival signals for chondroprogenitors once they have condensed.

Differences along the proximodistal axis

Our early viral infections direct *Noggin* misexpression before condensation of both zeugopodial and autopodial elements. The different results in the autopod versus the zeugopod could be explained as a difference in the extent of *Noggin* misexpression, or differences in the nature of the proteins involved in chondrogenesis in these two regions. With respect to the first point, a function for BMPs in the aggregation of chondroprogenitors in the zeugopod, or in their differentiation into chondrocytes in the autopod could exist, and may be unmasked by modifying the timing of *Noggin* misexpression. With respect to the second point, it is interesting that activin and TGFβ2 appear to play a role in chondrogenesis of the autopod but not of the zeugopod (Merino et al 1998, 1999b).

Clinical implications

Noggin can inhibit cartilage formation after formation of immature chondrocytes, suggesting that it may be clinically useful for the treatment of fibrodysplasia ossificans progressiva (FOP) (Kaplan et al 1998). This syndrome is characterized by ectopic chondrogenesis, followed by bone formation. This can occur in response to trauma, including medical treatment such as surgery or immunization. This *de novo* chondrogenesis is thought to correlate with aberrant production of BMP by lymphocytes at the site of injury (Shafritz et al 1996). Although Noggin action on other cell types would need to be investigated, treatment of FOP patients with Noggin protein might reverse the devastating effects of this disease.

Acknowledgements

We thank F. Luyten, W. Upholt and J. Andrews for *in situ* probes. This work was supported by Human Frontiers Science Program (S.P.), NIH (L.N.) and by the MSKCC Support Grant. S.P. and L.N. are, respectively, a Research Associate and an Assistant Investigator of the Howard Hughes Medical Institute.

References

Aulthouse AM, Solursh M 1987 The detection of a precartilage, blastema-specific marker. Dev Biol 120:377–384

Brunet LJ, McMahon JA, McMahon AP, Harland RM 1998 Noggin, cartilage morphogenesis, and joint formation in the mammalian skeleton. Science 280:1455–1457

Capdevila J, Johnson RL 1998 Endogenous and ectopic expression of *noggin* suggests a conserved mechanism for regulation of BMP function during limb and somite patterning. Dev Biol 197:205–217

Chang C, Hemmati-Brivanlou A 1999 *Xenopus* GDF6, a new antagonist of noggin and a partner of BMPs. Development 126:3347–3357

Chen WS, Manova K, Weinstein DD et al 1994 Disruption of the HNF-4 gene, expressed in visceral endoderm, leads to cell death in embryonic ectoderm and impaired gastrulation of mouse embryos. Genes Dev 8:2466–2477

Dunn NR, Winnier GE, Hargett LK, Schrick JJ, Fogo AB, Hogan BLM 1997 Haploinsufficient phenotypes in *Bmp4* heterozygous null mice and modification by mutations in *Gli3* and *Alx4*. Dev Biol 188:235–247

Erlacher L, McCartney J, Piek E et al 1998 Cartilage-derived morphogenetic proteins and osteogenic protein-1 differentially regulate osteogenesis. J Bone Miner Res 13:383–392

Fell HB, Canti RG 1934 Experiments on the development *in vitro* of the avian knee-joint. Proc R Soc 116:316–351

Francis-West PH, Abdelfattah A, Chen P et al 1999 Mechanisms of GDF-5 action during skeletal development. Development 126:1305–1315

Gañan Y, Macias D, Duterque-Coquillaud M, Ros MA, Hurle JM 1996 Role of TGFβs and BMPs as signals controlling the position of the digits and the areas of interdigital cell death in the developing chick limb autopod. Development 122:2349–2357

Grüneberg H, Lee AJ 1973 The anatomy and development of brachypodism in the mouse. J Embryol Exp Morphol 30:119–141

Hall BK, Miyake T 1995 Divide, accumulate, differentiate: cell condensation in skeletal development revisited. Int J Dev Biol 39:881–893

Hamburger V, Hamilton HL 1992 A series of normal stages in the development of the chick embryo. Dev Dyn 195:231–272

Hogan BLM 1996 Bone morphogenetic proteins: multifunctional regulators of vertebrate development. Genes Dev 10:1580–1594

Johnson RL, Tabin CJ 1997 Molecular models for vertebrate limb development. Cell 90:979–990

Kaplan FS, Delatycki M, Gannon FH, Rogers JG, Smith R, Shore EM 1998 Fybrodysplasia ossificans progressiva. In: Emery AEH (ed) Neuromuscular disorders: clinical and molecular genetics. Wiley, Chichester, p 289–321

Katagiri T, Boorla S, Frendo JL, Hogan BL, Karsenty G 1998 Skeletal abnormalities in doubly heterozygous *Bmp4* and *Bmp7* mice. Dev Genet 22:340–348

Kent J, Wheatley SC, Andrews JE, Sinclair AH, Koopman P 1996 A male-specific role for SOX9 in vertebrate sex determination. Development 122:2813–2822

Macias D, Gañan Y, Rodriguez-Leon J, Merino R, Hurle JM 1999 Regulation by members of the transforming growth factor beta superfamily of the digital and interdigital fates of the autopodial limb mesoderm. Cell Tissue Res 296:95–102

Merino R, Gañan Y, Macias D, Economides AN, Sampath KT, Hurle JM 1998 Morphogenesis of digits in the avian limb is controlled by FGFs, TGFβs, and noggin through BMP signaling. Dev Biol 200:35–45

Merino R, Macias D, Gañan Y et al 1999a Expression and function of *Gdf-5* during digit skeletogenesis in the embryonic chick leg bud. Dev Biol 206:33–45

Merino R, Macias D, Gañan Y et al 1999b Control of digit formation by activin signalling. Development 126:2161–2170

Nah H-D, Rodgers BJ, Kulyk WM, Kream BE, Kosher RA, Upholt WB 1988 *In situ* hybridization analysis of the expression of the type II collagen gene in the developing chicken limb bud. Coll Relat Res 8:277–294

Ng L-J, Wheatley S, Muscat GEO et al 1997 SOX9 binds DNA, activates transcription, and coexpresses with type II collagen during chondrogenesis in the mouse. Dev Biol 183:108–121

Pizette S, Niswander L 1999 BMPs negatively regulate structure and function of the limb apical ectodermal ridge. Development 126:883–894

Pizette S, Niswander L 2000 BMPs are required at two steps of limb chondrogenesis: formation of prechondrogenic condensations and their differentiation into chondrocytes. Dev Biol 219:237–249

Roark EF, Greer K 1994 Transforming growth factor-β and bone morphogenetic protein-2 act by distinct mechanisms to promote chick limb cartilage differentiation *in vitro*. Dev Dyn 200:103–116

Shafritz AB, Shore EM, Gannon FH et al 1996 Overexpression of an osteogenic morphogen in fibrodysplasia ossificans progressiva. N Engl J Med 335:555–561

Storm EE, Kingsley DM 1996 Joint patterning defects caused by single and double mutations in members of the bone morphogenetic protein (BMP) family. Development 122:3969–3979

Storm EE, Kingsley DM 1999 GDF5 coordinates bone and joint formation during digit development. Dev Biol 209:11–27

Storm EE, Huynh TV, Copeland NG, Jenkins NA, Kingsley DM, Lee SJ 1994 Limb alterations in *brachypodism* mice due to mutations in a new member of the TGFβ-superfamily. Nature 368:639–643

Stringa E, Tuan RS 1996 Chondrogenic cell subpopulation of chick embryonic calvarium: isolation by peanut agglutinin affinity chromatography and *in vitro* characterization. Anat Embryol (Berl) 194:427–437

Zhang H, Bradley A 1996 Mice deficient for BMP2 are nonviable and have defects in amnion/chorion and cardiac development. Development 122:2977–2986

Zimmerman LB, De Jesús-Escobar JM, Harland RM 1996 The Spemann organizer signal noggin binds and inactivates bone morphogenetic protein 4. Cell 86:599–606

Zou H, Wieser R, Massagué J, Niswander L 1997 Distinct roles of type I bone morphogenetic protein receptors in the formation and differentiation of cartilage. Genes Dev 11:2191–2203

DISCUSSION

Hall: How is joint specification related to condensation? Do you think that we are getting a little closer to answering this?

Pizette: Not really. Some results that I did not present here suggest that the prechondrogenic condensations are necessary for the joints to form. This is based on the fact that in the autopod of embryos infected with the Noggin virus there are no condensations, no expression of the joint marker Gdf5, nor any morphological signs of joint formation or initiation. In addition, the expression of *Gdf5* in a normal context is somewhat puzzling. Part of its expression pattern is clearly in transverse stripes in the prechondrogenic condensations, delineating the future joint regions. However in the zeugopod, *Gdf5* expression is a little different. Although it is also expressed in a transverse stripe between the stylopodial and the zeugopodial condensations, there is this additional longitudinal expression between the radius and ulna and tibia and fibula condensations, which has not been described before. What was described was that *Gdf5* is expressed throughout the whole condensation that will later give rise to radius and ulna and tibia and fibula. It is thus unclear to me what the relationship between the condensation and the joint forming region as marked by *Gdf5* expression is in the zeugopod.

Hall: The radius and ulna begin as a single condensation, which then has to separate. It looked as if what you had with your *Gdf5* expression was that single condensation in the middle that hadn't yet separated.

Pizette: That is possible. There is some interesting older work on joint formation in the knee of the chick (Fell & Canti 1934). In this paper they studied the region where the knee joint was coming from, looking at different stages. Early on, the region that will give rise to the knee joint maps the whole condensation, and then it reduces in size and approximately maps to what is called the opaque patch. There is a striking correlation between the expression of *Gdf5* at these stages, the joint forming region and the opaque patch. Somehow, I feel that there may be some apoptosis going on. This might be the mechanism by which the initial condensation is split. To me it is not clear whether or not there is real branching to form the radius and ulna, or if in the beginning there is an initial mass and the expression of *Gdf5* is restricted and maps the region of cell death, which would then help to split this condensation.

Blair: Is *Gdf5* expression maintained later on after the beginning of epiphyseal bone formation?

Pizette: It is still expressed in the cavitating joint.

Kingsley: Most of the *in situ* hybridizations showed the expression dropping down at later stages. We have been doing some experiments using an approach that attempts to bridge early patterning events to these more local events that are defining skeletal elements by doing promoter analysis. If we can figure out why *Gdf5* is turning-on in the stripes we might be able to link the stripy joint formation to earlier patterning events. One of the things we found in making various constructs that have *Gdf5* regulatory elements hooked up to the *lacZ* reporter gene, is that we can detect expression along the surface of joints at much later stages than we were ever able to by *in situ* hybridization, including articular surfaces of joints way after birth and into adulthood.

Mundlos: Did you say that type II collagen is up-regulated in the overexpression experiments?

Pizette: Yes.

Mundlos: Then I have a problem: my understanding was that BMP is upstream of collagen type II and can initiate chondrogenesis. Inhibiting BMPs should down-regulate collagen type II expression. This doesn't make sense.

Pizette: It doesn't make sense at all, especially because I looked at the expression of *Sox9*, which has been shown to regulate collagen II expression, and it is down-regulated. This is pretty much in agreement with BMPs being upstream of *Sox9* and collagen type II. If you compare expression of the BMP receptor 1B and collagen II, BMPR1B is slightly ahead in its expression in prechondrogenic condensations. I was expecting to see no or very little collagen II in the zeugopod region, because collagen II is expressed in prechondrogenic condensations. We were very surprised when we saw this huge up-regulation and ectopic expression. The next thing I want to do is to find out whether these regions that have ectopic collagen II expression are condensed or not. My initial feeling is that they are not.

Mundlos: If they express type II collagen, you should be able to see on a single histological section whether this is condensed.

Pizette: I haven't seen condensations on the histological sections, except for the region where *Gdf5* is ectopically expressed. I'm now doing PNA (peanut agglutinin) staining.

Karsenty: Type II collagen is expressed before the condensation. There is no condensation at Day 7.5, when we start to see type II collagen.

Poole: Are we talking about collagen type IIa or IIb? This is an important distinction.

Pizette: IIa.

Newman: A peculiarity of collagen type II relates to its regulation at the post-transcriptional level. It seems to be almost constitutively expressed in early mesenchyme, and then it comes up a little bit more in the condensing regions. Unlike the proteoglycan core protein, for example, which seems to be stringently regulated, with no sign of the mRNA until differentiation is about to happen, type II collagen is expressed much more broadly and much earlier (Kosher et al 1986a,b).

Hall: Including in places such as basal membranes where you normally would not expect to find type II collagen.

Sandrine Pizette, have you had any of your embryos go much longer to see whether those cells that express collagen type II do anything?

Pizette: They don't form chondrocytes.

Hall: But do they do anything else?

Pizette: I have no idea. On E5, in the limbs there is a huge domain of *Gdf5* expression, but there is not much muscle because of a defect in muscle formation. There is some tendon, but this is all I can say. I let the embryos carry on to E6 or E7, but they usually die by this stage. Have you any ideas about what these cells might do?

Hall: There are a number of possibilities. A late phase of cartilage might appear; another possibility is that collagen II tends to be expressed quite ubiquitously, and it may later be turned off: those cells might then become soft tissue, or die and disappear.

Pizette: I haven't seen cell death at E4 or E5. 24 h after this up-regulation we looked to see whether the levels of collagen type II were still high. It is still expressed in the zeugopod in a strange pattern that has nothing to do with where we would expect to see condensation or even chondrocytes. Compared with wild-type embryos, the expression level is lower than what one would expect in chondrocytes at the same stage.

Kronenberg: You mentioned that motivation behind the Noggin experiments was to sort out how much of the action of BMP is on the differentiation of cells, how much is on proliferation and how much is on cell death. You showed us data that suggested that there is a bit of cell death here and there that might turn out be relevant. You also showed us a bromodeoxyuridine experiment that indicated that proliferation may have a role. But it looked to me as if most of what you were showing were things that couldn't easily be explained simply by changing the cell numbers. What was your conclusion?

Pizette: Because there is no cell death it looks like BMPs are not required as a survival signal. However, because there is this defect in proliferation, BMPs may well have a role in promoting proliferation of cells. As I have shown you, once the condensation is established, the mitotic index decreases. If one wants to argue that BMPs are involved in stimulating proliferation, they would have to postulate that this is prior to condensation. It looks as if BMPs and GDF have a role in promoting chondrocyte differentiation. In the zeugopod BMPs may have a different role. They may be needed to provide a survival signal for cells once they have condensed. The other possibility is that they may be involved in repressing the joint fate, maintaining a limited number of cells which are going to acquire a joint fate.

Reddi: Have you looked at endogenous expression of *Noggin*? Does it play a role? It seems to me that it is very important to look at this.

Pizette: I forgot to mention the results of the *Noggin* knockout, which was done by Richard Harland (Brunet et al 1998). It results in the overgrowth of the cartilage elements. However, he has looked at the stage of the prechondrogenic condensed mesenchyme, and doesn't see any differences in the size of the condensations. This indicates that there is no effect at that stage. He also looked at proliferation of the growth plates, and didn't see any effect on proliferation either. The conclusion was that the BMPs may act at later stages through appositional growth and the recruitment of cells to the perichondria. In the embryo, *Noggin* is expressed in the prechondrogenic condensed mesenchyme, and also in immature and prehypertrophic chondrocytes. The striking thing is that if you compare the *Noggin* expression pattern with that of BMP receptor 1b in the prechondrogenic condensations of the digits, *Noggin* is expressed just in the centre of this domain of BMP receptor 1b expression, where it could potentially block all cells in the condensation from receiving BMP signalling. The knockout experiments suggested that *Noggin* was limiting the effect of BMPs. Here, if BMPs and GDFs are involved in forming the condensations, maybe *Noggin* would need to be expressed not in the middle of the condensation but at the edges to prevent more cells from being recruited. So the expression pattern we see is really puzzling.

Newman: This is one place where the dynamical models may be helpful, because in those models the source of the activator and the source of the inhibitor are in exactly the same place. Then it's just the relative levels of the proteins and their relative diffusion coefficients that determine what the spacing will be. This is the kind of pattern that you would expect. Now you just have to see the protein profiles to verify it.

Mundlos: Another possible explanation is that if you have expression of the inhibitor in the middle, it indicates that this expression is in the more mature cells. In contrast, those cells on the outside of the condensation are still recruiting and proliferating new chrondrocytes. The cells at the centre may have reached a certain level of differentiation and now this whole system is being switched off.

Pizette: I agree that this would make sense and you are right in that the cells in the centre of the condensations are the first cells that are going to start to form the chondrocytes. However this explanation conflicts with my results, since these results show that BMPs and GDFs are needed by the cells of the pre-chondrogenic condensed mesenchyme to make chondrocytes, those same cells which are in the centre of the condensation.

Mundlos: You misexpressed it where it is not usually expressed, meaning that you interfered at a position where it's not usually there, and then it does make sense.

Hall: Is the expression pattern of BMP the same as the pattern of expression of the receptor?

Pizette: Not at all: the BMPs are not expressed in the condensations, but instead in between them in the interdigit regions.

Burger: We should bear in mind that the skeleton is not just cartilage: it is also all the structures that link these cartilage elements. The continuity of the prechondrogenic condensation could also mean that the whole structure is laid down, possibly also including the muscles, tendons and ligaments, and it is only later that the foci of cartilage develop.

Hall: There are multiple roles of BMP in different tissues.

Pizette: There is a big effect in muscle and tendon formation.

Chen: Is the phenotype from your *Noggin* overexpression chicks the same as the dominant-negative BMP receptor chicks?

Pizette: Some features are comparable, but in the dominant-negative BMP receptor infected chicks it is mainly the distal tips of the digits that are missing. The phenotypes are pretty close; there is just a difference in efficiency.

Morriss-Kay: I would like to broaden this discussion to include another family of players—the retinoid receptors. I was struck by the similarity between the gene expression patterns illustrated by Sandrine and those of retinoic acid receptor (RAR)γ, which is expressed in pre-chondrogenic condensations and the cartilaginous bone models, and RARβ, which is expressed in the interdigital regions. The complementary expression patterns shown by RARγ and RARβ is similar to the expression patterns that we have just seen for GDF5 and BMPs in the interdigital regions, and Noggin and the BMP receptor 1B in the condensations. Is anything is known about the ways in which the RARs interact with the molecular elements that we've just heard about?

Underhill: It looks like RARγ will support the formation of cartilage, whereas RARβ seems to inhibit this process. Some of these signalling pathways appear to be functioning downstream of BMP signalling in regulating chondroblast differentiation (Weston et al 2000).

Pizette: There is a paper showing that retinoic acid signalling is upstream of BMP in the interdigits of the chick embryo and that it is required for apoptosis (Rodriguez-Leon et al 1999).

Hall: One of the studies Juan Hurle et al (1989) did was to show that in the interdigital region of the chick limb, if you take off this piece of epithelium, you can induce a little ectopic area of cartilage between the digits. Epithelium is inhibiting chondrogenesis. I think they then did an experiment where the epithelium was left intact, a bead of BMP was implanted and they obtained ectopic chondrogenesis when the epithelium was still present (Ganan et al 1998, Merino et al 1998). This makes me wonder about the role of the epithelium in the BMP story, as a sort of global inhibitor of chondrogenesis around the developing limb bud.

Pizette: That is one of the main problems in the chick. The model that is used involves the interdigit region, and people tend to put beads soaked in different factors in the interdigital region to see whether this promotes ectopic cartilage formation. I don't know whether this is such a good model. As you said, simple removal of the epithelium will induce ectopic cartilage formation. If you just put in BMP beads without removing anything, this will not induce ectopic cartilage formation but more cell death. The system seems to be already a little biased in the sense that there are things in the interdigit region that seem to be able to respond to BMPs only as transducing a cell death signal.

My other point has to do with the still unpublished work by Richard Maas on the *Msx1*/*Msx2* double knockout. BMPs are expressed in the ectoderm and in the mesenchyme. Expression in the ectoderm may regulate expression in the mesenchyme. In this double knockout it seems that *Msx* expression is lost in both compartments, as is BMP expression. Cell death is also lost. If you add back some BMP protein, you can still get apoptosis. The conclusion was that *Msx* expression is not needed for cell death, but just to maintain BMP expression. This goes back to the ectodermal control in the interdigital region where there is a signal in the ectoderm that tells the mesenchyme to die. When you put various factors into the interdigit region you might be competing in some way with this signalling.

Morriss-Kay: Juan Hurle would give credit to Michael Solursh for the interpretation of his first experiments, which is based on the fact that the surface ectoderm is secreting hyaluronan, which inhibits chondrogenesis (Solhursh 1984).

Hall: He also made the important point that these cells in the interdigital region are not naïve prechondrogenic mesenchyme; they are a separate population of cells. Whenever we are looking at these systems, we need to know something about cells that we're actually looking at, as they may respond quite differently to the same factors.

Tickle: What is the evidence that joint cells actually come from the prechondrogenic condensations? What is their origin?

Pizette: There is an old paper by Honor Fell on specification of the avian knee joint *in vitro*. The conclusion was that joint precursors are at least at some point present in the prechondrogenic condensations.

Kingsley: That is one of the classical studies. They essentially tried to culture the region where the joint was thought to form. They then asked whether it would still appear if they took the cells out, or transplanted the presumptive joint region into a new area to see whether a joint formed. The low resolution comes from the fact that the number of markers for identifying a presumptive joint region was small. All the transfers involved surgically removed cells, and it's not at the level of cell by cell resolution. There is always the question as to whether the source of the cells within the condensation is itself heterogeneous. There is no way to resolve that in this sort of experiment. My own feeling is that the best hope for that comes from a system where there are fewer cells to keep track of. For example, in the zebrafish joint there are two orders of magnitude fewer cells involved than in the mouse and chick joints. Chuck Kimmel's lab is interested in doing a much more sophisticated cell-by-cell lineage tracing of where the cells are coming from.

Tickle: There is a similar problem with the muscle mass splitting: how the initial, muscle mass is split into discrete regions that then go and form individual muscles. It is not clear here whether there is cell death or cells coming in from the outside.

References

Brunet LJ, McMahon JA, McMahon AP, Harland RM 1998 Noggin, cartilage morphogenesis, and joint formation in the mammalian skeleton. Science 280:1455–1457

Fell HB, Canti RG 1934 Experiments on the development in vitro of the avian knee joint. Proc R Soc 116:316–351

Ganan Y, Macias D, Basco RD, Merino R, Hurle JM 1998 Morphological diversity of the avian foot is related with the pattern of *msx* gene expression in the developing autopod. Dev Biol 196:33–41

Hurle JM, Ganan Y, Macias D 1989 Experimental analysis of the *in vivo* chondrogenic potential of the interdigital mesenchyme of the chick leg bud subjected to local ectodermal removal. Dev Biol 132:368–374

Kosher RA, Gay SW, Kamanitz JR et al 1986a Cartilage proteoglycan core protein gene expression during limb cartilage differentiation. Dev Biol 118:112–117

Kosher RA, Kulyk WM, Gay SW 1986b Collagen gene expression during limb cartilage differentiation. J Cell Biol 102:1151–1156

Merino R, Ganan Y, Macias D, Economides AN, Sampath KT, Hurle JM 1998 Morphogenesis of digits in he avian limb is controlled by FGFs, TGFβ2 and Noggin through BMP signaling. Dev Biol 200:35–45

Rodriguez-Leon J, Merino R, Macias D, Ganan Y, Santesteban E, Hurle JM 1999 Retinoic acid regulates programmed cell death through BMP signalling. Nat Cell Biol 1:125–126

Solursh M 1984 Cell and matrix interactions during limb chondrogenesis *in vitro*. In: Trelstad RL (ed) The role of extracellular matrix in development. Alan R Liss, New York
Weston AD, Rosen V, Chandraratnas RAS, Underhill TM 2000 Regulation of skeletal progenitor differentiation by the BMP and retinoid signaling pathways. J Cell Biol 148:679–690

General discussion I

Meikle: All the genes that are involved in patterning the vertebrate limb are also involved in patterning the teeth. Paul Sharpe and his colleagues (Tucker et al 1998) have done some interesting experiments in which they used Noggin-soaked beads to change the patterning of incisor teeth, to turn them into molars. I guess it is a question of location and what stage of development we are looking at.

Reddi: I want to raise a general issue about diffusion. On the one hand we talk about the important role of diffusion in creating peaks and valleys, and in morphogenesis. On the other hand, the powerful tool that Cheryll Tickle has developed using implanted beads soaked in various factors involves studying what in effect is a single source at very high concentrations. Many developmental phenomena involving bone morphogenetic proteins (BMPs) are critically dependent on thresholds. So how does one interpret these bead implantation experiments that involve a single source of a big bolus of a morphogen?

Tickle: Originally, Gregor Eichele and I used beads to release retinoic acid into the limb bud (Tickle et al 1985). At that time, we did many experiments to characterize the concentrations of retinoids that were released into the tissue and so on. In more recent experiments with beads soaked in proteins, we have very little idea about the release characteristics.

Reddi: Yet we strongly believe in the absence of an apical ectodermal ridge (AER), if we implant a bead containing fibroblast growth factor (FGF)4 or FGF8 it substitutes for the AER.

Kingsley: There are many cases where the bead experiments will give an effect with a molecule that is present endogenously at concentrations that might be too low to do this. But when the molecules are put in at a high enough level they will stimulate receptors that they normally wouldn't. The experiments are very powerful for identifying families of molecules that play an important role. They are less powerful, because of the uncertainties about concentrations, for identifying which particular member of a family is involved.

Ornitz: People have made efforts to label growth factors on beads with fluorescent tags and then have looked at their diffusion properties.

Morriss-Kay: We have used digoxygenin in labelling for looking at the time-course of diffusion from the bead. We use beads containing FGF2, and have used digoxygenin-labelled FGF2 which stops diffusing out from the bead within 48 h. It takes 24 h to diffuse enough cell diameters to get to the target tissue. People are doing these time-course studies.

Kronenberg: But that doesn't take away from David Kingsley's point, which is that we don't know the meaningful concentrations of any of these molecules.

Morriss-Kay: You're right, but we can actually see the time course of diffusion.

Tickle: I think the point about AER substitution is that we are discussing a generic effect of FGF. FGF2, FGF4 and FGF8 have the same effect: you can rescue limb development with all of these.

Ornitz: They definitely have similar specificity towards mesenchymal FGF receptor splice forms.

M. Cohn: When we did our original limb-induction experiments with FGFs, we worked through quite a few members of the FGF family and found that additional limbs could be induced with FGF1, FGF2 and FGF4. The utility of that experiment was really the identification of FGFs as a family that can induce limb formation, which was later confirmed by the genetic experiments in mouse. From our experiments and the expression profiles of FGFs known in 1994, it was not possible to identify the particular member of the FGF family which is the endogenous limb-inducing factor, but those experiments did identify the FGFs as candidates for the limb-inducing gene. In my opinion, that is the power of the bead technology.

Tickle: We have tried to look at the diffusion of FGF, but we found this to be very difficult. A short pulse of FGF is required for limb induction: within an hour we appear to trigger something which then sets the whole thing in motion.

Ornitz: Are you triggering the formation of the AER?

M. Cohn: FGF induces development of a limb bud at least a day before AER formation. The ectopic limb bud develops its own AER within 48 hours of FGF application, and it can develop autonomously into a complete limb.

Ornitz: You could be activating FGF10 expression in lateral plate mesoderm.

Wilkins: I would like to raise what I think is a general analytical and interpretative problem. It is clear that a lot of the responses in these bone-forming systems reflect complicated quantitative and qualitative balances of many factors. Each lab tends to specialize in studying one or two factors. Particularly in overexpression experiments, one can have artefacts or effects that don't have readily interpretable biological significance. This field needs to move towards systems of analysis that track all variables, comparable to these multi array treatments that follow transcription of all the genes. I'm not sure how that can be achieved, but I think it's something that people in the field need to think about.

Bard: The key point here is, what do you mean by 'all'? Do we know we've got all the factors?

Wilkins: No, but even of the ones so far identified, many of them are not followed in particular experiments, and yet may be doing interesting and significant things. For instance, when we raise Noggin expression and inhibit BMPs we get effects on limb development but we don't know what's happening to the retinoic acid receptors and so on.

Bard: The obvious approach to use here is microarray analysis. Here, you can use the technology to elicit the average expression patterns of a large number of genes in the sampled cells, unfortunately it is very much harder to discover in any quantitative way any non-uniform patterns of spatial expression across a cellular domain.

References

Tickle C, Lee J, Eichele G 1985 A quantitative analysis of the effect of all-*trans*-retinoic acid on the pattern of chick wing development. Dev Biol 109:82–95

Tucker AS, Matthews KL, Sharpe PT 1998 Transformations of tooth type induced by inhibition of BMP signalling. Science 282:11361–138

Developmental mechanisms of vertebrate limb evolution

Martin J. Cohn

Division of Zoology, School of Animal and Microbial Sciences, University of Reading, Whiteknights, Reading RG6 6AJ, UK

Abstract. Over the past few years, our understanding of the evolution of limbs has been improved by important new discoveries in the fossil record. Additionally, rapid progress has been made in identifying the molecular basis of vertebrate limb development. It is now possible to integrate these two areas of research in order to identify the molecular developmental mechanisms underlying the evolution of paired appendages in vertebrates. After the origin of paired appendages, several vertebrate lineages reduced or eliminated fins and limbs and returned to the limbless condition. Examples include eels, caecilians, snakes, slow worms and several marine mammals. Analyses of fossil and extant vertebrates show that evolution of limblessness frequently occurred together with elongation of the trunk and loss of clear morphological boundaries in the vertebral column. This may be suggestive of a common developmental mechanism linking these two processes. We have addressed this question by analysing python embryonic development at tissue, cellular and molecular levels, and we have identified a developmental mechanism which may account for evolution of limb loss in these animals.

2001 The molecular basis of skeletogenesis. Wiley, Chichester (Novartis Foundation Symposium 232) p 47–62

Skeletal morphology has two histories; one evolutionary and one developmental. These histories are intimately linked, but the details of this relationship have not been understood until very recently. This new focus owes largely to a number of factors coming together at approximately the same time, including (a) extremely rapid progress in the area of limb developmental genetics, (b) new palaeontological discoveries catalysing an interest in mechanisms of morphological development, and (c) a resurgence of interest in the relationship between embryonic development and evolution (a field now commonly referred to as 'evo-devo'). This integration of developmental and evolutionary biology has been bolstered by quite significant communication, including collaborative research, between developmental biologists and palaeontologists (e.g. Coates & Cohn 1998, Shubin et al 1997, Smith et al 1994).

One area which has seen considerable progress in recent years is the limb skeleton. Thanks to a rich fossil record, the evolutionary history of vertebrate limbs has become much clearer over the past decade. For example, work on Devonian amphibian fossils, such as *Acanthostega gunnari*, has changed our view of the fin-to-limb transition by raising strong evidence that polydactyly, rather than pentadactyly, is the primitive condition for tetrapod limbs (Coates & Clack 1990). Developmental genetics of limb development has contributed a mechanistic perspective to the study of limb evolution. Among these genetic studies, none has generated more discussion about developmental pathways of limb evolution than the work on *Hox* genes. The *Hox* complex is an evolutionarily ancient family of transcription factors which play fundamental roles in patterning the bodies of animal embryos (for review see Akam 1998). These homeobox-containing genes are arranged in clusters along the chromosome, and are best known for their roles as 'selector genes' which confer identity to cells (Rijli et al 1998). Quantitative and qualitative differences in *Hox* gene expression can, for example, determine whether a group of cells will form an antenna or a leg in flies, or a thoracic or cervical vertebra in vertebrates. *Hox* genes play very important roles in limb development. During early stages of development they are involved in specifying the position at which limbs will bud along the trunk (Cohn et al 1997, Rancourt et al 1995) and, at later stages of limb development, they govern cell identity, proliferation, adhesion and growth (Davis et al 1995, Duboule 1995, Yokouchi et al 1995). *Hox* gene expression is highly dynamic, and spatiotemporal changes in *Hox* expression domains correspond to the progression of limb development (Davis et al 1995). The discovery that late phases of *Hox* gene expression in the limbs control formation of digits (Dollé et al 1993) set the scene for some very exciting work in comparative developmental biology that led to the first suggestion of a specific molecular mechanism for a major transition in vertebrate limb evolution. Duboule and colleagues set-out to test the hypothesis that the late phase of *Hox* gene expression in the distal aspect of the limb bud was involved in specification of digits during the fin-to-limb transition. Their comparative analysis of *Hox* gene expression in mice and zebrafish revealed an intriguing difference in the dynamics of *Hox* expression in fins and limbs; zebrafish fins, which lack digits, also lack the late phase of distal *Hox* gene expression (Sordino et al 1995). This work gave rise to a model which suggested that changes in *cis*-regulation of *Hox* gene expression in distal fin/limb buds may have been a key step in the evolution of digits (reviewed in Zákány & Duboule 1999). This interesting study is an example of how one can relate the evolutionary and developmental histories of the skeleton to one another through an experimental approach.

Vertebrate limbs have diversified into an impressive range of anatomical patterns. In many cases, such as birds, salamanders, horses and sloths, these changes have involved reductions in the number of digits. More extreme examples of limb reduction include animals which have dispensed with limbs altogether, such as snakes. While snakes are probably the most widely known case of secondary limb loss, limblessness has evolved on many independent occasions in different vertebrate classes. Although the extent of limb reduction and the order in which limbs have been lost varies in different vertebrate lineages, evolution of limblessness frequently occurs together with elongation of the trunk and loss of clear axial regionalization of the vertebral column. This could suggest a common developmental basis of limb loss and homogenization of the axial skeleton. These anatomical changes, like those discussed above, have their bases in embryonic development, when the body plan is laid out. We took an experimental approach to try to identify the molecular mechanisms which may have generated the snake body plan.

For this study we focused on pythons, a primitive group of snakes which lack all traces of forelimbs, but have retained very small rudiments of the hindlimbs. When we began to analyse the skeletal anatomy of different python species, it became clear that the subdivisions of the vertebral column common to almost all tetrapods—cervical, thoracic, lumbar, sacral and caudal—were impossible to identify (Fig. 1A). Posterior to the atlas, all of the vertebrae looked similar down to the level of the hindlimb rudiment. The vertebral bodies showed very little regionalization, and each possessed a pair of true ribs, giving the appearance of a long series of thoracic vertebrae. Experimental data from a variety of organisms has demonstrated that vertebral identity is controlled by differential expression of *Hox* genes in paraxial mesoderm (from which vertebrae develop) along the primary body axis of the embryo. In order to determine whether development of vertebrae with thoracic morphology along most of the axial skeleton was associated with changes in *Hox* gene expression, we examined the distribution of three HOX proteins; HOXC6 and HOXC8 which, in other tetrapods, are restricted to thoracic somites, and HOXB5 which is expressed in all somites. In python embryos, we found that both thoracic markers, HOXC6 an HOXC8, were expressed over a broad domain extending from the first somite posteriorly to the level of the hindlimb buds, where a posterior boundary of expression was detected. This is in stark contrast to the general tetrapod condition, in which these genes are expressed in a domain restricted to the thorax (Fig. 1B). These gene expression patterns co-localize with the region of the python trunk that will form rib-bearing, or thoracic, vertebrae. The posterior boundary of expression lies at the posterior limit of the thoracic series, where there is an abrupt transition in vertebral identity from vertebrae with true ribs to vertebrae with short, forked fused ribs known as lymphapophyses (Fig. 1C). Thus, extension of thoracic identity along the python axial skeleton is associated with extension of *Hoxc6* and *Hoxc8* expression domains. Expansion of these domains in transgenic mice result in

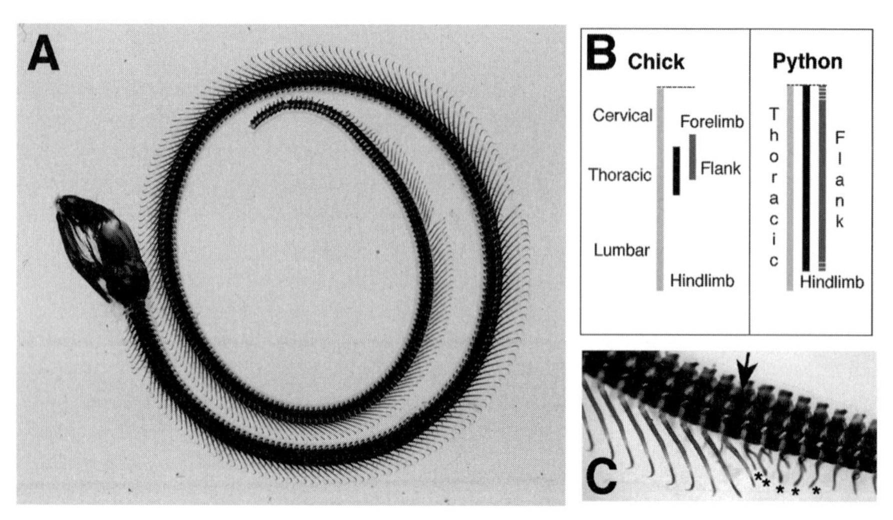

FIG. 1. Morphological and molecular regionalization of the python axial skeleton. Anterior to left in (A) and (C). (A, C) Alcian blue- and alizarin red-stained skeletal preparation of python embryo at 24 days of incubation. (A) Lateral view of complete skeleton. Note homogeneity of vertebrae, almost all of which bear ribs and have a thoracic appearance. (B) Schematic diagram comparing expression domains of HOXB5 (light grey), HOXC8 (black) and HOXC6 (dark grey) in chick and python embryos. Broken line at anterior and posterior extremes of red line indicates lack of certainty about precise limits of HOXC6 expression. Note that expansion of HOXC8 and HOXC6 domains in python correlates with expansion of thoracic identity in axial skeleton and flank identity in lateral plate mesoderm. (C) High magnification view of cloacal region of embryo shown in (A). Arrow indicates position of the hindlimb (removed) relative to axial skeleton. Hindlimb position corresponds to a transitional vertebra with intermediate morphology (arrow), separating vertebrae with large, movable ribs (left) from vertebrae with lymphapophyses in cloacal region (right, with asterisks).

expansion of ribs along the mouse axial skeleton (Jegalian & De Robertis 1992, Pollock et al 1995), and therefore, there may be a causal relationship in snakes. *Hoxb5* is also expressed over a broad anteroposterior domain, but this is true of all tetrapods examined. An interesting difference, however, is seen in lateral plate mesoderm, the tissue which will give rise to the limbs and body wall. In other tetrapods, *Hoxb5* is expressed in the proximal, anterior part of the forelimb, where it plays a role in determining forelimb position (Rancourt et al 1995). In python embryos, we were unable to detect this regionally specific pattern of expression; instead we saw widespread expression of *Hoxb5* throughout the lateral plate mesoderm. Loss of regionally specific expression in lateral plate mesoderm is associated with loss of forelimb specification, and in the context of the altered forelimb position seen in *Hoxb5* mutants, may underlie the failure of forelimb specification in python embryos.

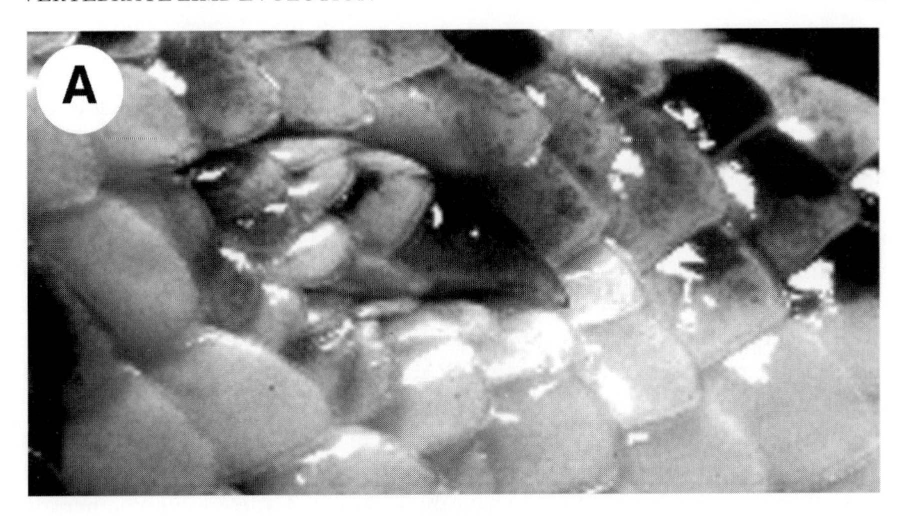

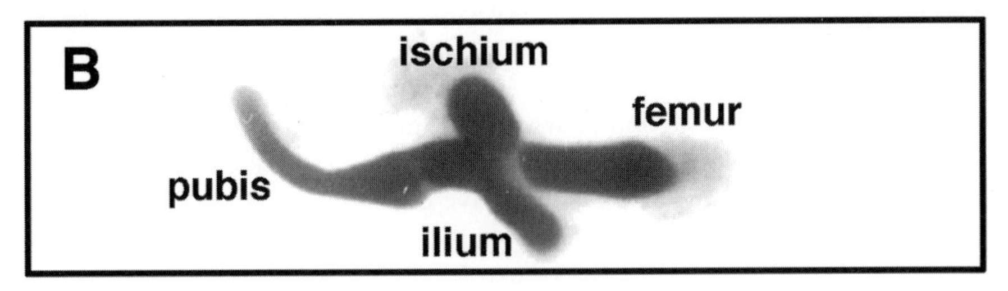

FIG. 2. Rudimentary hindlimbs of pythons. (A) Lateral view of left hindlimb protruding from the body wall of adult Python regius. (B) Skeletal preparation of hindlimb and associated pelvis dissected from Burmese python embryo at 14 days incubation. Femur and all three elements of pelvic girdle are present (pubis, ilium and ischium).

Pythons, unlike more derived snakes, have partially developed hindlimbs, known as spurs (Fig. 2A). The truncated limb skeleton consists of all three elements of the pelvic girdle, and a severely stunted femur (Fig. 2B). Early development of the hindlimb during embryogenesis appears to be normal, as a pair of well-formed limb buds emerge from lateral plate mesoderm on either side of the cloaca. Shortly after initiation of limb budding, bud outgrowth arrests. Outgrowth of the tetrapod limb skeleton is controlled by the apical ectodermal ridge (AER), a specialized epithelial ridge which runs along the distal edge of the limb bud (Cohn & Bright 1999). Surgical removal of this ridge from early limb buds of chick embryos results in the arrest of limb outgrowth and loss of distal skeletal structures. To determine whether hindlimb development arrests in python embryos due to a failure of apical ridge function, we analysed early limb buds for morphological and molecular evidence of an AER. Scanning electron microscopy showed a relatively smooth ectodermal jacket covering the limb bud, in contrast to the chick limb bud in which the AER is clearly visible. Immunohistochemical analysis also failed to reveal expression of genes associated with AER function in the python limb bud ectoderm. These results suggested that hindlimb bud outgrowth arrests in python embryos because they lack an AER.

The AER maintains another signalling region in the limb bud, known as the zone of polarizing activity (ZPA) or polarizing region. The polarizing region is a specialized group of mesenchymal cells at the posterior margin of the limb bud which controls patterning of the limbs along the anterior to posterior (thumb to small finger) axis. These cells express a gene called Sonic hedgehog (*Shh*), which mediates the polarizing activity of the ZPA. Maintenance of *Shh* expression, and signalling activity of ZPA cells, depends on fibroblast growth factors secreted by the apical ectodermal ridge. We were interested in determining whether pythons had retained any evidence of a polarizing region from their limbed ancestry, and what effect the lack of an AER might have on these cells. To determine whether any cells in python hindlimb buds have molecular characteristics of ZPA cells, we examined the distribution of SHH protein in limb bud-stage python embryos. We were unable to detect any SHH in the hindlimb buds, although strong expression was seen in the notochord and in the floor plate of the neural tube. Thus, in the absence of an AER, the underlying mesenchymal cells fail to express *Shh*. We next tested whether these cells have the ability to polarize a limb bud. Mesenchymal cells were transplanted from the posterior and anterior margins of python hindlimb buds to the anterior margin of chick wing buds. Surprisingly, cells from both positions induced mild digit duplications in chick wings, indicating that they have retained polarizing potential even though they do not express *Shh*. When we assayed the transplanted posterior cells for *Shh* expression, we found that SHH could be detected in the python cells after they were grafted under a functional chick apical ridge. This indicated that python hindlimb cells have retained the potential to express *Shh* and polarize a limb in the presence of AER signals. Moreover, polarizing potential extends into the anterior part of the limb bud. This differs from the condition found in other vertebrate embryos, in which polarizing activity is always confined to the posterior margin of the limb bud. The anteroposterior extent of polarizing potential in mouse lateral plate mesoderm is related to the extent of *Hoxb8* expression, and, as such, expansion of this potential in python lateral plate mesoderm may be related to expansion of *Hox* gene expression domains.

We next turned our attention to the apical ridge to investigate the basis of failed ridge formation. In the chicken *limbless* mutant, limb outgrowth fails because apical ridge formation fails. It is also known that dorsoventral polarity is lost in limb ectoderm of these mutants, which is significant because dorsoventrally polarized expression of genes such as *Radical Fringe* and *Wnt3a* is needed for normal ridge formation (Kengaku et al 1998, Laufer et al 1997, Zeller & Duboule 1997). When we examined dorsoventral gene expression in python hindlimb buds, we found that both *Engrailed* (an ventral ectodermal marker) and *Wnt7a* (a dorsal mesenchymal marker) were expressed in their normal positions. These results demonstrate that the mechanism underlying limb truncation in pythons differs from that which affects *limbless* mutants.

These experiments allowed us to eliminate a number of possibilities for the basis of hindlimb truncation, but the nature of the mechanism underlying failure of ridge formation remained unclear. During normal tetrapod limb development, the ridge is induced in apical ectoderm by a signal from underlying mesenchymal cells. The ectoderm responds to that signal by activating expression of genes such as *Fgf4* and organizing itself into a pseudo-stratified, columnar epithelium. Failure of AER formation could be due to a deficiency in the inductive signal or in the response to such a signal. To determine whether python limb mesenchyme is competent to produce a ridge-inducing signal, we transplanted python limb bud mesoderm under the non-ridge ectoderm of a chick wing and then monitored expression of *Fgf8* (a marker for apical ridge cells). We found that python cells were able to extend the domain of *Fgf8* expression into ectoderm overlying the graft, indicating that a functional ridge-inducing signal was produced by the python limb mesenchyme. This suggests that the deficient tissue in python hindlimbs could be the ectoderm, although this hypothesis will require further testing by recombining python limb bud ectoderm with chick limb bud mesoderm.

Our findings uncover a remarkable amount of the limb development program intact in pythons, which is surprising given that digits were probably last present in snakes during the Cretaceous. This retention of signalling potential and molecular polarity of python limb buds suggests that limb outgrowth could be rescued if an apical ridge could be restored. Because fibroblast growth factors (FGFs) mediate the signalling activity of the AER, we grafted FGF-loaded carrier beads to the distal margin of python limb buds. Although technical difficulties associated with *in ovo* operations prevented us from maintaining the embryos beyond 24 h after surgery, we did observe a dramatic increase in the proximodistal outgrowth of FGF-treated limb buds within the first day, suggesting that FGF can sustain python limb development beyond the normal stage of arrest. This is somewhat similar to the observations of Raynaud et al (1995), who demonstrated that outgrowth of slow worm hindlimb buds in culture can be stimulated by addition of FGF to the culture media. Whether replacement of a single growth factor will be sufficient to fully restore limb development in pythons is unclear at present, but the ability of FGFs to catalyse complete limb development in the flank (inter-limb) region of avian embryos suggests that autonomous limb development can be initiated by a single molecular switch.

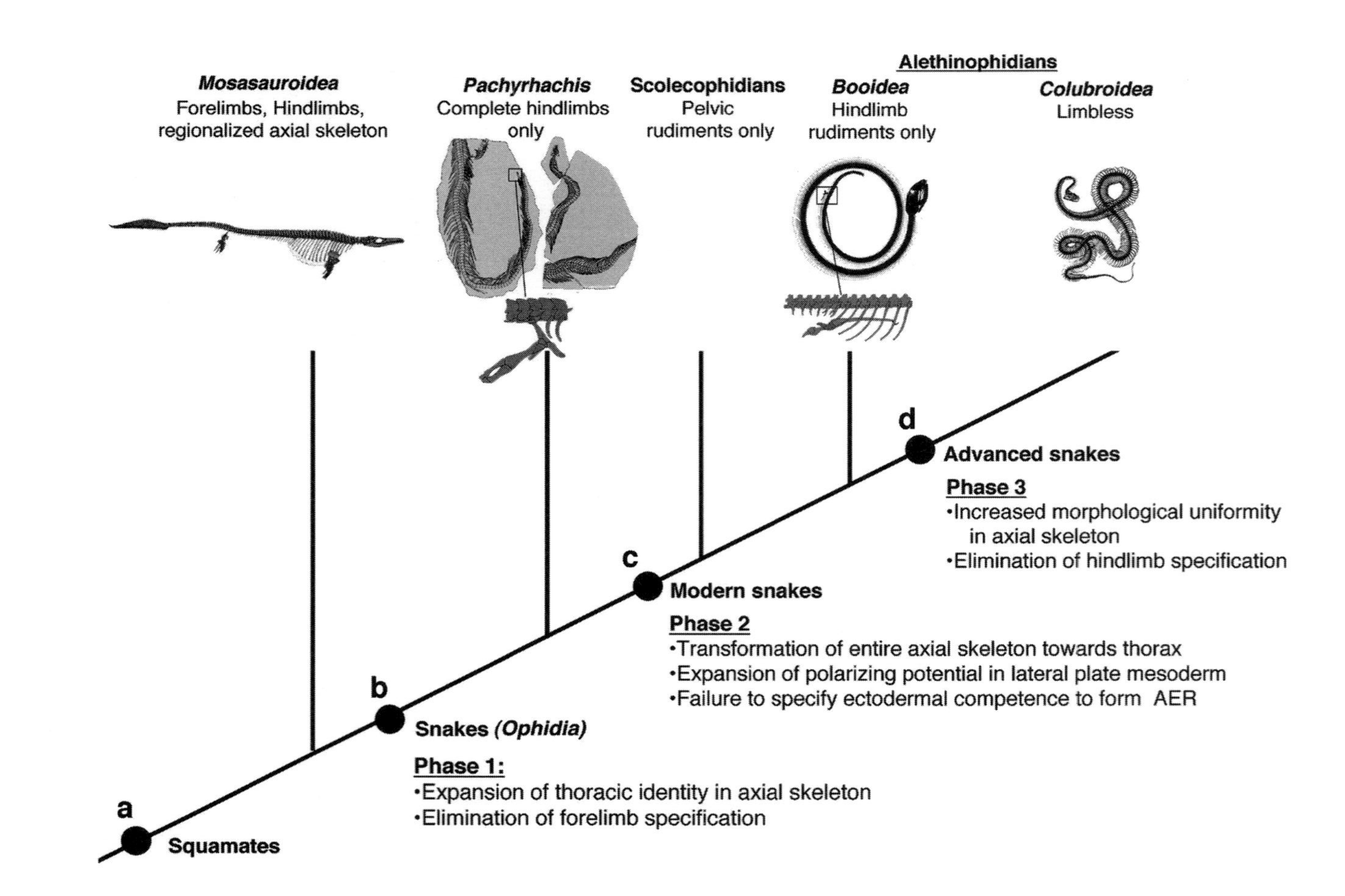

Alethinophidians
Mosasauroidea
Forelimbs, Hindlimbs, regionalized axial skeleton
Pachyrhachis
Complete hindlimbs only
Scolecophidians
Pelvic rudiments only
Booidea
Hindlimb rudiments only
Colubroidea
Limbless
a
Squamates
b
Snakes (Ophidia)
Phase 1:
•Expansion of thoracic identity in axial skeleton
•Elimination of forelimb specification
c
Modern snakes
Phase 2
•Transformation of entire axial skeleton towards thorax
•Expansion of polarizing potential in lateral plate mesoderm
•Failure to specify ectodermal competence to form AER
d
Advanced snakes
Phase 3
•Increased morphological uniformity in axial skeleton
•Elimination of hindlimb specification

On the basis of the above results, our current view is that loss of limbs and axial regionalization in snakes may stem from changes in the regulation of *Hox* gene expression along the primary body axis. Both limb position and axial skeletal identity are regulated by these factors, and as such, they are good candidates for coordinating changes to the axial and appendicular skeletons. This does not necessarily imply that *Hox* gene expression in paraxial and lateral plate mesoderm are co-regulated by the same *cis*-acting elements; this linkage may occur at the level of secondary or tertiary signalling between paraxial and lateral plate mesoderm, or via *trans*-acting factors which operate on global *Hox* expression. Our model suggests that the major morphological transitions in snake evolution can be accounted for by several phases of expansion of *Hox* gene expression domains along the anteroposterior axis of the trunk (Fig. 3). While some of these hypotheses concerning fossil taxa are not directly testable by an experimental approach, we can test predictions at the top and bottom of the tree by expanding our comparative analysis of development. For example, experiments are currently underway to test the hypothesis that more derived snakes, which lack limbs

FIG. 3. Developmental model for the evolution of snakes. Tree shows evolutionary relationships among the following: *Colubroidea* (advanced snakes) which lack both forelimbs and hindlimbs and have a large number of nearly-identical vertebrae; *Booidea* (including pythons and boas) which lack forelimbs, but have rudimentary hindlimbs and a large number of morphologically uniform vertebrae with few or equivocal regional differences; Scolecophidians, which have pelvic rudiments and a large number of morphologically uniform vertebrae; the primitive snake *Pachyrhachis problematicus*, which lacks forelimbs, but has complete (or nearly-complete) hindlimbs and a large number of similar vertebrae which nonetheless have identifiable regional differences; and mosasaurs, which have a morphologically regionalized axial skeleton and complete, normally polarized forelimbs and hindlimbs. According to this model, progressive expansion of Hox gene expression domains can account for loss of forelimbs, hindlimbs and regional identity in the axial skeleton. Additionally, the increase in vertebral number would have required continuous production of mesoderm for axial elongation, and this could have been achieved by sustained growth of the tail bud and movement of mesoderm through the primitive streak (Wilson & Beddington 1997). Node '**a**' indicates origin of squamates. (**b**) *Hox* expansion initiated prior to the divergence of the *Pachyrhachis* lineage could have lead to reduction of regional differentiation in the axial skeleton and elimination of forelimb specification, with hindlimb development remaining unaffected. (**c**) Continued expansion of *Hox* domains after the divergence of the *Pachyrhachis* lineage could have lead to transformation the entire axial skeleton (anterior to the tail) towards thoracic identity and to reduction of hindlimb development by eliminating ectodermal competence to form an apical ridge and expanding polarizing potential (competence to express *Shh*). This condition is retained in scolecophidians and in modern pythons, which together with boas comprise the *Booidea*. (**d**) Further homogenization of *Hox* gene expression domains is predicted to have lead to the origin of advanced snakes/*colubroidea*. (Phylogenetic relationships among these taxa based on Caldwell & Lee 1997; Figures modified from Caldwell & Lee 1997, Carroll 1988, Gasc 1976.) Reproduced with permission from Cohn & Tickle (1999).

completely, will show less regionalization of *Hox* expression domains than do pythons.

How broadly applicable are the principles that we have discovered in python embryos? While it is tempting to speculate that secondary limb loss in other lineages may have been driven by the same developmental mechanisms, such speculation is avoidable when these hypotheses can be tested with relative ease in the laboratory. Analyses similar to the one we have described above can be performed using phylogenetically relevant taxa to determine whether independent evolution of limblessness in different vertebrate lineages may stem from similar developmental mechanisms.

Acknowledgements

I thank the Novartis Foundation, Gillian Morriss-Kay and Adam Wilkins for the invitation to participate in this symposium. I am grateful to Cheryll Tickle, with whom the work on limblessness was performed, for encouragement and for many interesting discussions, and to Ketan Patel for critical reading of the manuscript. I thank the Welsh Mountain Zoo, Drayton Manor Zoo Park, London Zoo, Edinburgh Zoo, Jason Fletcher and the Reptile Trust for generously donating fertile eggs. Our work on snake embryos was funded by the BBSRC.

References

Akam M 1998 Hox genes, homeosis and the evolution of segment identity: no need for hopeless monsters. Int J Dev Biol 42:445–451

Caldwell MW, Lee MS 1997 A snake with legs from the marine Cretaceous of the Middle East. Nature 386:705–709

Carroll R 1988 Vertebrate paleontology and evolution. Freeman, New York

Coates MI, Clack JA 1990 Polydactyly in the earliest tetrapod limbs. Nature 347:66–69

Coates MI, Cohn MJ 1998 Fins, limbs and tails: outgrowth and patterning in vertebrate evolution. Bioessays 20:371–381

Cohn MJ, Bright PE 1999 Molecular control of vertebrate limb development, evolution and congenital malformations. Cell Tissue Res 296:3–17

Cohn MJ, Tickle C 1999 Developmental basis of limblessness and axial patterning in snakes. Nature 399:474–479

Cohn MJ, Patel K, Krumlauf R, Wilkinson DG, Clarke JD, Tickle C 1997 *Hox9* genes and vertebrate limb specification. Nature 387:97–101

Davis AP, Witte DP, Li Hsieh H, Potter SS, Capecchi MR 1995 Absence of radius and ulna in mice lacking Hoxa-11 and Hoxd-11. Nature 375:791–795

Dollé P, Dierich A, LeMeur M et al 1993 Disruption of the Hoxd-13 gene induces localized heterochrony leading to mice with neotenic limbs. Cell 75:431–441

Duboule D 1995 Vertebrate Hox genes and proliferation: an alternative pathway to homeosis? Curr Opin Genet Dev 5:525–528

Gasc JP 1976 Les rapportsanatomiques du membre pelvien vestigial chez les squamates serpentiformes. Bull Mus Nat Hist 2e Ser 38:99–110

Jegalian BG, De Robertis EM 1992 Homeotic transformations in the mouse induced by overexpression of a human *Hox3.3* transgene. Cell 71:901–910

Kengaku M, Capdevila J, Rodriguez-Esteban C et al 1998 Distinct WNT pathways regulating AER formation and dorsoventral polarity in the chick limb bud. Science 280:1274–1277
Laufer E, Dahn R, Orozco OE et al 1997 Expression of *Radical fringe* in limb-bud ectoderm regulates apical ectodermal ridge formation. Nature 386:366–373 (erratum: 1997 Nature 388:400)
Pollock RA, Sreenath T, Ngo L, Bieberich CJ 1995 Gain of function mutations for paralogous *Hox* genes: implications for the evolution of *Hox* gene function. Proc Natl Acad Sci USA 92:4492–4496
Rancourt DE, Tsuzuki T, Capecchi MR 1995 Genetic interaction between Hoxb-5 and Hoxb-6 is revealed by nonallelic noncomplementation. Genes Dev 9:108–122
Raynaud A, Kan P, Bouche G, Duprat AM 1995 Fibroblast growth factors (FGF-2) and delayed involution of the posterior limbs of the slow-worm embryo (*Anguis fragilis* L.). C R Acad Sci Iii 318:573–578
Rijli FM, Gavalas A, Chambon P 1998 Segmentation and specification in the branchial region of the head: the role of the *Hox* selector genes. Int J Dev Biol 42:393–401
Shubin N, Tabin C, Carroll S 1997 Fossils, genes and the evolution of animal limbs. Nature 388:639–648
Smith M, Hickman A, Amanze D et al 1994 Trunk neural crest origin of caudal fin mesenchyme in the zebrafish *Brachydanio rerio*. Proc R Soc Lond Ser B Biol Sci 256:137–145
Sordino P, van der Hoeven F, Duboule D 1995 Hox gene expression in teleost fins and the origin of vertebrate digits. Nature 375:678–681
Wilson V, Beddington R 1997 Expression of T protein in the primitive streak is necessary and sufficient for posterior mesoderm movement and somite differentiation. Dev Biol 192:45–58
Yokouchi Y, Nakazato S, Yamamoto M et al 1995 Misexpression of Hoxa-13 induces cartilage homeotic transformation and changes cell adhesiveness in chick limb buds. Genes Dev 9:2509–2522
Zákány J, Duboule D 1999 *Hox* genes in digit development and evolution. Cell Tissue Res 296:19–25
Zeller R, Duboule D 1997 Dorso-ventral limb polarity and origin of the ridge: on the fringe of independence? Bioessays 19:541–546

DISCUSSION

Kronenberg: What abnormalities are you expecting to find (or have you found) in the *Hox* gene pattern in the python mesenchyme that might correlate with the expansion of the thoracic vertebrae?

M. Cohn: We know that the somites, which form the vertebrae, are regionalized by nested domains of *Hox* gene expression. In transgenic mouse experiments in which the *Hoxc6* and *Hoxc8* genes are mis-expressed and expanded anteroposteriorally, there is an expansion of rib-bearing vertebrae beyond the normal thorax. This might mimic, to some degree, what has happened in snakes. The lateral plate mesoderm, which forms the limbs and body wall, is also regionalized by *Hox* gene expression. This is something that Cheryll Tickle and I showed a couple of years ago (Cohn et al 1997). Molecular regionalization of the lateral plate is an important step in determining the position at which limbs develop relative to the main body axis. The same principle of regionalization by differential *Hox* gene expression is operating in the axial skeleton and in the lateral plate mesoderm. I'm arguing that, in snakes, expression of some *Hox* genes has been homogenized in both of these tissues. This might account for expansion of thorax in the axial skeleton, and expansion of flank — the limbless part of the body wall — in the lateral plate. The latter would posteriorize the positional identity of cells that would otherwise form forelimbs.

Kronenberg: That's the disappearance of the anterior limb. But it looked as if the hindlimb ended up in the right place, so why don't you get a normal hindlimb *Hox* pattern?

M. Cohn: There is a posterior boundary of *Hoxc8* expression precisely at the level of the hindlimb bud. Hindlimb position is specified and budding is initiated. Our data suggest that we're seeing a downstream effect in the hindlimb, in which formation of the apical ridge is affected. There is no evidence that ridge formation is related to *Hox* gene expression, even in the model systems, but I think that this is an interesting possibility.

Kronenberg: Why would whales be different? Is there anything you know yet about whales that explains the evolution of its thorax and limbs?

M. Cohn: Eocene whales like *Basilosaurus* challenge our idea that limb reduction and axial regionalization have to be wrapped-up in the same developmental package. They have elongated and homogenized the posterior part of the axial skeleton, yet they make hindlimbs which are complete all the way to the toes.

Wilkins: In principle there can be independent selection for retention of the limbs. One just needs genetic changes that uncouple the initiation of the bud development from these earlier signals that they used to be linked to. I don't think that this pre-historic whale disproves your idea.

Ornitz: In light of that, have you looked to see whether FGF10 is expressed in lateral plate mesoderm in the python? Is FGF receptor 2 expressed in ectoderm in the ridge?

M. Cohn: No, but I like the FGF receptor 2 idea. This could explain failure of the interaction between limb bud mesenchyme and the overlying ectoderm, which should normally result in formation of an AER.

Morriss-Kay: It would be good to look at this, because in the early mouse limb bud, there is no morphological AER — in this sense it resembles python more than chick

Have you looked at *Patched* (*Ptc*) expression in the limb? Although you have shown there is no *Shh* there, the limb clearly has the potential to turn *Shh* on. The limb also has a much broader region of polarizing activity, which in a minor way (since it doesn't form any digits) is reminiscent of the *Doublefoot* mutant, in which the whole limb mesenchyme acts as a polarizing region (Hayes et al 1998). Is there any *Ptc* expression in the python limb bud?

M. Cohn: I haven't looked at *Ptc* or BMPs in the python.

Ornitz: Have you looked in lateral plate mesoderm for *Tbx4* or *Tbx5* expression?

M. Cohn: No. It seems likely that *Tbx4* must be expressed, since the hindlimb field is specified. This is an interesting question with respect to the forelimbs: Is a forelimb field specified in python embryos? From the *Hox* pattern that we see, I would predict that there is not, but we should certainly look at *Tbx* genes.

Beresford: Is the assumption that if you carried out an experiment in which you grafted a chick AER onto the python mesenchyme, you would get a limb? That is, is the assumption that there is nothing intrinsically wrong with that mesenchyme, and that nothing has been lost over evolutionary time?

M. Cohn: This is an experiment we have tried for many years. The more we learn about python hindlimb bud mesenchyme, the more reason we have to believe that we can rescue limb outgrowth with an AER signal. I have been treating python embryos with FGF, which is the key ridge factor, by grafting FGF-loaded beads to the hindlimb bud. So far, what I have seen is that in the first 24 h, proximo-distal outgrowth is increased by about 30%. FGF can, therefore, sustain outgrowth over a longer period. However, we ran into technical problems doing these experiments *in ovo* because of the soft nature of snake eggs; they tend to collapse after being opened and there are problems with infection. The next step is to start doing this experiment in explant cultures, which I plan to start next season.

Newman: The fact that the python mesenchyme retains the ability to make *Shh* suggests two things. First is that *Shh* might be induced by FGFs in other regions of the embryo and, second, there probably isn't a limb-specific promoter for *Shh* induction.

M. Cohn: That is an interesting point. If *Shh* and FGF were always acting in a feedback loop in the embryo, then there would be strong selection to maintain this circuit. While they are often co-expressed in the embryo, this is not always the case. I would say that it does seem likely that *Shh* expression in the limb is controlled by some FGF-responsive mechanism, but we don't yet know whether there is a limb-specific hedgehog regulatory element.

Blair: Is there a theoretical basis for the uncoupling of the limb formation from the axial skeleton?

M. Cohn: There is an experimental basis for it. The morphological evidence suggests that the position of the forelimb in vertebrates tends to co-localize with the junction between thoracic and cervical vertebrae, but during development they can be uncoupled experimentally. One of our findings when we were looking at *Hox* regulation in lateral plate mesoderm was that FGF can reprogramme the *Hox* code and induce ectopic limb formation in the lateral plate without altering *Hox* expression or identity in the axial skeleton.

Blair: What I find confusing is that the *HoxC8* boundary still exists in the snake, but the hind limb rudiments are completely dissociated from the axial skeleton.

M. Cohn: The disarticulation of the pelvic girdle from the column could be an epiphenomenon that happens late, during differentiation or growth, and not during specification of pattern. These tissues are patterned by *Hox* domains that lie adjacent to one another.

Mundlos: Could you not explain this simply by the loss of *Shh* expression? If Sonic is lost in the mouse, there is also truncation of limbs, and in the hindlimb there is exactly the same situation as you have seen in the python: there is a femur and nothing else. Perhaps loss of *Hox* expression leads to the regional turn off of *Shh* in the digit.

M. Cohn: We did notice that the extent of hindlimb outgrowth in pythons is almost the same as the extent of outgrowth seen in the *Shh* knockout mice. This is an interesting point. Our observations that python mesenchymal cells are competent to express *Shh* in the presence of a functional AER, and that the ridge is normally absent from these limb buds, suggests that absence of *Shh* is due to absence of an AER.

Meikle: How long ago did the pythons loose their hind limbs? There is a limit to the amount of time which a gene can be silenced without undergoing some kind of mutational degradation.

Ornitz: But *Shh* is still used by pythons in other places. Perhaps there is a tissue-specific enhancer that is lost.

Kingsley: This is the part that isn't known: how modular are the controls that build these limbs? Once the limbs are gone you should lose selection to maintain those modules that are limb specific. If there are limb-specific regulatory elements that are no longer used in making a useful limb, one would expect to see secondary changes in these.

Wilkins: In the python the limbs are still useable to some extent, for certain functions. So in this case there would be some selection for retention of some of these regulatory elements.

M. Cohn: There is an evolutionary principle known as 'Dollo's Law', which says that evolution cannot reverse itself. More specifically, once a structure is lost from an animal, it can not re-evolve. This 'law' has to be re-examined in light of molecular developmental biology. If there is selection to maintain a developmental cassette like FGFSHH in another part of the embryo, then there should be no great difficulty in re-activating that cassette in the limb. But if development is truly modular and there is, for example, a transcriptional enhancer required for *Shh* expression in the limb, then I would agree that once limbs are lost, genetic drift could eventually make it difficult, if not impossible, for them to re-appear.

Wilkie: Do you know what underlies the changes in distribution of *Hox* gene expression? Is *Hox* gene organization altered, are there differences in particular *Hox* promoters, or is this all driven by another gene that's switching on *Hox*—which is itself relatively unchanged in gene organization ?

M. Cohn: We have no idea—we don't even know how many *Hox* genes these animals have. However, this is a key question. These developmental changes could be associated with gene loss. Puffer fish, for example, have stripped-down their axial skeletons and lost a set of fins, and this is associated with loss of several *Hox* genes.

Hall: Now that you have done this quite prodigious study on a snake that retains hindlimb elements, it would be nice to do the whole thing all over again with a snake that completely loses hindlimb elements, or some of the legless lizards, because this might help you to get at the modularity and the connection to the *Hox* genes.

M. Cohn: We are, in fact, now looking at corn snakes and king snakes, both of which lack limbs entirely.

Hall: One of the reasons the limb buds regress in legless lizards is because the somitic cells move into the limb and inhibit cell death. Do you know anything about the somitic contributions coming into the python's limbs? There are clearly muscles there, so there must be a normal somitic contribution.

M. Cohn: I'm glad that you brought this up, because I was going to ask you about it! You are correct that there is muscle in these rudimentary limbs. One idea proposed by Raynaud several years ago is that limbless lizards have reduced limbs because of a quantitative deficiency in the number of somitic processes that invade and somehow 'stimulate' the lateral plate. Additionally, explant experiments in mouse suggested that a somitic contribution is needed for limb outgrowth. The data from chick somite-removal experiments, however, show that one can get perfectly good, but muscle-free, limbs without somites. In light of this, I wonder whether there may have been some misinterpretation of the mouse and legless lizard work. By removing or separating paraxial from lateral plate mesoderm, one might inadvertently remove the adjacent intermediate mesoderm. The intermediate mesoderm expresses Fgf8, and there are data which suggest that it is required for limb development. Somites, however, do not seem to be required for limb outgrowth. I have not looked for somitic processes in pythons primarily because I think that the somitic contribution to the limb bud is myogenic rather than stimulatory.

Hall: Jonathan Bard and I were chatting yesterday about the derivatives of the intermediate mesoderm, such as the mesonephros and the role that it plays in development.

There is a notion that the ability to be competent to make an AER is specified by the mesoderm. I had a feeling it must be known whether the initial ability to become AER is ectoderm specific or mesoderm specific in the chick.

Ornitz: In the mouse *Fgf10* knockout there is a very transient limb bud but the AER is not formed. *Fgf10* is a mesoderm-expressed gene.

Pizette: If you put FGF7 beads in the mesenchyme of the back, you can actually induce AER-specific gene expression in the overlying ectoderm (Yonei-Tamura et al 1999). This indicates that the competence to form an AER is in the mesenchyme.

Hall: Unless that back ectoderm is competent to respond, then what you probably mean is making it competent to respond.

Tickle: In those experiments, there's a difference in the competence of the ectoderm at the dorsal midline versus the competence of the ectoderm on the sides of the body. The competence dorsally extended much further anteriorly then it did on the lateral side of the body.

References

Cohn MJ, Patel K, Krumlauf R, Wilkinson DG, Clarke JD, Tickle C 1997 *Hox9* genes and vertebrate limb specification. Nature 387:97–101

Hayes C, Lyon MF, Morriss-Kay GM 1998 Morphogenesis of *Doublefoot (Dbf)*, a mouse mutant with polydactyly and craniofacial defects. J Anat 193:81–91

Yonei-Tamura, Endo T, Yajima H, Ohuchi, Ide H, Tamura K 1999 FGF7 and FGF10 directly induce the apical ectodermal ridge in chick embryos. Dev Biol 211:133–143

Regulation of chondrocyte growth and differentiation by fibroblast growth factor receptor 3

David M. Ornitz

Department of Molecular Biology and Pharmacology, Washington University School of Medicine, St Louis, MO 63110, USA

Abstract. Both gain-of-function and loss-of-function mutations in fibroblast growth factor receptor 3 (*Fgfr3*) have revealed unique roles for this receptor during skeletal development. Loss-of-function alleles of *Fgfr3* lead to an increase in the size of the hypertrophic zone, delayed closure of the growth plate and the subsequent overgrowth of long bones. Gain-of-function mutations in *Fgfr3* have been genetically linked to autosomal dominant dwarfing chondrodysplasia syndromes where both the size and architecture of the epiphyseal growth plate are altered. Analysis of these phenotypes and the biochemical consequences of the mutations in FGFR3 demonstrate that FGFR3-mediated signalling is an essential negative regulator of endochondral ossification.

2001 The molecular basis of skeletogenesis. Wiley, Chichester (Novartis Foundation Symposium 232) p 63–80

Skeletal growth is regulated by the process of endochondral ossification, a developmental program that occurs at the ends of growing long bones and in vertebrae. During endochrondral ossification, chondrocytes differentiate through a series of well-defined morphological zones within the epiphyseal growth plate. The proliferative zone provides a renewable source of chondrocytes for longitudinal bone growth. After exiting the cell cycle maturing chondrocytes secrete a specialized extracellular matrix that includes chondroitin-sulfate proteoglycans and type II collagen. Encapsulated in this matrix, the chondrocytes undergo hypertrophy and subsequently express type X collagen and alkaline phosphatase. Hypertrophic chondrocytes undergo an apoptotic death as their surrounding matrix is mineralized and replaced by trabecular bone (Caplan & Pechak 1987).

FGF receptors (*Fgfrs*) 1 and 3 are both expressed in the epiphyseal growth plate. *Fgfr3* is expressed in proliferating chondrocytes, whereas *Fgfr1* is expressed in hypertrophic chondrocytes (Peters et al 1993, Deng et al 1996). *Fgfr3* is also expressed in the cartilage of the developing embryo, prior to formation of ossification centres. This expression pattern suggests a direct role for FGFR3 in regulating chondrocyte proliferation and possibly differentiation (Peters et al 1993, Delezoide et al 1998, Naski et al 1998). In contrast, *Fgfr1* is prominently expressed in hypertrophic chondrocytes (Peters et al 1992, Delezoide et al 1998) suggesting a role for *Fgfr1* in maintaining the hypertrophic phenotype of these cells (cell survival), in regulating the production of unique extracellular matrix products of hypertrophic chondrocytes or in signalling their eventual apoptotic death.

Dwarfing chondrodysplasias

FGFR3 is an essential regulator of endochondral bone growth. When the process of endochondral bone growth is disrupted by mutations in *Fgfr3* mild to severe dwarfism ensues. The dwarfing chondrodysplasias, including hypochondroplasia (HCH) (Bellus et al 1995), achondroplasia (ACH) (Rousseau et al 1994, Shiang et al 1994, Ikegawa et al 1995, Superti-Furga et al 1995) and thanatophoric dysplasia (TD) (Rousseau et al 1995, 1996, Tavormina et al 1995a,b) are caused by mutations in the gene encoding *Fgfr3*. HCH is a mild and relatively common skeletal disorder with clinical features similar to that of ACH. ACH is the most common form of genetic dwarfism. ACH is characterized by shortening of the proximal, and to a lesser extent, distal long bones. The cranium of ACH patients is characterized by frontal bossing, and the face is characterized by a depressed nasal bridge. Rare homozygous cases of ACH usually result in neonatal lethality (Stanescu et al 1990). These individuals have features similar to that of TD (Rimoin & Lachman 1993). TD results from several dominant mutations in the *Fgfr3* gene. TD is the most common lethal-neonatal skeletal disorder and is clinically similar to homozygous ACH (Stanescu et al 1990).

The dwarfing chondrodysplasias display a graded spectrum of phenotypic severity which result from distinct missense mutations in *Fgfr3* (Fig. 1). TD, characterized by neonatal lethality and profound dwarfism, is the result of mutations including an R248C substitution in the extracellular domain or a K650E substitution in the tyrosine kinase (TK) domain (Tavormina et al 1995b). ACH, which is non-lethal and presents less severe dwarfism, results almost exclusively from a G380R substitution in the transmembrane domain (Rousseau et al 1994, Shiang et al 1994). The mild proportional dwarfism characteristic of HCH results from a N540K mutation in the proximal tyrosine kinase domain of *Fgfr3* (Bellus et al 1995).

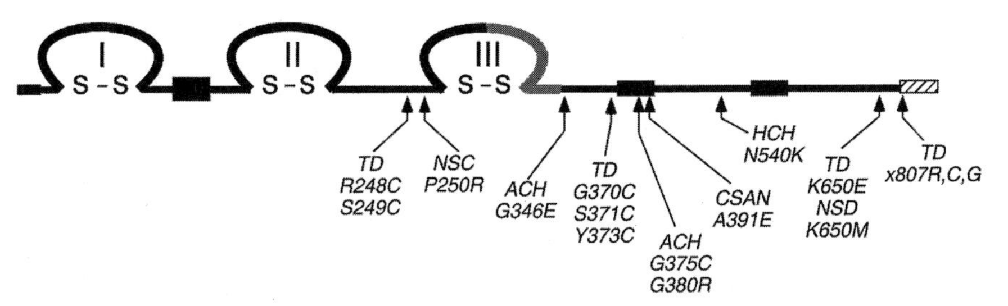

FIG. 1. Mutations that have been identified in the *FGFR3* gene in humans. Mutations responsible for achondroplasia (ACH), thanatophoric dysplasia (TD), hypochondroplasia (HCH), Crouzon syndrome associated with acanthosis nigricans (CSAN) and a non-syndromic craniosynostosis (NSC) are mapped onto the structure of the full length FGFR3. The full-length receptor has three immunoglobulin (Ig)-like domains. The shaded region beginning in Ig-like domain III is subject to alternative splicing. The hatched line at the C-terminus of FGFR3 represents an extension of the protein resulting from mutations in the stop codon of the receptor. The numbers represent the position of the amino acid in the coding sequence of the human receptor. Amino acids are abbreviated using standard single letter codes. (Adapted from Naski & Ornitz 1998).

Mixed cases of chondrodysplasia and craniosynostosis

A few rare cases have been described in which both chondrodysplasia and craniosynostosis phenotypes occur within a single patient. In one case a patient with an FGFR3 N540K mutation presented with hypochondroplasia and a clover leaf skull (Angle et al 1998). Additionally, in patients with an FGFR3 R248C mutation (TD type 1), the clover leaf skull deformity is observed in about half of the cases and is likely to be due to variable expressivity of the mutation. In patients with an FGFR3 K650E mutation (TD type II), the cloverleaf skull deformity occurs often (Tavormina et al 1995b). The vast majority of craniosynostosis syndromes are caused by mutations in *Fgfr2* (Cohen 1997, Webster & Donoghue 1997, Wilkie 1997). Although these syndromes are often associated with syndactly, dwarfism is not a common feature. It is therefore possible that the mechanism by which FGFR3 causes craniosynostosis is, directly or indirectly, activating FGFR2 in cranial sutures. Expression studies have localized *Fgfr1*, *2* and *3* in close proximity in the cranial suture (Kim et al 1998, Iseki et al 1999) suggesting that interactions between these receptors are possible.

Mutations in FGFR3 activate signalling

All of the mutations identified in FGFRs are autosomal dominant and frequently arise sporadically. The great majority of these disorders result from point mutations in the coding sequence of the FGFR that result in a single amino acid substitution.

Biochemical studies have shown that mutations causing ACH and TD are gain-of-function mutations resulting in increased receptor tyrosine kinase activity (Naski et al 1996, Webster et al 1996, Webster & Donoghue 1996, Li et al 1997) (Fig. 2). The G380R mutation in the transmembrane domain of FGFR3 (responsible for most cases of ACH) partially activates receptor signalling (Naski et al 1996). The basal mitogenic activity of this receptor (assayed as a chimeric receptor containing the tyrosine kinase domain from FGFR1) could be augmented by the addition of ligand. The dose response curve suggested that this receptor has a similar ligand binding affinity to that of the wild-type receptor. Additionally, studies of receptor tyrosine phosphorylation showed ligand independent receptor autophosphorylation. The K650E and R248C mutations of TD are also activating mutations. These mutations result in ligand independent receptor activation as evidenced by ligand independent cell proliferation and receptor tyrosine phosphorylation. Significantly, the mutations causing TD were more strongly activating than the mutation causing ACH. This suggested a correlation between the degree of receptor activation and the severity of the dwarfing chondrodysplasia. This study also demonstrated that the FGFR3 R248C mutation constitutively activated the receptor by forming a disulfide-linked receptor homodimer in which an unpaired cysteine residue in the extracellular domain of the receptor forms an intermolecular disulfide linkage (Naski et al 1996). The FGFR3 K650E mutation occurs in a highly conserved lysine residue in the activation loop of the receptor (Hanks et al 1988, Mohammadi et al 1996). This mutation results in a constitutively active tyrosine kinase, presumably by altering the structure of the activation loop. Unlike the R248C mutation which showed constitutive activation matching that of maximally stimulated wild-type receptor, the K650E mutant receptor can be activated by ligand to a level greater than that of the wild-type receptor. These observations are consistent with the observed covalent homodimerization consequence of the R248C mutation and the deregulation of the kinase domain by the K650E mutation.

Functional and structural studies on the FGFR have localized a ligand binding domain to the second and third Ig domains and the intervening linker (Johnson et al 1990, Wang et al 1995, Chellaiah et al 1999, Plotnikov et al 1999). Interestingly, many of the mutations that affect FGFR activity (such as FGFR3:R248C) are localized to three highly conserved amino acid residues (RSP) within the linker sequence between Ig domain II and III (Fig. 3). The conserved RSP motif is itself embedded within a highly conserved sequence that is thought to function as a receptor dimerization domain (Wang et al 1997) or ligand binding surface (Plotnikov et al 1999). The biochemical and structural data suggest that mutations in this sequence either affect ligand binding or receptor dimerization or both.

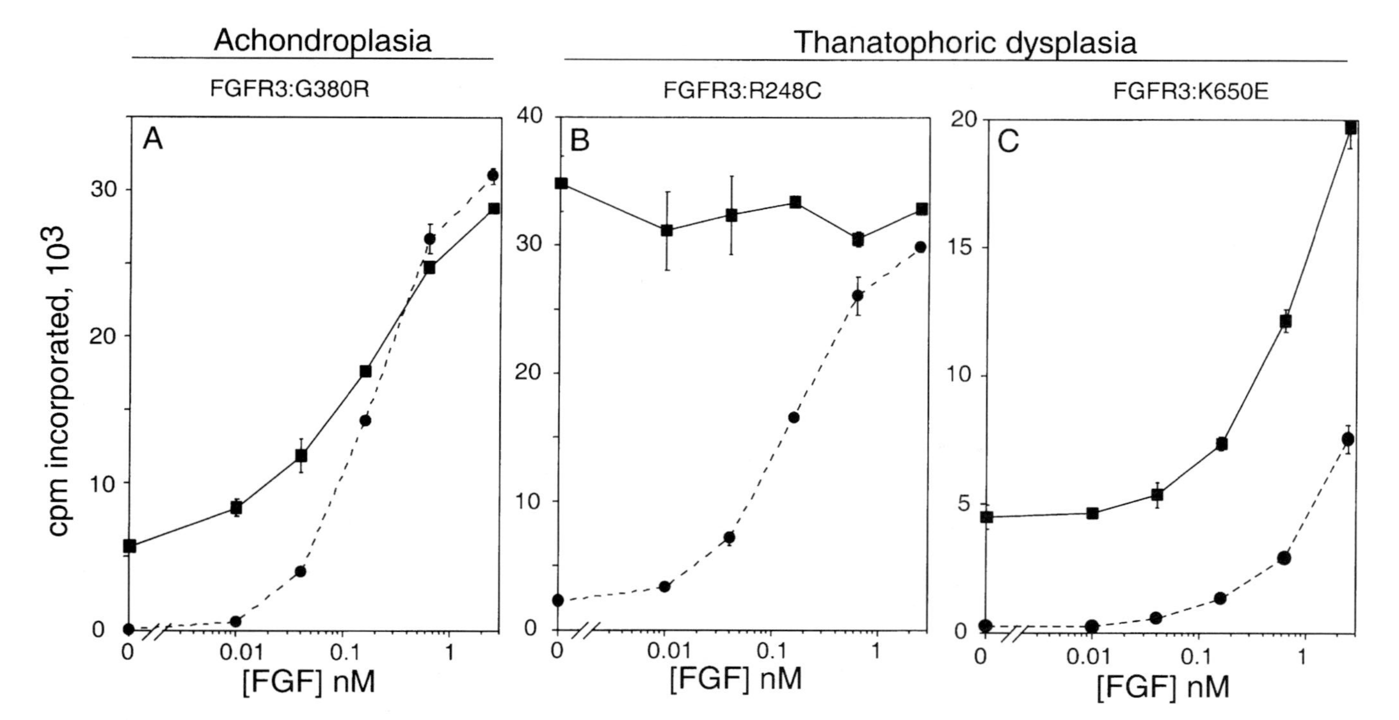

FIG. 2. Proliferation of BaF3 cells expressing wild-type and mutant FGFR3 in response to FGF1. Mitogenic response curves of BaF3 clones expressing: (A) wild-type or FGFR3:G380R chimeric receptors encoding the IIIb isoform of the FGFR3 ligand binding domain and the FGFR1 tyrosine kinase domain; (B) wild-type or FGFR3:R248C chimeric receptors encoding the IIIc isoform of FGFR3 and the FGFR I tyrosine kinase domain; or (C) wild-type or full length FGFR3:K650E. The dose response of a representative wild-type (circle) and mutant (square) clone are shown. The level of receptor expression was determined to be comparable between the clones expressing the wild-type or mutant receptors by either measuring the binding of [125I]-FGF1 or Western blotting for the receptor with the antibody AB6 (Adapted from Naski et al 1996).

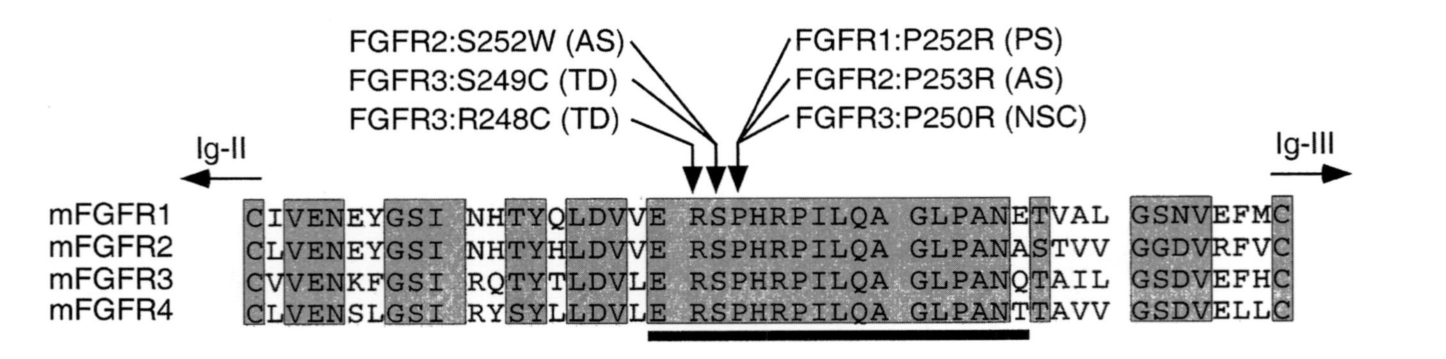

FIG. 3. Conserved sequence in the linker region between Ig-like domain II and III of FGFRs 13. The relative position of mutations occurring in FGFRs 1–3 clustered within three conserved amino acid residues are indicated. The conserved region thought to serve as a receptor dimerization domain is underlined (Wang et al 1997). Apert syndrome (AS), thanatophoric dysplasia (TD), Pfeiffer syndrome (PS), non-syndromic craniosynostosis (NSC), immunoglobulin loops II and III (IgII, IgIII) are indicated. (Adapted from Ornitz 2000.)

Loss-of-function mutations in FGFR3 in mice

Mice homozygous for null alleles of *Fgfr3* exhibit skeletal overgrowth (Colvin et al 1996, Deng et al 1996) (Table 1). The contrasting phenotypes between the *Fgfr3*$^{-/-}$ mice and the human dwarfing conditions resulting from mutations in *Fgfrs* suggest that the mutations causing dwarfism are gain-of-function alleles, a hypothesis strongly supported with biochemical data (Naski et al 1996, Webster et al 1996, Webster & Donoghue 1996). These gain-of-function and loss-of-function studies and the biochemical data define *Fgfr3* as a negative regulator of endochondral bone growth.

Studies of *Fgfr3* null mice show prolonged expression of markers for cell proliferation (Deng et al 1996) and over-expression of activated FGFR3 (achondroplasia mutation) in a chondrocytic cell line or in the growth plate of transgenic mice results in diminished cell proliferation (Naski et al 1998, Henderson et al 1999). In addition to affecting chondrocyte proliferation, FGFR3 may also regulate chondrocyte differentiation. Significantly, differentiation of cultured chondrocytes is inhibited by the addition of FGF2 (Kato & Iwamoto 1990). Histological studies of biopsies from individuals with achondroplasia show either extensive or focal disorganization of the growth plate (Ponseti 1970, Rimoin et al 1970, Briner et al 1991). Furthermore, *Fgfr3*$^{-/-}$ mice have an expanded zone of hypertrophy in the epiphyseal growth plate (Colvin et al 1996, Deng et al 1996) and mice over-expressing activated FGFR3 in the growth plate show decreased numbers of cells in the prehypertrophic and hypertrophic zones (Naski et al 1998).

Fgf and *Fgfr* expression in the growth plate

The physiologic FGF ligand(s) for FGFR3 in the epiphyseal growth plate is not known. FGF2 is abundantly expressed in growth plate chondrocytes

TABLE 1 Comparison of bone length in wild-type and *Fgfr*$^{-/-}$ mice

Bone	*Bone length (percentage of control)*[a]	*P value*[b]	*Body weight (percentage of control)*
Femur	115	0.0001	59
Tibia	107	0.0006	59
Humerus	115	0.0001	68

[a]Comparison of *Fgfr3*/ mice both to wild-type and heterozygous mice.
[b]$n = 58$ mice for each genotype.
Adapted from Colvin et al (1996).

(Gospodarowicz & Mescher 1977, Klagsburn et al 1977, Twal et al 1994) and for many years was thought to be a candidate for physiologic ligand in the growth plate. Furthermore, FGF2 is a potent mitogen for growth plate (Quarto et al 1997) or articular (Jones & Addison 1975) chondrocytes in culture and FGF2 is a ligand for the mesenchymal splice forms of both FGFR1, 2 and 3 (Ornitz et al 1996). Surprisingly, however, mice lacking *Fgf2* have no overt skeletal defects (Dono et al 1998, Tobe et al 1998, Zhou et al 1998). This suggests that FGF2 is not the physiological ligand for FGFR3 in the growth plate. Alternatively, FGF2 may cooperate with other members of the family to regulate chondrogenesis.

The expression of FGF8 and FGF17, two ligands for FGFR3, has recently been observed in developing bone in the mouse between E14.5 and E16.5 (Xu et al 1999). Both ligands were expressed in dorsal costal cartilage and in costal perichondrium. *Fgf8* expression was observed in the osteoblast compartment of calvarial bone, in the mandible, in cortical bone and in the growth plate of long bones. *Fgf17* expression was not detected in long bones but was detected in some intramembranous bone such as the maxilla and the scapula. After E16.5, expression of these ligands decreased in the growth plate. It is not yet known whether they are re-expressed in the growth plate of long bones after birth. Besides these ligands, other FGFs can function as ligands for FGFR3 *in vitro*, yet have not been rigorously examined for expression in the growth plate of developing bones. Because the ligand for FGFR3 appears to be limiting, the regulation of such a ligand, in terms of its expression or bioavailability, will be an important component of the complex regulatory cascades controlling longitudinal bone growth.

Mouse models for achondroplasia

Transgenic mice in which an *Fgfr3* cDNA containing the G380R mutation (ACH) is over-expressed in the growth plate or knocked into the *Fgfr3* genomic locus develop a phenotype resembling achondroplasia (Naski et al 1998, Wang et al 1999). Mice in which the K650E (TD type II) mutation is introduced into the *Fgfr3* genomic locus by homologous recombination showed a similar phenotype (Li et al 1999). These studies demonstrate that activated FGFR3 signalling inhibits chondrocyte proliferation and slows chondrocyte differentiation *in vivo*. Studies in which tissue from TD type I patients was examined showed that the constitutive activation of FGFR3 (R248C) does not prevent chondrocyte proliferation but rather alters their differentiation by triggering premature apoptosis through activation of the STAT1 signalling pathway (Legeai-Mallet et al 1998). Similarly, FGFR3 containing the K650E mutation has constitutive tyrosine kinase activity which can specifically activate the transcription factor STAT1 leading to growth arrest *in vitro* (Su et al 1997, Sahni et al 1999) and *in vivo* (Li et al 1999).

Why does FGF signalling inhibit chondrocyte growth?

Classically, FGFs are considered powerful mitogens for many cell types including primary chondrocytes (Gospodarowicz & Mescher 1977, Klagsburn et al 1977, Basilico & Moseatelli 1992). It is therefore surprising and provocative that FGFR3 signalling inhibits chondrocyte proliferation *in vivo* in the growth plate. This paradoxical activity may be a unique property of FGFR3 or may result from a unique response of the growth plate chondrocyte to an FGFR-derived signal. Interestingly, *in vitro*, FGFR3 activation by ligand or by activating mutations results in a poor mitogenic response compared to FGFR1 (Lin et al 1996, 1998, Naski et al 1996). Recent *in vitro* studies, in which a constitutively active FGFR3 (containing the K650E mutation) was over-expressed in a non-chondrocytic cell line, showed decreased cell proliferation and a coincident increase in STAT1 activity suggesting that activate FGFR3 may stimulate growth inhibitory pathways (Su et al 1997, Sahni et al 1999).

The effects of an activating mutation in FGFR3 on the proliferation and differentiation of epiphyseal chondrocytes are now well described. However the signalling pathways downstream of FGFR3 that negatively regulate chondrocyte proliferation are poorly defined. Furthermore, it is not known whether these pathways are specific to FGFR3 (versus FGFR1) or whether this response is unique to the chondrocyte.

What is the rate-limiting step in endochondral ossification?

The physiological FGF ligand that regulates endochondral ossification is not known. However, gain-of-function and loss-of-function mutations in *Fgfr3* and comparison to transgenic mice that over-express an FGF in the growth plate suggest that an FGF ligand is the rate-limiting signal in the FGF pathway regulating endochondral ossification. Transgenic mice that over-express wild-type FGFR3 in the growth plate do not show signs of abnormal chondrocyte proliferation or abnormal growth plate morphology (Naski et al 1998). However, postnatal transgenic mice over-expressing activated FGFR3 or ectopic FGF2 in proliferating chondrocytes develop a dwarfing condition similar to that of achondroplasia (Coffin et al 1995, Naski et al 1998). These observations suggest that in the postnatal growth plate, the concentration of an, as yet unidentified, FGF ligand is limiting relative to that of FGFR3.

In contrast, during embryonic development the activated FGFR3 transgene had no effect on the proliferation of chondrocytes and loss of FGFR3 results in a broadened growth plate. This suggests that during embryonic development chondrocyte proliferation is insensitive to FGFR signalling either because the concentration of FGF is in excess, saturating FGFR signalling pathways, or that other mitogens function dominantly during embryonic life and mask the growth inhibitory effects of FGFR3.

How FGF signalling pathways interact with other signalling pathways to coordinate bone chrondrogenesis

Along with FGFs, endochondral bone growth is regulated by many signalling molecules, including growth hormone, insulin-like growth factor I (IGF-I), parathyroid hormone related protein (PTHrP), Indian hedgehog (Ihh) and bone morphogenetic proteins (BMPs) (Reddi 1994, Erlebacher et al 1995). A feedback loop has been identified in which Ihh and PTHrP interact to coordinate chondrocyte differentiation (Lanske et al 1996, Vortkamp et al 1996). The relationship between the Ihh/PTHrP and FGF signalling pathways has been investigated by examining the expression of *Ihh* and its receptor, patched, and of *PTHrP* in the growth plate of mice that are either lacking *Fgfr3* or that overexpress *Fgfr3* containing the activating mutation, G380R. In mice that overexpress *Fgfr3* G380R, *Ihh* expression and signalling, as assessed by examining *patched* expression, is inhibited. Furthermore, expression of *Bmp4*, a molecule thought to be downstream of *Ihh*, is also suppressed in both growth plate cartilage and in the surrounding perichondrium (Naski et al 1998). In contrast, in mice lacking *Fgfr3*, the expression of these molecules is increased relative to wild-type mice (Naski et al 1998). These data suggest that FGFR3 signalling is genetically upstream of the hedgehog, BMP and PTHrP signalling pathways and that FGFR3 signalling may serve to globally coordinate endochondral bone growth.

Conclusion

To investigate the genetic and biochemical mechanisms used by FGFR3 to regulate chondrocyte growth and differentiation, we have constructed transgenic mice that either lack a functional *Fgfr3* gene or that express the activating FGFR3 mutation (G380R) that causes achondroplasia. The effects on proliferation and differentiation of epiphyseal chondrocytes showed that activation of FGFR3 dramatically inhibits both chondrocyte proliferation and differentiation. The consequence of this effect on chondrogenesis is a histologically shortened growth plate and a gross phenotype resembling the human skeletal disorder, achondroplasia. In contrast, loss of FGFR3 activity results in an opposite phenotype in which proliferation is increased, the size and longevity of the growth plate is extended and skeletal overgrowth ensues. Examination of signalling pathways that regulate chondrocyte differentiation showed that FGFR3 signalling inhibits Ihh signalling and BMP4 expression in cartilage and perichondrium. These data place *Fgfr3* genetically upstream of *Ihh* and suggest that FGFR3 and its endogenous ligand may globally regulate chondrogenesis and osteogenesis.

Acknowledgements

This work was supported by grants HD35692 and CA60673 from the National Institutes of Health, USA.

References

Angle B, Hersh JH, Christensen KM 1998 Molecularly proven hypochondroplasia with cloverleaf skull deformity: a novel association. Clin Genet 54:417–420

Basilico C, Moscatelli D 1992 The FGF family of growth factors and oncogenes. Adv Cancer Res 59:115–165

Bellus GA, McIntosh I, Smith EA et al 1995 A recurrent mutation in the tyrosine kinase domain of fibroblast growth factor receptor 3 causes hypochondroplasia. Nat Genet 10:357–359

Briner J, Giedion A, Spycher MA 1991 Variation of quantitative and qualitative changes of enchondral ossification in heterozygous achondroplasia. Pathol Res Pract 187:271–278

Caplan AI, Pechak DG 1987 The cellular and molecular embryology of bone formation. In: Peck WA (ed) Bone and mineral research, vol 5. Elsevier Science, New York, p 117–183

Chellaiah A, Yuan W, Chellaiah M, Ornitz DM 1999 Mapping ligand binding domains in chimeric fibroblast growth factor receptor molecules. Multiple regions determine ligand binding specificity. J Biol Chem 274:34785–34794

Coffin JD, Florkiewicz RZ, Neumann J et al 1995 Abnormal bone growth and selective translational regulation in basic fibroblast growth factor (FGF-2) transgenic mice. Mol Biol Cell 6:1861–1873

Cohen MM Jr 1997 Short-limb skeletal dysplasias and craniosynostosis: what do they have in common? Pediatr Radiol 27:442–446

Colvin JS, Bohne BA, Harding GW, McEwen DG, Ornitz DM 1996 Skeletal overgrowth and deafness in mice lacking fibroblast growth factor receptor 3. Nat Genet 12:390–397

Delezoide AL, Benoist-Lasselin C, Legeai-Mallet L et al 1998 Spatio-temporal expression of FGFR 1, 2 and 3 genes during human embryo-fetal ossification. Mech Dev 77:19–30

Deng C, Wynshaw-Boris A, Zhou F, Kuo A, Leder P 1996 Fibroblast growth factor receptor 3 is a negative regulator of bone growth. Cell 84:911–921

Dono R, Texido G, Dussel R, Ehmke H, Zeller R 1998 Impaired cerebral cortex development and blood pressure regulation in FGF-2-deficient mice. EMBO J 17:42134–225

Erlebacher A, Filvaroff EH, Gitelman SE, Derynck R 1995 Toward a molecular understanding of skeletal development. Cell 80:371–378

Gospodarowicz D, Mescher AL 1977 A comparison of the responses of cultured myoblasts and chondrocytes to fibroblast and epidermal growth factors. J Cell Physiol 93:117–127

Hanks SK, Quinn AM, Hunter T 1988 The protein kinase family: conserved features and deduced phylogeny of the catalytic domains. Science 241:42–52

Henderson JE, Naski MC, Stregger S et al 1999 Inhibition of cell growth in chondrocytes expressing FGFR3*Ach* is linked to disruption of signaling through $\alpha 5\beta 1$ integrin. J Bone Miner Res, in press

Ikegawa S, Fukushima Y, Isomura M, Takada F, Nakamura Y 1995 Mutations of the fibroblast growth factor receptor-3 gene in one familial and six sporadic cases of achondroplasia in Japanese patients. Hum Genet 96:309–311

Iseki S, Wilkie AO, Morriss-Kay GM 1999 *Fgfr1* and *Fgfr2* have distinct differentiation- and proliferation-related roles in the developing mouse skull vault. Development 126:5611–5620

Johnson DE, Lee PL, Lu J, Williams LT 1990 Diverse forms of a receptor for acidic and basic fibroblast growth factors. Mol Cell Biol 10:4728–4736

Jones KL, Addison J 1975 Pituitary fibroblast growth factor as a stimulator of growth in cultured rabbit articular chondrocytes. Endocrinology 97:359–365

Kato Y, Iwamoto M 1990 Fibroblast growth factor is an inhibitor of chondrocyte terminal differentiation. J Biol Chem 265:5903–5909

Kim HJ, Rice DP, Kettunen PJ, Thesleff I 1998 FGF-, BMP- and Shh-mediated signalling pathways in the regulation of cranial suture morphogenesis and calvarial bone development. Development 125:1241–1251

Klagsburn M, Langner R, Levenson R, Smith S, Lillehei C 1977 The stimulation of DNA synthesis and cell division in chondrocytes and 3T3 cells by a growth factor isolated from cartilage. Exp Cell Res 105:99–108

Lanske B, Karaplis AC, Lee K et al 1996 PTH/PTHrP receptor in early development and Indian hedgehog-regulated bone growth. Science 273:663–666

Legeai-Mallet L, Benoist-Lasselin C, Delezoide AL, Munnich A, Bonaventure J 1998 Fibroblast growth factor receptor 3 mutations promote apoptosis but do not alter chondrocyte proliferation in thanatophoric dysplasia. J Biol Chem 273:13007–13014 (erratum: 1998 J Biol Chem 273:19358)

Li CL, Chen L, Iwata T, Kitagawa M, Fu XY, Deng CX 1999 A Lys644Glu substitution in fibroblast growth factor receptor 3 (FGFR3) causes dwarfism in mice by activation of STATs and ink4 cell cycle inhibitors. Hum Mol Genet 8:35–44

Li Y, Mangasarian K, Mansukhani A, Basilico C 1997 Activation of FGF receptors by mutations in the transmembrane domain. Oncogene 14:1397–1406

Lin HY, Xu JS, Ornitz DM, Halegoua S, Hayman MJ 1996 The fibroblast growth factor receptor-1 is necessary for the induction of neurite outgrowth in PC12 cells by aFGF. J Neurosci 16:4579–4587

Lin HY, Xu JS, Ischenko I, Ornitz DM, Halegoua S, Hayman MJ 1998 Identification of the cytoplasmic regions of fibroblast growth factor (FGF) receptor 1 which play important roles in the induction of neurite outgrowth in PC12 cells by FGF-1. Mol Cell Biol 18:3762–3770

Mohammadi M, Schlessinger J, Hubbard SR 1996 Structure of the FGF receptor tyrosine kinase domain reveals a novel autoinhibitory mechanism. Cell 86:577–587

Naski MC, Ornitz DM 1998 FGF signaling in skeletal development. Front Biosci 3:D781–D794

Naski MC, Wang Q, Xu J, Ornitz DM 1996 Graded activation of fibroblast growth factor receptor 3 by mutations causing achondroplasia and thanatophoric dysplasia. Nat Genet 13:233–237

Naski MC, Colvin JS, Coffin JD, Ornitz DM 1998 Repression of hedgehog signaling and BMP4 expression in growth plate cartilage by fibroblast growth factor receptor 3. Development 125:4977–4988

Ornitz DM 2000 Fibroblast growth factors, chondrogenesis and related clinical disorders. In: Canalis E (ed) Skeletal growth factors. Lippincott Williams & Wilkins, Philadelphia, PA, p 197–209

Ornitz DM, Xu J, Colvin JS et al 1996 Receptor specificity of the fibroblast growth factor family. J Biol Chem 271:15292–15297

Peters KG, Werner S, Chen G, Williams LT 1992 Two FGF receptor genes are differentially expressed in epithelial and mesenchymal tissues during limb formation and organogenesis in the mouse. Development 114:233–243

Peters K, Ornitz DM, Werner S, Williams L 1993 Unique expression pattern of the FGF receptor 3 gene during mouse organogenesis. Dev Biol 155:423–430

Plotnikov AN, Schlessinger J, Hubbard SR, Mohammadi M 1999 Structural basis for FGF receptor dimerization and activation. Cell 98:641–650

Ponseti IV 1970 Skeletal growth in achondroplasia. J Bone Joint Surg (Am) 52:701–716

Quarto R, Campanile G, Cancedda R, Dozin B 1997 Modulation of commitment, proliferation, and differentiation of chondrogenic cells in defined culture medium. Endocrinology 138:4966–4976

Reddi AH 1994 Bone and cartilage differentiation. Curr Opin Genet Dev 4:737–744

Rimoin DL, Lachman RS 1993 Genetic disorders of the osseous skeleton. In: McKusick's Heritable disorders of connective tissue, 5th edn. Mosby-Year Book, St Louis, p 557–689

Rimoin DL, Hughes GN, Kaufman RL, Rosenthal RE, McAlister WH, Silberberg R 1970 Endochondral ossification in achondroplastic dwarfism. N Engl J Med 283:728–735

Rousseau F, Bonaventure J, Legeai-Mallet L et al 1994 Mutations in the gene encoding fibroblast growth factor receptor-3 in achondroplasia. Nature 371:252–254

Rousseau F, Saugier P, Le Merrer M et al 1995 Stop codon FGFR3 mutations in thanatophoric dwarfism type 1. Nat Genet 10:11–12

Rousseau F, el Ghouzzi V, Delezoide AL et al 1996 Missense FGFR3 mutations create cysteine residues in thanatophoric dwarfism type 1 (TD1). Hum Mol Genet 5:509-512

Sahni M, Ambrosetti DC, Mansukhani A, Gertner R, Levy D, Basilico C 1999 FGF signaling inhibits chondrocyte proliferation and regulates bone development through the STAT-1 pathway. Genes Dev 13:1361–1366

Shiang R, Thompson LM, Zhu YZ et al 1994 Mutations in the transmembrane domain of *FGFR3* cause the most common genetic form of dwarfism, achondroplasia. Cell 78:335–342

Stanescu R, Stanescu V, Maroteaux P 1990 Homozygous achondroplasia: morphologic and biochemical study of cartilage. Am J Med Genet 37:412–421

Su WCS, Kitagawa M, Xue NR et al 1997 Activation of Stat1 by mutant fibroblast growth-factor receptor in thanatophoric dysplasia type II dwarfism. Nature 386:288–292

Superti-Furga A, Eich G, Bucher HU et al 1995. A glycine 375-to-eysteine substitution in the transmembrane domain of the fibroblast growth factor receptor-3 in a newborn with achondroplasia. Eur J Pediatr 154:215–219

Tavormina PL, Rimoin DL, Cohn DH, Zhu YZ, Shiang R, Wasmuth JJ 1995a Another mutation that results in the substitution of an unpaired cysteine residue in the extracellular domain of FGFR3 in thanatophoric dysplasia type 1. Hum Mol Genet 4:2175–2177

Tavormina PL, Shiang R, Thompson LM et al 1995b Thanatophoric dysplasia (types I and II) caused by distinct mutations in fibroblast growth factor receptor 3. Nat Genet 9:321–328

Tobe T, Ortega S, Luna JD et al 1998 Targeted disruption of the FGF2 gene does not prevent choroidal neovascularization in a murine model. Am J Pathol 153:1641–1646

Twal WO, Vasilatos-Younken R, Gay CV, Leach RM Jr 1994 Isolation and localization of basic fibroblast growth factor-immunoreactive substance in the epiphyseal growth plate. J Bone Miner Res 9:1737–1744

Vortkamp A, Lee K, Lanske B, Segre GV, Kronenberg HM, Tabin CJ 1996 Regulation of rate of cartilage differentiation by Indian hedgehog and PTH-related protein. Science 273:613–622

Wang F, Kan M, Yan G, Xu J, McKeehan WL 1995 Alternately spliced NH2-terminal immunoglobulin-like loop I in the ectodomain of the fibroblast growth factor (FGF) receptor 1 lowers affinity for both heparin and FGF-1. J Biol Chem 270:10231–10235

Wang F, Kan M, McKeehan K, Jang JH, Feng SJ, McKeehan WL 1997 A homeo-interaction sequence in the ectodomain of the fibroblast growth factor receptor. J Biol Chem 272:23887–23895

Wang YC, Spatz MK, Kannan K et al 1999 A mouse model for achondroplasia produced by targeting fibroblast growth factor receptor 3. Proc Natl Acad Sci USA 96:4455–4460

Webster MK, Donoghue DJ 1996 Constitutive activation of fibroblast growth factor receptor 3 by the transmembrane domain point mutation found in achondroplasia. EMBO J 15:520–527

Webster MK, Donoghue DJ 1997 FGFR activation in skeletal disorders: too much of a good thing. Trends Genet 13:178–182

Webster MK, D'Avis PY, Robertson SC, Donoghue DJ 1996 Profound ligand-independent kinase activation of fibroblast growth factor receptor 3 by the activation loop mutation responsible for a lethal skeletal dysplasia, thanatophoric dysplasia type II. Mol Cell Biol 16:4081–4087

Wilkie AOM 1997 Craniosynostosis — genes and mechanisms. Hum Mol Genet 6:1647–1656

Xu J, Lawshé A, MacArthur CA, Ornitz DM 1999 Genomic structure, mapping, activity and expression of fibroblast growth factor 17. Mech Dev 83:165–178

Zhou M, Sutliff RL, Paul RJ et al 1998 Fibroblast growth factor 2 control of vascular tone. Nat Med 4:201–207

DISCUSSION

Ornitz: I would like to add that our recent unpublished work shows that ectopic expression of chimeric FGFRs in the joint space blocks development of the joint.

Kronenberg: I wondered whether any of the differences between what happens in the joint and what happens in the proliferative zone might have to do with the patterns of endogenous receptor expression. That is to say, with your type I chimera near the joint surface, not only is it being highly expressed there, but so also is normal type I, and type III isn't there, so you have lots of homodimers there. Whereas if you express FGF3R there, you also are expressing the endogenous type I receptor near the joint surface. When you make a transgenic, you have type I and type III receptor there, so that the type III receptor might be only in heterodimers near the joint surface, and I'm postulating that this might work less well that the type I which is expected to form only homodimers.

Ornitz: Certainly heterodimers can form. It will also depend on what ligands are present. We still have the FGFR3 extracellular domain that has different specificities for ligand than FGFR1, which is normally expressed on the periarticular surface. We still don't know what the endogenous ligand is in the growth plate. The chimeric *Fgfr3* transgene also has the achondroplasia mutation so it is activated to some extent in a ligand independent fashion. The chimeric receptor is going to be active regardless of ligand to some extent at the periarticular surface, and it is clearly dose-dependent since we have this difference between the homozygous (fused joints) and heterozygous animals (normal joints).

Kingsley: You mentioned a couple times the idea of coupling between the BrdU rates in the perichondrium and the growth plate.

Ornitz: This is a very striking observation.

Kingsley: Do you think any of that could come from the expression of the transgene itself?

Ornitz: The transgene uses the collagen II promoter, which is tightly restricted to the chondrocyte lineage; there's no evidence of expression of the transgene in the perichondrium. Expression of the collagen II promoter in the transgenic construct is shown clearly by the *in situ* hybridizations. Therefore, there has to be a diffusible molecule that can signal between the growth plate and the perichondrium. Obvious candidates are IHH and PTHrP, and possibly BMPs.

Meikle: I don't know whether this is relevant to what you're saying, but if you take long bones and transplant them into a non-functional environment, the long bones will continue to grow longitudinally, but there is a cessation of periosteal chondrogenesis. There is quite a marked demarcation between the two. This suggests that functional activity plays an important role in growth in width of the epiphyseal plate and hence the bone itself.

Russell: Is there any human clinical counterpart of the failure to form joint spaces, and what happens in the homozygous achondroplasia offspring which occasionally occur?

Ornitz: I'm not aware of any joint problems in achondroplasia patients but endogenous FGFR3 is not expressed in tissue adjacent to where the joint will form. What we're seeing is due ectopic expression of the chimeric receptor, but it does say that these cells are responsive to an FGFR signal, and suggests that FGF signals, possibly through FGFR1, may be involved in regulating the formation of the joint space.

Russell: In achondroplasia, it is striking how just this one mutation occurs sporadically and so consistently. What's the current thinking about why there aren't more mutations?

Ornitz: This mutation weakly activates the receptor. If you have a mutation that strongly activates the receptor, it would be lethal and you would not see survival of the embryo. The actual sequence where the achondroplasia mutation forms is one of the most common spontaneous mutation spots in the human genome. This has something to do with the content of the actual DNA sequence.

Wilkie: I don't buy that as the mechanism. It is not just FGFR3, but also FGFR1 and FGFR2, which are in completely different sequence contexts in different parts of the genome, all of which have elevated frequencies for very specific missense mutations. The other point is that, in the case of Apert, Crouzon and Pfeiffer syndrome mutations in *FGFR2* and the achondroplasia mutation in *FGFR3*, the parental origin of these mutations has been looked at and they're exclusively paternal in origin, so essentially they're arising only in sperm (Moloney et al 1996, Wilkin et al 1998, Glaser et al 2000). A hypothesis that I think is more attractive relates to the fact that FGFR signalling is important in gametogenesis (van Dissel-Emiliani et al 1996, Resnick et al 1998). Perhaps these mutations, which are all activating in one way or another (Wilkie et al 2000, this volume), can confer a selective advantage to those germ cells in which they arise. This would result in a higher birth prevalence for the mutations than you would normally expect for a point mutation.

Russell: You could presumably look at that directly by just looking at sperm.

D. Cohn: Some of those kinds of studies have been done. When you look at single sperm in males who have not fathered children, there is a proportion of sperm that carry that mutation. And this proportion appears to increase at a steady rate with age, which goes along with the paternal age effect seen in achondroplasia. However, the population to really look at is the population of normal fathers who have fathered children with achondroplasia, to see if they really represent a subset in which an early mutation or selection has occurred such that they are mosaic at a relatively high level.

Wilkie: The important point is the fact that you *can* identify these mutations in sperm from normal males who haven't fathered children with achondroplasia. The mere fact that you can detect them at the same sorts of rates at which you would identify live-born achondroplastics indicates that males who have fathered children with achondroplasia aren't very different from other males: they just happen to be the people whose rare FGFR3 mutant sperm have by chance fertilized an egg, thus giving rise to an achondroplastic offspring. So my own hunch is that the majority of fathers who have had an achondroplastic child don't have a higher rate of the mutation in their sperm than in other men.

D. Cohn: The experiments haven't been done, but among males who have not fathered achondroplastic children, there is quite a range of variation in the proportion of sperm carrying the mutation. This suggests there may be a subset of males with a higher proportion of mutated sperm. Whether that proportion is a stochastic effect or is related to this possibility of selection is not yet known.

Kingsley: You can look for the selection experimentally in chimeras of mouse mutants made between cells carrying a reconstruction of the relevant allele, and wild-type cells, because the prediction from the model would be that in the chimera the germ cells of the mutant would completely take over the germline.

Ornitz: Since we have done our transgenic study, the achondroplasia mutation has been knocked in to the endogenous *Fgfr3* locus. As far as I know, there is a Mendelian pattern frequency of inheritance in those mice.

Kingsley: You would have to make the chimeric mouse to test this, because if all the cells carry the mutation then there is no selective advantage to having it.

Wilkie: In Apert syndrome, which is due to mutations in *FGFR2*, one of the clinical features is that the number of phalangeal elements is reduced from three to two (Wilkie et al 2000, this volume). This presumably reflects the fact that one of the joints hasn't formed. I would like to ask a question that leads on from this. Is there any evidence that the likelihood of forming a joint may depend in some way on the overall mass or length of a condensing mesenchymal element? What I'm suggesting here is a model whereby if you shorten elements down to below a critical size, then the joint doesn't form. That is, there may be some sort of measurement process, and it may not be something specific to receptor signalling.

Ornitz: I would say probably not, because in achondroplasia, where there is very severe dwarfism, joints are formed normally.

Kingsley: An important question concerns where the shortening occurs. The type of mechanism that you are mentioning could be relevant, if there is drastic shortening of early condensation. I don't know what the clinical cause of the shortening is in the human example you just mentioned, but it could be that the early condensations are completely normal.

Kronenberg: With regard to the BMP4 effect, I recall that in the *Ihh* knockout mouse, BMP4 in the growth plate is normal (A. McMahon, unpublished data). This would argue that the dramatic decrease you're seeing might either be an effect of the FGFR itself on BMP4, or some other indirect effect, rather than due to the decrease in IHH.

Ornitz: The difference between that result and our result is in development. The *Ihh* knockout mice die in late embryogenesis, so St-Jacques et al (1999) are looking at BMP4 in late embryogenesis. The stage at which we looked at BMP4 is at 14 days postnatally. If we look at BMP4 in the achondroplasia mice early on, there's no difference. The effect we're seeing is later in development. Andy McMahon's lab is planning to conditionally knockout *Ihh* postnatally to test this idea.

Kronenberg: The decrease in *Ihh* and, in particular, in *Patched*, is very striking. Could the thinness of the line of prehypertrophic cells, and the thinness of the line of hypertrophic cells, be a passive consequence of the lack of proliferation of the precursor cells in those compartments? That is, if there are fewer proliferating cells and everything else stays normal, the expectation is the thinner line. I don't think the thinner line needs a unique explanation.

Hall: The area of perichondrium in which you saw the reduced proliferation was in the distal zone around the epiphysis, the zone that is called perichondrial ring, from which cartilage, fibrous tissue and bone comes. It is an area that we call perichondrium but which actually does a number of different things, some of which are functionally dependent, so it will be interesting to see what was developing in that zone.

Ornitz: We haven't looked at any markers. I don't know what markers to use to distinguish different populations of cells in the perichondrial ring.

Wilkie: You said that at E18 there was no difference in the amount of BrdU labelling in your mutants and wild-type. Are you implying that it's really an early embryonic phenotype, and that later on there are no measurable functional differences?

Ornitz: In activating mutations in *Fgfr3*, the phenotype begins postnatally. But in the loss-of-function *Fgr3* knockout animals we detect a histologic phenotype in which the size of the hypertrophic and proliferating zone are increased at late embryonic stages (E16–18). I should add that the collagen II promoter FGFR3G380R transgene is expressed early, it is just that the cells are not sensitive to increased FGFR3 activity early in development. The presumption would be that chondrocytes are maximally stimulated by ligand in that tissue and the receptor only becomes rate-limiting postnatally.

Morriss-Kay: Have you looked at the development of the skull in either of your two mouse models in relation to both the tissue developmental functional aspects and in terms of the other *Fgfr* gene expression patterns?

Ornitz: We have not looked at other *Fgfr* gene expression patterns in these animals. The facial bones are smaller, but we haven't actually looked at anything in great detail.

Morriss-Kay: And not the skull vault?

Ornitz: Fgfr3 is expressed around the sutures; there could be a subtle phenotype that we have missed. We haven't looked that closely.

Kingsley: Where would the collagen II transgene drive expression in the skull?

Ornitz: There is a band of collagen underlying the suture.

Hall: But you don't see fusion of the sutures in the skull?

Ornitz: No. There is no cranial synostosis.

Hall: There's no connection between joint changes that occur in the long bones and the changes in the sutures?

Ornitz: In the homozygous animals there is no suture phenotype.

Morriss-Kay: You would expect the opposite, because in achondroplasia the skull is enlarged.

Ornitz: In achondroplasia, I have heard some reports that there may be an increased brain size, and *Fgfr3* is expressed throughout the CNS, and probably in glial cells. It is therefore conceivable that part of the frontal bossing seen in achondroplasia patients could be secondary to a larger brain.

References

Moloney DM, Slaney SF, Oldridge M et al 1996 Exclusive paternal origin of new mutations in Apert syndrome. Nat Genet 13:48–53

Glaser RL, Jiang W, Boyadjiev SA et al 2000 Paternal origin of FGFR2 mutations in sporadic cases of Crouzon and Pfeiffer syndrome. Am J Hum Genet 66:768–777

St-Jacques B, Hammerschmidt M, McMahon AP 1999 Indian hedgehog signaling regulates proliferation and differentiation of chondrocytes and is essential for bone formation. Genes Dev 13:2072–2086

Wilkin DJ, Szabo JK, Cameron R et al 1998 Mutations in fibroblast growth factor receptor 3 in sporadic cases of achondroplasia occur exclusively on the paternally derived chromosome. Am J Hum Genet 63:711–716

van Dissel-Emiliani FMF, de Boer-Brouwer M, de Rooij DG 1996 Effect of fibroblast growth factor-2 on Sertoli cells and gonocytes in coculture during the perinatal period. Endocrinology 137:647–654

Resnick JL, Ortiz M, Keller JR, Donovan PJ 1998 Role of fibroblast growth factors and their receptors in mouse primordial germ cell growth. Biol Reprod 59:1224–1229

Wilkie AOM, Oldridge M, Tang A, Maxson RE Jr 2001 Craniosynostosis and related limb anomalies. In: The molecular basis of skeletogenesis. Wiley, Chichester (Novartis Found Symp 232) p 122–143

Defects of human skeletogenesis — models and mechanisms

Stefan Mundlos

Universitätsklinikum der Humboldt-Universität zu Berlin, Campus Charité Mitte, Institut für Medzinische Genetik, Schumannstrasse 20/21, 10117 Berlin, Germany

Abstract. Heritable diseases of the skeleton are a highly complex group of genetic disorders. Skeletal morphogenesis involves, in principle, four distinct developmental processes: patterning, organogenesis, growth and homeostasis. Defects in patterning affect the number and shape of bones and will result in dysostosis. Organogenesis involves the formation of bone and cartilage as an organ. Defects in growth plate function lead to abnormal proliferation and/or differentiation of chondrocytes resulting in dwarfism and dysplasia. Bone mass, shape and strength are maintained in equilibrium throughout development and adulthood (homeostasis). Animal studies are providing good correlations between specific embryological events and gene function, and consequently a framework for understanding the fundamental pathways that build and pattern bone. Based on the remarkable conservation of basic developmental mechanisms between animal species, connections to human disorders are frequently possible. As examples for recent advances in our understanding of the processes that underlie skeletal pathology, the molecular basis of a patterning defect, synpolydactyly, and a defect of organogenesis, cleidocranial dysplasia, will be presented and discussed.

2001 The molecular basis of skeletogenesis. Wiley, Chichester (Novartis Foundation Symposium 232) p 81–101

Skeletal morphogenesis involves, in principle, four distinct developmental processes: patterning, organogenesis, growth, and homeostasis. Patterning describes functions that play a role in the eventual size, shape (gestalt) and number of individual skeletal elements. Genes effecting patterning are expressed in progenitor cells long before overt skeletogenesis. Accordingly, defects in these genes will not affect the skeleton as a whole but can be expected to result in an altered number and shape of individual skeletal elements. Organogenesis involves the differentiation of progenitor cells into chondrocytes or osteoblasts, and the subsequent formation of bone and cartilage as an organ. All longitudinal growth occurs in the growth plates, highly specialized regions of cartilage in which chondrocytes proliferate and differentiate, and where cartilage is ultimately transformed into bone. Defects in growth plate function result in altered growth rates and consequently in dwarfism. Bone mass, shape and strength are maintained in equilibrium throughout development and adulthood by the compensating forces of bone resorption (osteoclasts) and bone formation (osteoblasts).

The study of genetic diseases of bone offers the unique opportunity to identify new genes that have important roles during skeletal development and/or maintenance. At the same time, the patients' phenotype can give important clues to the possible function of the mutated gene. However, frequently the function of a disease-causing gene cannot be investigated in humans. This is particularly true for developmental genes that act during the first weeks after conception. Model organisms are needed to elucidate the function of these genes. Animal studies are providing a good correlation between specific embryological events and gene function, and consequently a framework for understanding the fundamental pathways that build and pattern bone. Based on the remarkable conservation of basic developmental mechanisms between animal species, connections to human disorders and vice versa are frequently possible.

Patterning the limb

The appendicular skeleton originates from a dual contribution of the lateral plate and the somitic mesoderm. After the initial outgrowth of the limb bud, cells from the lateral edges of nearby somites migrate into the limb muscles, nerves and vasculature. All other limb tissues, including the skeletogenic mesenchyme, are derived from the lateral plate mesoderm. Skeletal elements are laid down as cartilaginous templates in a proximal to distal fashion, so that the humerus forms first, followed by radius and ulna and lastly the digits.

Cells of the limb bud are embedded in a dynamic three-dimensional structure controlled by multiple signalling molecules that ultimately determine their fate. Hox genes of the A and D clusters play an important role in this process. They are expressed in a stage-dependent pattern that determines the shape and identity of the individual elements. How this works and what the downstream targets of the Hox genes are is not clear, but it seems that they function in a dose-dependent fashion regulating proliferation and cell adhesion. Congenital limb defects and knock out experiments in the mouse provide a first glance at the mechanisms by which these genes control limb development.

HOXD13 is mutated in synpolydactyly

Syndactyly (the osseous or cutaneous fusion of fingers or toes) and polydactyly (extra fingers or toes) are some of the most common malformations in humans. In many instances they occur together resulting in a complex malformation of the limb. Synpolydactyly, or syndactyly type II (SPD), comprises syndactly

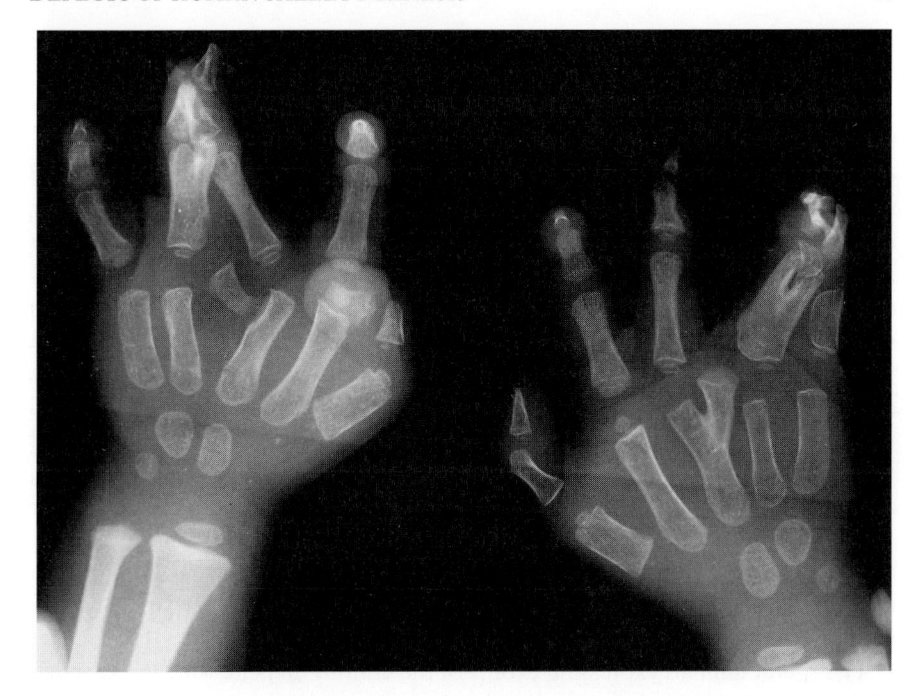

FIG. 1. Typical case of synpolydactyly in a 3 year old patient. Note V-shaped metacarpal III, duplication of phalanges and bony fusion with phalanges of finger IV.

between the third and fourth fingers with an additional finger in the syndactylous web (Fig. 1). The feet are similarly affected with postaxial polydactyly (extra digit at the lateral side) and a variable syndactyly of toes III to VI. The condition is inherited as a dominant trait with variable expressivity and penetrance. Usually all four limbs are involved, but many individuals have both normal and affected hands/feet indicating the presence of intra-individual variability.

Synpolydactyly is caused by mutations in *HOXD13* (Muragaki et al 1996). The mutations lead to an in frame expansion of a 15-residue polyalanine tract, encoded by an imperfect trinucleotide repeat sequence by 7, 8, 9, 10, or, in one family, 14 additional residues (Goodman et al 1997). Expansions of a polyalanine tract represent a new mechanism of mutation that had not been described before. Subsequently, similar expansions were observed in other transcriptions factors such as *CBFA1* (Mundlos et al 1997) supporting the concept that alanine expansions may cause disease. Further support came from the discovery of a mouse mutant that represents an exact pheno- and genocopy of human SPD. This mutant has been named synpolydactyly homologue (spdh). The underlying mutations was identified as an expansion of alanines from 15 to 22 residues (Johnson et al 1998).

The polyalanine expansions described here are reminiscent of trinucleotide expansions observed in neurodegenerative diseases. However, in these conditions, the CAG trinucleotide repeat sequence encodes for glutamine stretches of variable length (Huntington disease, spinocerebellar ataxia, and others). Expansion of the CAG repeat beyond a certain threshold (37–40 residues) leads to disease. Whereas CAG trinucleotide repeats are invariably unstable when inherited from one generation to the next and frequently show varying degrees of somatic mosaicism, polyalanine coding repeats are generally not polymorphic and are very stable, when passed on from one generation to the next. In *HOXD13* the trinucleotide repeat encodes for a series of alanines located within the N-terminal part of the protein. Such alanine repeats are extremely common among Hox genes and other transcription factors. Their specific function is unknown but removal results in a loss of function, as demonstrated for the *Drosophila* even-skipped protein (Han & Manley 1993). Interestingly, *in vitro* experiments have demonstrated that the expansion of an alanine tract beyond 21 repeat units results in the complete loss of transactivation activity (Lanz et al 1995). This is in agreement with the clinical data where a total length of 22 alanines (additional 7 repeats) is the shortest disease causing repeat. Thus, like in the glutamine repeat diseases, there seems to be a threshold.

Transgenic animals expressing exon 1 of the huntingtin gene with an expanded glutamine repeat under the huntingtin promotor exhibit a progressive neurological phenotype most likely caused by neuronal intranuclear inclusions containing the proteins huntingtin and ubiquitin (Mangiarini et al 1996, Davies et al 1997). Thus, the expansion of the polyglutamine stretch within huntingtin results in the accumulation of degraded protein in the nucleus, a mechanism that may lead to a gain of function explaining the dominant negative effect of the mutation.

Whether the *HOXD13* mutation in SPD works in a similar way is unknown. Like in the glutamine expansions, this mutation is unlikely to be a simple loss of function, because mice with targeted disruption of *Hoxd13* do not show an SPD-like phenotype (Zákány et al 1997). One explanation for the weak phenotype in *Hoxd13* knockouts is that other Hox genes expressed in the limb such as *Hoxd11*, *Hoxd12* and *Hoxa13* compensate for the loss of *Hoxd13*. In contrast, the mutant HOXD13 protein, having the largest expression domain, may interfere with other Hox genes and partially inactivate them.

This concept is supported by knockout experiments performed in D. Duboule's lab. The inactivation of *Hoxd12*, *Hoxd11* and *Hoxd13* results in a phenotype that is strikingly similar to human SPD (Zákány & Duboule 1996). In these mice the overall size of the limb buds was reduced but the strongest defects were seen in the digits, where polydactyly, fusions and webbing were observed in all homozygotes. Simple *Hoxd13* mutant mice showed a much milder phenotype, whereas mice mutated for either *Hoxd11* or *Hoxd12* were normal. This illustrated the functional cooperation between these genes. Since this triple knockout phenocopies human SPD, it seems likely that the alanine expansion in HOXD13 works through the functional neutralization of several other Hox proteins.

In a study by Goodman et al (1997), we have analyzed a large number of families with SPD (a total of 20 pedigrees and 99 affected individuals) for their clinical phenotypes and the corresponding mutations in *HOXD13*. The mildest phenotype was found to be an isolated fourth finger brachydactyly or a fifth finger camptodactyly. Syndactyly III/IV without digit duplication was also observed and occurred in approximately one-third of affected hands of 7- and 8-alanine expansions carriers. In contrast, there was a duplicated digit in the syndactylous web in virtually all affected hands of 9-, 10- and 14-alanine expansion carriers. Interestingly, penetrance was also related to the length of the alanine tract. Close to 40% of individuals with 7-alanine expansions had no hand involvement. In contrast, all of the individuals with expansions of 8 and more alanines had one or more hands involved. Penetrance and expressivity was clearly related to expansion size.

One individual had a different phenotype, which first suggested a different diagnosis. However, analysis of her pedigree revealed that her parents were first cousins with her mother having an SPD-like phenotype, and she had two affected sons that had typical SPD. Mutational analysis demonstrated a homozygous expansion by 10 alanines in this patient. Her phenotype is interesting in that it is substantially different from the heterozygous phenotype. Her hands were very small and there was syndactyly between fingers III, IV and V. Her feet were short and had only three toes (Fig. 2). The metacarpals were extremely short and had the appearance of carpal bones. Similar changes were observed in the feet. The dominant feature in this case is the brachydactyly/hypodactyly which stand in contrast to the polydactyly characteristic for heterozygous SPD. It seems difficult to reconcile how polydactyly and hypodactyly can be part of a continuous spectrum.

Experiments with knock out mice may provide some explanation. As mentioned above, inactivating *Hoxd11*, *Hoxd12* and *Hoxd13* causes a SPD-like phenotype. Removing one more of the limb-expressed Hox genes, in this case *Hoxa13*, causes severe brachydactyly and hypodactyly. Zákány et al (1997) describe a dose-dependent mechanism by which Hox genes of the 11, 12, and 13 group control both the size and the number of digits. Thus, a progressive reduction in the dose of Hox gene products leads first to postaxial polydactyly, then synpolydactyly and finally cumulates in oligodactyly/adactyly. The phenotype of mice lacking all *Hoxd11–13* genes and one allele of *Hoxa13* resembles in many ways the phenotype of the above described homozygous patient. It thus seems conceivable that the dosage of mutant alanine expansions as well as their size control the phenotype in a dose-dependent manner. Figure 2 demonstrates this concept.

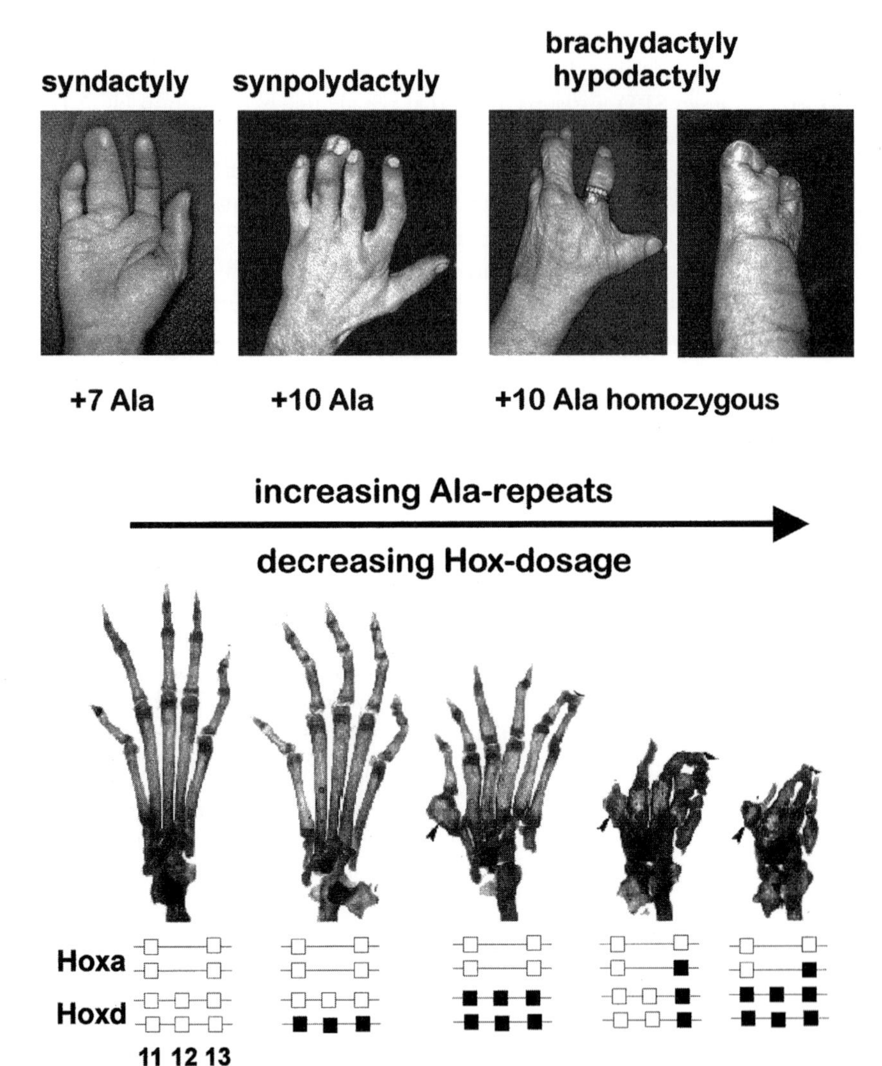

FIG. 2. Increasing alanine repeats and decreasing dosage of *Hox* genes cause similar defects. Top panel from left to right shows hands from: a patient with a +7 alanine expansion and syndactyly of fingers III/IV, the mildest form of SPD, a patient with a +10 alanine expansion showing the full clinical picture with syndactyly III/IV (surgically removed) and polydactyly in the syndactylous web, mother of this patient with a homozygous +10 alanine expansion showing severe brachydactyly, syndactyly of fingers III/IV and oligodactyly of feet. Bottom panel shows dose dependent variation of dactyly in feet of mice with varying *Hox* genotypes. Genes from the *Hoxd* and *Hoxa* clusters were inactivated as schematized below with the genes whose functions were removed in black. The progressive inactivation goes together with a transition from pentadactyly (left) to polydactyly (middle) to oligodactly (right). From Zákány et al (1997), with permission.

Organogenesis of cartilage and bone

Organogenesis involves the formation of bone and cartilage as an organ (skeleton). Condensation of undifferentiated cells follows a pattern set by the early patterning genes. Differentiation of condensed cells into chondrocytes or osteoblasts follows, and these cells start producing a bone- or cartilage-specific extracellular matrix (a process referred to as histogenesis). These aspects of early skeletogenesis, the 'anlage', represent the outlines of the future skeletal elements. Bones then develop either directly from osteoblastic progenitor cells (desmal or intramembranous) or within a temporary framework of hyaline cartilage (endochondral). Recently, disease related genes have been characterized that are involved in the regulation of organogenesis. As expected, mutations in these genes result in altered shapes and/or numbers of certain skeletal elements since they affect the differentiation of progenitor cells and hence the formation of the anlage.

One example of such a gene is the recently identified transcription factor *Cbfa1*. *Cbfa1* belongs to the core binding factor (Cbf) transcription factors, a small family of heterodimeric proteins of two unrelated subunits comprising a DNA binding α subunit and a non-DNA binding β subunit. The mammalian Cbfα subunits are encoded by three distinct genes, *Cbfa1*, *Cbfa2* and *Cbfa3*. They share a conserved 128 amino acid domain, called the runt domain because of its homology to the *Drosophila* pair-rule gene *runt*.

CBFA1 is mutated in cleidocranial dysplasia

CBFA1 is mutated in cleidocranial dysplasia (CCD), a dominantly inherited skeletal malformation syndrome (Mundlos et al 1997, Lee et al 1997, Quack et al 1999). Patients with CCD exhibit a number of typical clinical findings including (i) hypoplasia/aplasia of the clavicle, (ii) abnormal craniofacial growth, (iii) supernumerary teeth, (iv) short stature, and (v) several other minor skeletal changes (for review see Mundlos 1999) (Fig. 3).

Most mutations in *CBFA1* are expected to result in heterozygous loss of function indicating that haploinsufficiency is sufficient to cause disease. Deletions of the entire gene, splice site mutations within the first exons, insertions resulting in frame shifts and missense mutations are common (Mundlos et al 1997). Two nonsense mutations located within the DNA binding runt domain were shown to interfere with DNA binding (Lee et al 1997). Other point mutations located in the region coding for the nuclear localization signal apparently inactivate this signal resulting in a protein that stays in the cytoplasm instead of being transported into the nucleus (Quack et al 1999).

Mutation analysis in a wide spectrum of individuals has helped to define and extend the phenotypic spectrum associated with mutations in *CBFA1*.

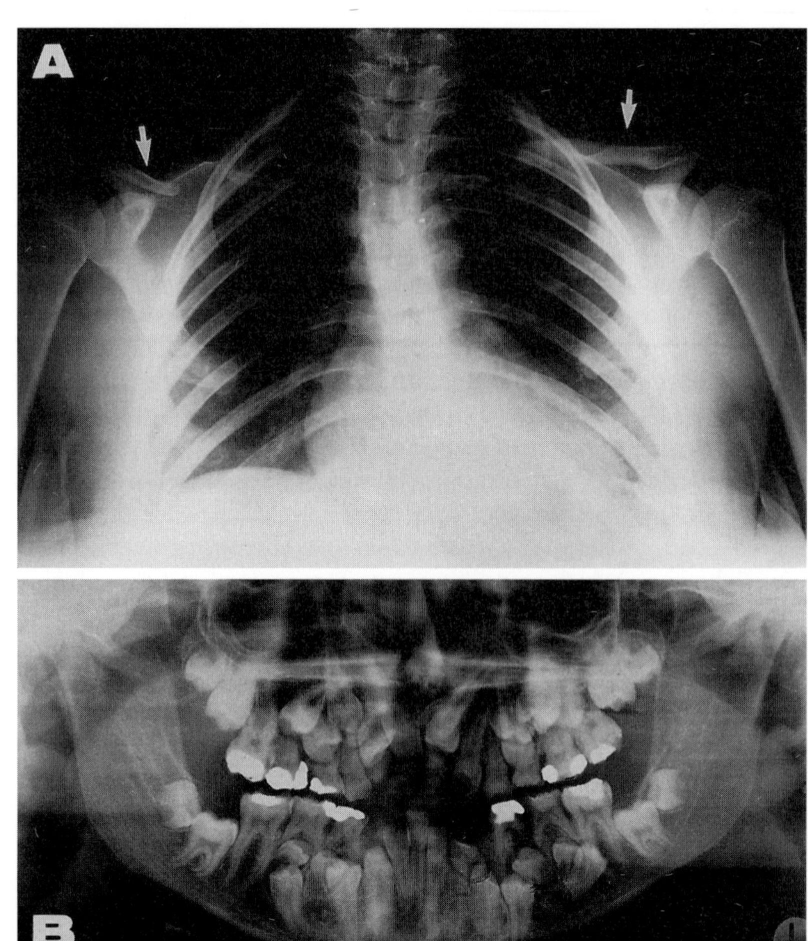

FIG. 3. Radiological findings in cleidocranial dysplasia. (A) Chest radiograph demonstrating cone-shaped thorax, low set shoulders and bilateral clavicular hypoplasia (right > left, arrows). (B) Panoramic radiograph of teeth from an affected patient aged 15 years showing multiple impacted teeth and partly persistent deciduous dentition.

Supernumerary teeth may be the only finding, indicating that mutations in *CBFA1* can be associated exclusively with a dental phenotype. One patient with a frame shift mutation 3′ of the runt domain coding region had, in addition to severe CCD, congenital osteoporosis, severe skoliosis and recurrent fractures. At age 26 this patient had a bone density of −6.34 (DEXA scan of left forearm, value in SD with < -1.5 representing osteopenia) (Quack et al 1999). These results indicate that *CBFA1* has a role beyond development and is necessary for the

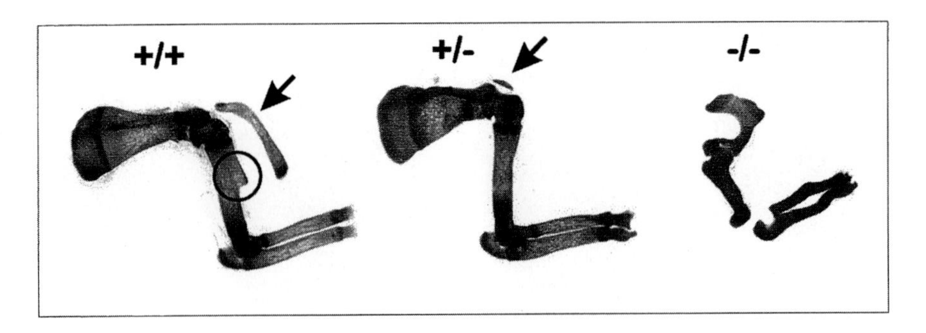

FIG. 4. *Cbfa1* in skeletal development. Alizarin red alcian blue stained skeletal preparations of fore limbs from wild-type ($^{+/+}$), *Cbfa1*$^{+/-}$ and *Cbfa1*$^{-/-}$ mice. Note the size of the clavicle (arrowed) in wild-type and heterozygous mice. The tuberositas humeri (circled) is absent in *Cbfa1*$^{+/-}$ mice. The skeleton of *Cbfa1*$^{-/-}$ mice is without signs of ossification and severely reduced in size. Note the defect in the scapula.

homeostasis of adult bone mass. This is in agreement with data generated from transgenic mice that express a dominant negative form of *Cbfa1* under the control of an osteoblast-specific promotor (Ducy et al 1999).

The role of *Cbfa1* in development has been elucidated by the generation of mutated mice in which the *Cbfa1* gene locus was targeted (Otto et al 1997, Komori et al 1997). Inactivation of one allele results in skeletal changes that are remarkably similar to those observed in CCD (Fig. 4). *Cbfa1*$^{-/+}$ mice have hypoplastic clavicles and open fontanelles as well as other minor skeletal abnormalities frequently observed in CCD patients such as changes in the pelvis, vertebrae and ribs (Mundlos et al 1996, Otto et al 1997). The development of the clavicle has been investigated in these mice in some detail. The study by Huang et al (1997) demonstrates that the clavicle forms by the condensation of mesenchyme as early as E13. Cells in the centre of the condensation differentiate into characteristic precursor cells that express markers of the chondrogenic (collagen type II) and osteoblastic (collagen type I) lineage. In *Cbfa1*$^{-/+}$ mice the condensation takes place, yet the differentiation into precursor cells and consequently into chondrocytes and osteoblasts is lacking. These studies suggested that CBFA1 is a crucial factor in skeletal development regulating the differentiation of mesenchymal stem cells in bone and cartilage precursors.

This concept is further substantiated by the phenotype of mice completely deficient for *Cbfa1*. Such mice die immediately after birth owing to a complete absence of bone (Otto et al 1997, Komori et al 1997). Further histological analysis showed an arrest in endochondral, as well as membranous ossification. *Cbfa1*$^{-/-}$ mice develop normal cartilage anlagen but differentiation of mesenchymal stem cells into osteoblasts is completely missing (Fig. 4). Hence,

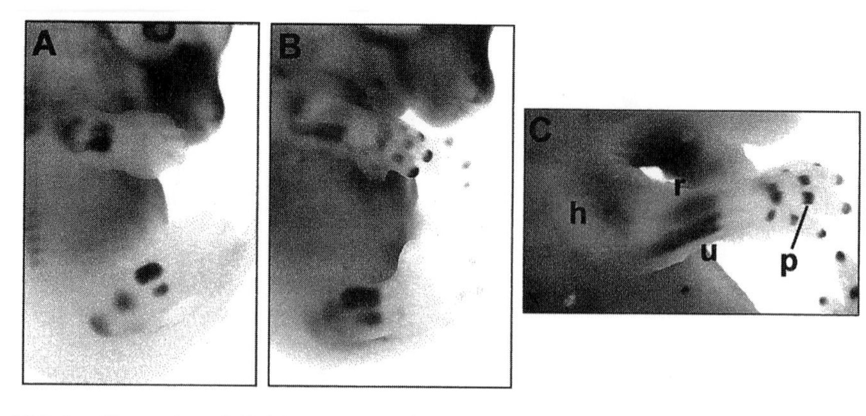

FIG. 5. Expression of *Cbfa1* in mouse embryos of stage (A) E12.5, (B) E13.5 and (C) E14.5. Note strong expression in the primordia of the facial bones (A, B) and in the cartilaginous anlagen of radius and ulna (r,u), humerus (h), and in phalanges (p). Ossification appears in humerus and radius/ulna at E14.5. With permission from Kim et al (1999).

structures such as the skull, the mandible and the maxilla do not develop, whereas the cartilaginous anlagen are present but are not replaced by bone. *Cbfa1* is thus essential for osteoblast differentiation.

More detailed investigations showed that *Cbfa1* is not only essential for osteoblast formation, but also a major regulator of chondrocyte differentiation. In *Cbfa1*$^{-/-}$ mice hypertrophy does not take place or is severely delayed (Kim et al 1999). Interestingly, the effect of *Cbfa1* on chondrocyte differentiation differs in the three fore limb segments. Whereas the humerus completely lacks hypertrophic cells, radius and ulna are less affected and will eventually show some hypertrophy and ossification. In the phalanges, hypertrophy is initiated but not maintained. A direct role for *Cbfa1* in chondrocyte differentiation is supported by its expression pattern. At E12.5 *Cbfa1* is strongly expressed in all cartilage anlagen long before ossification is present (Fig. 5). Besides in the perichondrium and in mature osteoblasts *Cbfa1* is expressed in hypertrophic and pre-hypertrophic chondrocytes. The expression in chondrocytes overlaps with that of collagen type X (exclusively expressed in hypertrophic cells) and that of indian hedgehog (expressed in pre-hypertrophic cells). A role for *Cbfa1* in chondrocyte differentiation can provide an explanation for the short stature of CCD patients.

References

Davies SW, Turmaine M, Cozens BA et al 1997 Formation of neuronal intranuclear inclusions underlies the neurological dysfunction in mice transgenic for the HD mutation. Cell 90: 537–548

Ducy P, Starbuck M, Priemel M et al 1999 A Cbfa1-dependent genetic pathway controls bone formation beyond embryonic development. Genes Dev 13:1025–1036

Goodman FR, Mundlos S, Muragaki Y et al 1997 Synpolydactyly phenotypes correlate with size of expansions in HOXD13 polyalanine tract. Proc Natl Acad Sci USA 94:7458–7463
Han K, Manley JL 1993 Transcriptional repression by the *Drosophila* even-skipped protein: definition of a minimal repression domain. Genes Dev 7:491–503
Huang L, Fukai N, Selby PB, Olsen BR, Mundlos S 1997 Mouse clavicular development: analysis of wild-type and cleidocranial dysplasia mutant mice. Dev Dyn 210:33–40
Johnson KR, Sweet HO, Donahue LR, Ward-Bailey P, Bronson RT, Davisson MT 1998 A new spontaneous mouse mutation of Hoxd13 with a polyalanine expansion and phenotype similar to human synpolydactyly. Hum Mol Genet 7:1033–1038
Kim IS, Otto F, Zabel B, Mundlos S 1999 Regulation of chondrocyte differentiation by Cbfa1. Mech Dev 80:159–170
Komori T, Yagi H, Nomura S et al 1997 Targeted disruption of Cbfa1 results in a complete lack of bone formation owing to maturational arrest of osteoblasts. Cell 89:755–764
Lanz RB, Wieland S, Hug M, Rusconi S 1995 A transcriptional repressor obtained by alternative translation of a trinucleotide repeat. Nucleic Acids Res 23:138–145
Lee B, Thirunavukkarasu K, Zhou L et al 1997 Missense mutations abolishing DNA binding of the osteoblast-specific transcription factor OSF2/CBFA1 in cleidocranial dysplasia. Nat Genet 16:307–310
Mangiarini L, Sathasivam K, Seller M et al 1996 Exon 1 of the HD gene with an expanded CAG repeat is sufficient to cause a progressive neurological phenotype in transgenic mice. Cell 87:493–506
Mundlos S 1999 Cleidocranial dysplasia: clinical and molecular genetics. J Med Genet 36:177–182
Mundlos S, Huang LF, Selby P, Olsen BR 1996 Cleidocranial dysplasia in mice. Ann NY Acad Sci 785:301–302
Mundlos S, Otto F, Mundlos C et al 1997 Mutations involving the transcription factor CBFA1 cause cleidocranial dysplasia. Cell 89:773–779
Muragaki Y, Mundlos S, Upton J, Olsen BR 1996 Altered growth and branching patterns in synpolydactyly caused by mutations in HOXD13. Science 272:548–551
Otto F, Thornell AP, Crompton T et al 1997 Cbfa1, a candidate gene for cleidocranial dysplasia syndrome, is essential for osteoblast differentiation and bone development. Cell 89:765–771
Quack I, Vonderstrass B, Sock M et al 1999 Mutation analysis of core binding factor A1 in patients with cleidocranial dysplasia. Am J Hum Genet 65:1268–1278
Zákány J, Duboule D 1996 Synpolydactyly in mice with a targeted deficiency in the HoxD complex. Nature 384:69–71
Zákány J, Fromental-Ramain C, Warot X, Duboule D 1997 Regulation of number and size of digits by posterior Hox genes: a dose-dependent mechanism with potential evolutionary implications. Proc Natl Acad Sci USA 94:13695–13700

DISCUSSION

Karsenty: You mentioned that CCD patients have a mild short stature. We know that in both mice and humans, *Cbfa1* expression and function is extremely superimposable. You showed that the effect on chondrocytes is mostly on the humerus. How can this explain the short stature phenotype?

Mundlos: It doesn't.

Karsenty: We showed that in *Cbfa1* mice overexpressing a dominant-negative form of *Cbfa1* after birth there is a normal number of osteoblasts, although there is less bone, as you showed in the patients. Do you have any histology or any other indication that CCD patients have a normal number of osteoblasts?

Mundlos: We have no histology. To me, that patient is somewhat in between the 100% and 50% gene loss. It looks as if he has 70–80% gene loss. This would correlate with your knockout mice. Perhaps there is some sort of threshold beyond that, and then there is an effect on the number of osteoblasts. This patient was born with an osteoporotic skeleton and has maintained this. He probably just started off with a lower number of osteoblasts.

Bard: You didn't touch on what I thought was the most extraordinary feature of the human phenotype — the increase in number of teeth. And yet the gene seemed work through a completely different system (exoskeleton, neural crest cells). Does the mouse model also have abnormal numbers of teeth?

Mundlos: No, but that can quite easily be explained, because mice have only one set of teeth. In humans, it's only the secondary teeth — the permanent teeth — that are affected. Why this happens remains a mystery.

Meikle: Most of those supernumerary teeth are premolars. In mice the whole premolar and canine series has been silenced.

Karsenty: One hypothesis could be that bone tissue prevents tooth formation, and by having less bone you have a permissive effect on tooth formation. This is a great hypothesis because it is not testable!

Ornitz: Is there any effect on fibroblast growth factor (FGF) receptor 3 expression?

Mundlos: I don't know. We looked at proliferation, and that does not seem to be the reason why they are short. In that sense it is likely that FGF receptor 3 will not be affected.

Wilkins: In genetics courses one is told that most loss-of-function mutations are recessive. It is my impression that many transcription factor loci are haplo-insufficient. Is there a compilation of data on this? Has anyone actually looked at the databases to see if this is a general property of transcription factor loci?

Mundlos: That is probably true. Interestingly, in mice this is not a general assumption. Mice frequently don't have a phenotype in the heterozygote, and you need to have the homozygous to see an effect. In this respect, *Cbfa1* is really quite an exception.

D. Cohn: Certainly for many structural proteins, haploinsufficiency in humans produces a phenotype.

Wilkins: I wasn't saying it would be diagnostic for or true only of transcription factor genes. But quite often it seems to be true for these genes; in *Drosophila* too, it's not just humans.

Mundlos: There seems to be a very important dosage effect.

Morriss-Kay: Why do you think the clavicle is singled out to be affected in the heterozygotes? Is it anything to do with the peculiar way in which the clavicle is made, being both a membrane bone and an endochondral bone, or is there some genetic precedent?

Mundlos: It is difficult to say. We have looked at the clavicles in these mice. It seems to be a bone which, from the developmental point of view, is very different to all the others. The precursor cells are unlike osteoblasts or chondrogenic precursor cells, and express either type II or type I collagen. Also, a condensation forms, but the precursor cells do not differentiate.

Morriss-Kay: The other special feature of the clavicle is its evolutionary origin from the dermal skeleton, which was added secondarily to the endoskeletal pectoral girdle (Goodrich 1958).

Hall: Therefore it has closer links to the teeth than it does to the endoskeleton, which is interesting in terms of the supernumerary teeth phenotype.

The cartilage on the mouse clavicle is secondary: it arises from the periosteum after bone formation. If you are losing the membrane bone first, but still getting the cartilage, is it coming from the periosteum?

Mundlos: It's not really secondary. In early condensations, there are large, peculiar precursor cells that build up in the middle, and around these cells, there are condensed cells. Some of these precursor cells express type I collagen, so they are more in the osteoblastic lineage, whereas the others express type II collagen, so they are more in the chondrogenic lineage. Then, at the same time there is bone formation and cartilage formation, and at the border between the two they form a growth plate.

Hall: There must be big differences between the human clavicle and the mouse clavicle.

Mundlos: A little bit of a compact bone is built at a later stage.

Kingsley: Cbfa1 is not specific to osteoblasts. In embryonic development it shows up earlier in both chondrogenic and osteoblast precursors. Is this the only known example of a phenotype in the cartilage-based element? Most of the phenotypes that are usually described are the dramatic loss of the alizarin red-stained bone throughout the skeleton.

Kronenberg: There are a lot of markers of differentiation that are missing, decreased or delayed in the chondrocytes. The chondrocytes do not appear to be normal. Nobody knows whether that's because the precursors which have reasonably high level *Cbfa1* expression have a delayed effect, or whether it's the low level of *Cbfa1* in those chondrocytes that none the less matters.

Mundlos: There is a cartilage phenotype in the heterozygotes, in the patients, because there are all these skeletal abnormalities on the X-ray. There is an abnormal metaphysis and epiphysis, and there is a small pelvis: this is all due to the abnormal growth of cartilage.

Wilkie: Have you considered the possibility that your patient with the *CBFA1* mutation and atypical, severe phenotype could have a second mutation somewhere else? In the mouse, how much does the genetic background affect either the phenotype or the detailed timing and location of gene expression?

Mundlos: We have not looked at the timed expression. We have crossed our mouse in a few different backgrounds, and there is absolutely no difference. The phenotype is very strong and very penetrant.

Wilkie: Going back to your *short-digit* mouse, the obvious question is, have you looked for a *sonic hedgehog* (*Shh*) mutation in type A1 brachydactyly? If it is not caused by an *Shh* mutation, do you think that the *short-digit* phenotype could be attributable to the deletion that you're seeing near to *Shh*? Is this deleting another gene as well as causing a position effect on *Shh*?

Mundlos: We have tested a family for linkage to the syntenic region, and it does not map there. If there was a *trans* effect, meaning there is a gene that maps close to it and which is either deleted or mutated or whatever, then the complementation study would have not worked. That is, if you were to cross them, then the double heterozygotes should not have the phenotype, but since they have exactly the same phenotype it is highly likely that we are actually talking about the same gene. It must therefore be something which regulates it, unless we have missed the mutation of the gene, which I hope we haven't.

Wilkie: How would you explain why a deletion in *cis* to *Shh* is associated with a more severe phenotype in the heterozygote than knocking out the gene? Is the gene 'knockout' not a true knockout, for instance?

Mundlos: That's a difficult one. It may be also that it is a factor of the level of expression. We know from some knockouts, for instance, that they're not complete knockouts: there is linkage of some normal allele which may account for that. I don't know. It could also be that within that division you have something which is regulatory plus another gene.

Kingsley: There could also be negative regulatory elements within the deletion. In this case not only would expression be reduced, but also ectopic expression might be produced by regulatory changes at the same time.

Mundlos: There was no expression.

Kingsley: Rob Krumlauf's group recently published a paper on a mouse mutant called *Sasquatch* (Sharpe et al 1999). They also proposed an effect on *Shh* regulatory elements. But, if I remember correctly, the postulated location of those elements was way on the other side of the genes. They have a transgene insertion that gives a limb phenotype that doesn't affect *Shh* coding region. He proposed it was affecting sonic regulatory region, but it was distal instead of proximal. This creates an unusual situation. Your allele gave essentially all the phenotypes that sonic did, which would suggest all the regulatory elements would have to be affected in your deletion, in order to recapitulate all of the phenotypes. Yet this other paper suggests that at least some of the regulatory elements are on the other side of gene. Have you any thoughts about how to reconcile these observations?

Mundlos: I recall the paper. Do they know where it is? It is a random insertion, and they have mapped this random insertion close to Sonic hedgehog but they can't really differentiate where it is.

Kingsley: I think there was an argument that it was distal.

Wilkie: As you know, in humans there are quite a lot of 7q translocations that are associated with human holoprosencephaly that don't actually disrupt the *Shh* gene itself (Belloni et al 1996). These led people slightly off-track at first during the search for the disease gene. They are all on the 5′ (telomeric) side of the gene. This is consistent with important regulatory elements lying upstream of the gene.

Mundlos: These translocations are also quite far away.

Wilkie: Yes, up to 450 kb according to the paper by Belloni et al (1996).

Kingsley: Your map looked like it was drawn with deletions proximal to the sonic locus. The centimorgans were labelled from the centromere.

Hall: The supernumerary teeth really are very striking. Whilst less bone may be a trigger, presumably you need factors other than just less bone to get supernumerary teeth from the permanent dentition. I assume that there is no basis normally in the permanent dentition for making additional teeth, so what else do you need to get them?

Blair: In osteogenesis imperfecta there are not supernumerary teeth but there is microdontia.

Hall: I'm presuming you need something that will continue to maintain a proliferative population at the base of these permanent teeth.

Meikle: You mentioned the importance of bone in this. We might well be looking at neural crest migration in the number of cells that are available for making teeth. In patients with hypodontia, they are missing teeth and they also have small jaws and relatively little dentoalveolar bone.

Hall: If it was such an early event, you would expect it to affect both the primary and secondary teeth. This seems to come in very late and only affects the permanent dentition.

Mundlos: The mechanisms of formation of primary and secondary teeth are quite different, though. The secondary tooth formation happens at an early stage but these are quiescent. The number of condensations that are giving rise to the primary teeth is something that the gene must affect. They are built after the primary teeth have been formed and stay quiescent.

Hall: So the gene is affecting the number of centres that are being formed.

Burger: I seem to remember that odontoblasts also express *Cbfa1*.

Newman: At the beginning of you paper you said that although we know the gene, we still don't understand the syndrome. Different parts of the body that one might expect to behave similarly behave differently; different individuals behave differently. One conclusion from this is that the gene may be the wrong level of description: instead, we should concentrate on processes and recognize that processes involve cooperation among many different gene products. But any given gene product won't necessarily be doing the exact same thing each time that process is occurring.

Mundlos: That is right. We see quite frequently that we have similar phenotypes, when we have effects in the same pathways. For instance if you have defects in somite differentiation, then there's a whole bunch of genes that regulate that. The phenotype will always be similar in that they have these segmentation defects. I think that's how it has to be defined. Similar diseases can be caused by defects in quite different genes, but they are phenotypically similar because these genes act on the same pathway or target. This makes life difficult for us.

Bard: Stuart Newman's last point has touched on something that has been going through my mind as I've been listening. The process that seems the most important in bone formation, at least in limbs, is size delimitation: how does the embryo know, for example, that the humerus is the right size, for example? This isn't merely the action of one gene. There is some process which is determining size here. It happens in the digits as the condensations fragment into the phalanges. In a sense, nothing I've heard today has touched on this process of spatial determination. We have some sense of the mechanisms of segmentation size for the vertebrae, but the equivalent processes in the limb remain obscure.

Newman: With analogy to other types of systems, if you have a container of water and you agitate it, this will cause waves of a certain size no matter whether you agitate it with a pencil or with a spoon. It is a characteristic of the system. Similarly with a string of a violin, the vibrations depend on the thickness, length and tension of the string, not how you excite it. In the end its characteristics will come through, even though they might not instantly, after the transients have subsided you get waves of a particular size.

Bard: I just don't see how we can bring size into this, even though it seems so integral to what's going on.

Newman: Don't these reaction–diffusion systems that we both study give rise to structures with characteristic sizes?

Bard: No one has looked at 3D simulations of reaction–diffusion models and they're much harder to interpret than 2D ones, because it's not clear how diffusion will affect the patterns.

Kingsley: In the old Oster et al (1988) paper that used essentially the formalism of the reaction–diffusion equations (although not in exactly that form), a combination of a local activator and a lateral inhibitor with the right constraints will generate a repeating pattern. The claim was made that the same sets of equations could be used either to branch the condensations or to segment the condensations.

Kingsley: The other point that was made in those papers is that although the equations do a good job of predicting the behaviour, the equations themselves are compatible with many different physical realizations that could either be mechanical or chemical or diffusing. Therefore I do still think there's a huge need to do the difficult embryology and gene hunting to try to put real molecules and processes on the limbs. Even if the behaviour fundamentally is based upon those properties, since the equations are compatible with different physical models, without some embryology and gene hunting we won't ever know whether any of them will match up.

Bard: The best the models can do is suggest plausibility, and then put the ball back into the court of the experimentalists. But, because it is extraordinarily difficult to devise experiments that might disprove such vaguely constructed models it is actually rather hard to show that the models are implausible. Although they are helpful metaphors, I remain dubious about them.

Kingsley: I agree.

Hall: Today we have heard that retinoic acid is distributed across the limb by a gene which is differentially distributed and which has an effect which relates to the specification of pattern. This results in digits 1,2,3 in the chick limb; I don't think it tells us anything about why digit 1 is a different size from digit 3. Clearly something like SHH or retinoic acid may be playing a role in specifying the size of those digits, as well as their polarity. This seems to be close to the sorts of things that Turing was looking for: the differential response to a single molecule.

Ornitz: Sure, but there is also differential expression of *Hox* genes D12 and D13.

Mundlos: The other problem is that there is no diffusion gradient.

Hall: Is that true for retinoic acid as well? I know there's been argument about this in the literature.

Morriss-Kay: There isn't any good evidence for a retinoic acid gradient, or even for a major role of retinoic acid in limb development. All we know is that in the chick, if you put a bead in, you will get a change in gene expression which leads to extra digit formation.

Tickle: I'm not sure that's entirely true. Thaller & Eichele (1987) could detect more retinoic acid posteriorly in the chick. Malcolm Maden has looked at this again recently, and he substantiates the idea that retinoic acid is enriched posteriorly (Maden et al 1998). It depends what you mean by a 'gradient': how many points you would require for a gradient?

Morriss-Kay: The trouble is that it is not possible to distinguish between free and bound (unavailable) retinoic acid.

Hall: So what is SHH or retinoic acid doing when you put it anteriorly in the limb bud?

Morriss-Kay: Retinoic acid up-regulates other genes local to the bead. There is also up-regulation of retinoic acid receptor (RAR)β, which has a retinoic acid response element in its promoter.

Kronenberg: A recent paper shows that SHH affects FGF synthesis through a complicated cascade (Zúñiga et al 1999). SHH stimulates the production of formin in the developing limb. Formin then stimulates the production of gremlin, a bone morphogenetic protein (BMP) agonist. Blockade of BMP signalling leads to induction of FGF4 synthesis. SHH doesn't move far but triggers a cascade. The diffusion of SHH is biologically regulated in an elaborate fashion, and it doesn't go very far — no more than a few cell layers. This then kicks off other things. It seems to be a pattern for SHH and Indian hedgehog (IHH) that they stimulate the production of things that have the ability to move further and faster.

Hall: So you get your cascade by sequentially turning on genes which have quite local actions, but progressing across the system.

Newman: There seem to be two independent phenomena. Everything that causes extra digits, in addition to inducing these genes, also expands the mesoblast. There is more mesenchyme to be organized. If all you had was more mesenchyme, you might expect to get just more indeterminate digits. But since you also affect these gradients of Hox proteins and other things, you get a polarity that makes any digits that arise have a particular character rather than a general character. It therefore seems that these duplication experiments are looking at two kinds of overlapping but distinct phenomena: first, more digits because of more mesenchyme and, second, particular kinds of digits because of these gradients that are induced by the retinoids or SHH.

M. Cohn: Addition of mesenchyme alone to the limb bud will not result in additional digits. For example, when tissue from the anterior part of one limb bud is grafted to the anterior part of a host limb bud, this will result in broadening of the host limb but not in the formation of extra digits unless the grafted cells have potential polarizing activity.

Pizette: The reason it doesn't is because in that case it is regulated by the zones of cell death. However, in the talpid mutants, the zones of cell death are lost and this may be the reason why there is more mesenchyme, which then results in extra digits that lack a true identity. The mesenchyme mass is indeed critical in at least setting the number of elements, but not their identity.

Hall: Presumably therefore the mesenchyme mass has some role in setting the size of those elements?

Pizette: I have been thinking about this. Digit 2 is obviously shorter in the forelimb. If you look at the overall shape of the forelimb bud, it is striking that on the anterior side there is much less mesenchyme than in the hind limb even. Perhaps small differences in how the apical ectodermal ridge (AER) extends anteriorly and in how big the zones of necrosis on the anterior side are, dictates the amount of mesenchyme available to form cartilage elements. Maybe the zone of cell death is even bigger on the anterior side of the wing than it is on the anterior side of the leg, which then leaves less mesenchyme for wing digit 2 to form.

M. Cohn: In limb buds of the short-digit mutant embryos, you showed that there is no detectable expression of *Hoxd11–13;* however it seemed like you were looking at relatively late stages of *Hox* expression, when posterior *Hox* expression domains are moving anteriorly through the autopod. What about the early stages of *Hox* gene expression in the limb buds? Have you detected *Hoxd11–13* in early bud or pre-bud stages?

Mundlos: We looked at Day 9, and there is no *Hox* expression. There is strong expression of *Fgf8* in AER at this stage.

Tickle: What about *Hoxd9*? This is expressed in the forelimb. It would be interesting to know if that was there.

Mundlos: At Day 11 there might be some weak expression.

Tickle: Martin Cohn, were you thinking of the limbless mutant, where there is early activation of *Hox* genes even in the absence of *Shh*?

M. Cohn: I was actually thinking of the early initiation of the limb bud, which we found involves expression of *Hoxd9*. That expression domain originates in the pre-limb lateral plate mesoderm at forelimb levels long before *Shh* expression. As the forelimb bud emerges, *Hoxd9* expression is maintained and more posterior *HoxD* genes are activated. I was leading to the question of whether your mutants have arrested the unravelling of the posterior *Hox* genes after limb bud initiation, or do you think that *HoxD* genes are never expressed in mutant limb buds?

Mundlos: I can't tell you whether there is expression at a very early stage. We tested the most posterior *Hox* genes and found no expression. If you lose *Hox* expression completely then these parts of the limb are just gone. If sonic regulates it either by maintaining it or initiating it, they're not there when they're actually needed at that later stage and this part of the skeleton is gone.

Hall: So the notion of a potential alternate pathway to sonic is something one should keep an open mind on I guess, both from the python work and from the sorts of studies that you are doing. Martin Cohn, I wasn't quite sure when you put your python tissue into the anterior part of the limb of chick, was it the python cells that were producing this sonic?

M. Cohn: From looking at sections of those limbs, it is the graft of python cells that is expressing *Shh*. There is also no experimental evidence from work on chick limb buds to indicate that transplanting cells with polarizing activity induces *Shh* in neighboring cells. In our experiments, *Shh* expression is induced in python cells that are transplanted under the chick apical ectodermal ridge, which is a source of FGF.

Perrin-Schmitt: I would like to propose a model. We are working on *M-twist*. In *twist* null heterozygous mice, there is a duplicated hallux (big toe), with three phalanges instead of two, all the other digits are normal. This is on a C57/Black 6 background. Thus we tried to build a model to understand what would happen at the gene expression level. When we look at *M-twist* expression by *in situ* hybridizations, we see a gradient in the forming limb bud. When the mesoderm is not differentiated, there is a high expression of *M-twist*, but as soon as mesoderm begins to differentiate, *M-twist* expression vanishes. At least at one stage there is a gradient with a higher *M-twist* expression in the posterior part of the limb, and later a higher expression in what will be the future anterior part. We are interested in what happens with the model of FGF expression, and have heard today that FGF and FGF receptor (FGFR) expression allows the limb bud to grow.

I would propose that the proximal/distal axis would be established through the FGF/FGFR signalling plus the *Hoxa9–Hoxa13* expression gradient; then the anterior/posterior gradient would be controlled by the zone of polarizing activity/SHH/retinoic acid signalling plus the *Hoxd9–Hoxd13* expression gradient.

We wonder whether the gradient of *Hox* gene expression is modified in *twist* null heterozygous mice so that instead of the correct differentiation of the hallux in the heterozygous mice, there is another type of differentiation at the opposite side from the *Shh* expression region (the *M-twist* gradient might also be affected). So we tried to test this hypothesis, using *Hox* mutants with a reduced number of digits, namely *Hoxa13*$^{-/-}$: if the gradient of *M-twist* expression is important for the differentiation of the anterior digits, we should be able to rescue this phenotype in double mutants by crossing *twist* null heterozygotes (carrying an additional hallux) with the *Hoxa* strain lacking the anterior digit.

Hall: But you don't have the double heterozygotes yet?

Perrin-Schmitt: We are constructing them now.

Hall: This nicely emphasizes that a cell which is sitting in a particular position is encountering a number of different gradients coming from different directions.

References

Belloni E, Muenke M, Roessler E et al 1996 Identification of *Sonic hedgehog* as a candidate gene responsible for holoprosencephaly. Nat Genet 14:353–356

Goodrich ES 1958 Studies on the structure and development of vertebrates. Dover, New York

Maden M, Sonneveld E, van der Saag PT, Gale E 1998 The distribution of endogenous retinoic acid in the chick embryo: implications for developmental mechanisms. Development 125:4133–4144

Oster GF, Shubin N, Murray JD, Alberch P 1988 Evolution and morphogenetic rules: the shape of the vertebrate limb in ontogeny and phylogeny. Evolution 42:862–884

Sharpe J, Lettice L, Hecksher-Sorensen J, Fox M, Hill R, Krumlauf R 1999 Identification of *sonic hedgehog* as a candidate gene responsible for the polydactylous mouse mutant Sasquatch. Curr Biol 9:97–100

Thaller C, Eichele G 1987 Identification and spatial distribution of retinoids in the developing chick limb bud. Nature 327:625–628

Zúñiga A, Haramis AP, McMahon AP, Zeller R 1999 Signal relay by BMP antagonism controls the SHH/FGF4 feedback loop in vertebrate limb buds. Nature 401:598–602

Genetic control of the cell proliferation–differentiation balance in the developing skull vault: roles of fibroblast growth factor receptor signalling pathways

G. M. Morriss-Kay*, S. Iseki*† and D. Johnson‡¶

**Department of Human Anatomy and Genetics, University of Oxford, South Parks Road, Oxford OX1 3QX, UK, †Department of Molecular Craniofacial Embryology, Tokyo Medical and Dental University, 1-5-45, Yushima, Bunkyo-ku, Tokyo 113-8549, Japan, ‡Institute of Molecular Medicine, John Radcliffe Hospital, Headington, Oxford OX3 9DS, and ¶Department of Plastic and Reconstructive Surgery and Oxford Craniofacial Unit, The Radcliffe Infirmary NHS Trust, Woodstock Road, Oxford OX2 6HE, UK*

Abstract. Activating mutations of genes encoding the transmembrane tyrosine kinase receptors fibroblast growth factor receptors (FGFRs)1–3, and haploinsufficiency of the transcription factor TWIST, cause human craniosynostosis syndromes that typically involve the coronal suture. We have investigated the functional roles of these genes in development of the coronal suture in mouse fetuses, and tested the effects of increasing FGFR signalling by applying exogenous FGF2 to the suture. The results indicate that the proliferation–differentiation balance in normal sutural development involves a gradient of extracellular FGF from the region of differentiation, in which *Fgfr1* is expressed, to the sutural mesenchyme, in which low levels of FGF are associated with *Fgfr2* expression in osteogenic stem cells. Experimental increase of sutural FGF levels leads to down-regulation of *Fgfr2*, up-regulation of *Fgfr1*, up-regulation of the osteogenic differentiation gene *Osteopontin*, and cessation of proliferation. *Twist* is expressed in the midsutural mesenchyme and is partially co-expressed with *Fgfr2*, consistent with the possibility that it is involved in maintaining proliferation through regulating transcription of *Fgfr2*.

2001 The molecular basis of skeletogenesis. Wiley, Chichester (Novartis Foundation Symposium 232) p 102–121

The pattern of growth in the developing skull vault

The vertebrate skull develops within a layer of mesenchyme between the embryonic brain and surface ectoderm. This mesenchymal layer gives rise to an outer skeletogenic membrane in which the skull bones form, and an inner meningeal layer from which the dura, arachnoid and pia mater layers form. The bones of the skull vault, like those of the face, are formed directly in mesenchyme (intramembranous ossification), in contrast to the cartilaginous origin of the bones of the skull base and occipital region (endochondral ossification).

The sutures between the membrane bones of the skull vault are growth centres in which proliferating osteogenic stem cells provide the source material for incremental growth of the calvarial bones. Osteogenic stem cells are located at the periphery of each membrane bone and also, in smaller numbers, on their inner and outer surfaces. Continued growth of the skull vault depends on maintenance of a balance between proliferation of the osteogenic stem cells and their differentiation to form new bone.

The developmental origin of the coronal suture is of particular interest in understanding mechanisms of skull growth. This bilateral, vertical suture is formed between the frontal and parietal bones; it is responsible for most of the growth of the skull in the fronto-occipital (front-to-back) plane. Interspecific grafting experiments in avian embryos suggested that both frontal and parietal bones have the same origin, either from mesoderm (Noden et al 1988) or neural crest (Couly et al 1993). However, investigation of a transgenic mouse with a permanent neural crest cell lineage marker suggests that the coronal suture is formed at the interface of a neural crest-derived frontal bone and a mesoderm-derived parietal bone (X. Jiang, H. Sucov & G. Morriss-Kay, unpublished work). This juxtaposition of tissues of different origins is likely to be functionally significant.

Fibroblast growth factor receptor signalling and the coronal suture

Fibroblast growth factor receptors (FGFRs) are transmembrane proteins that play major roles in skeletogenesis. Activating mutations of the human *FGFR1*, *FGFR2* and *FGFR3* genes cause premature fusion of the skull bones (craniosynostosis), as well as other skeletal defects.

The elucidation of these mutations, and study of their associated phenotypes, has provided important clues to the biological functions of FGFR signalling and of the mechanisms involved in normal sutural development. In craniosynostosis caused by mutation of the genes encoding FGFR1, -2 or -3, the coronal suture is typically involved, sometimes uniquely (Bellus et al 1996). The significance of this correlation is illustrated by the incidence of synostosis in different sutures: coronal synostosis (unilateral or bilateral) accounts for only 17–30% of all craniosynostosis cases, the most commonly synostosed suture being the sagittal (50–60%) (Wall 1997).

Simple coronal synostosis may be caused by an activating mutation of *FGFR3* (Pro250Arg; Muenke et al 1997) but a very similar phenotype can also result from

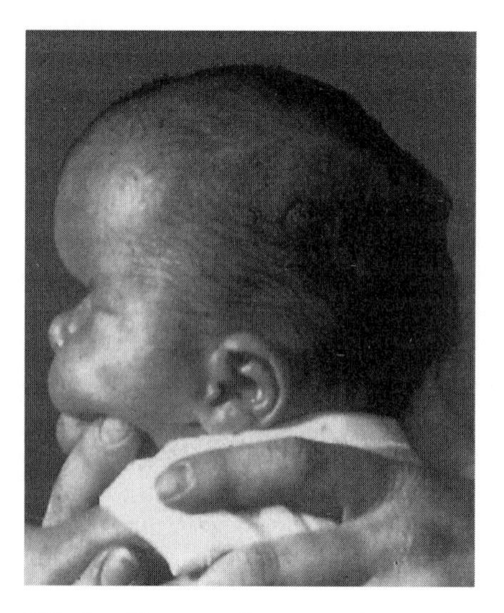

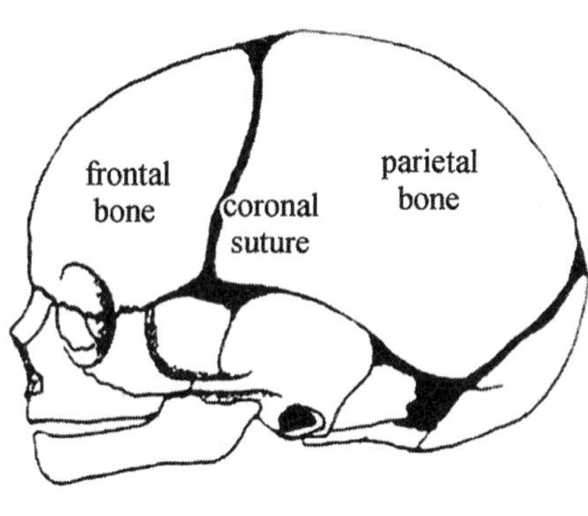

FIG. 1. The coronal suture and coronal synostosis. (*Left*) Two-week-old infant with Apert syndrome, in which all calvarial sutures eventually fuse but at this stage only the coronal is fused, inhibiting growth in the fronto-occipital plane but allowing the brain to expand into the region of the metopic and sagittal sutures. (*Right*) Outline of the human skull to approximately the same scale.

inactivating mutations and microdeletions of another gene, *TWIST* (el Ghouzzi et al 1997, Howard et al 1997, Paznekas et al 1998, Johnson et al 1998). *TWIST* encodes a basic helix-loop-helix transcription factor whose *Drosophila* homologue, *twist*, regulates transcription of *heartless* (*htl*, formerly *DFR1* or *DFGF-R2*), a homologue of vertebrate *FGFR* genes that is expressed in mesodermal tissues (Beiman et al 1996). Although a genetic association between vertebrate *TWIST* and *FGFR* genes has not yet been established, the phenotypic similarity between activating mutations of *FGFR* genes and inactivating mutations of *TWIST* is highly suggestive of a conserved pathway.

These observations suggest that the coronal suture is a particularly good site for studying the roles of FGFR signalling and TWIST function during sutural initiation, development, and maintenance. We have therefore investigated the functional roles of FGFRs during prenatal development of the coronal suture in the mouse, as a basis for understanding their roles in human sutural development, and to gain some insight into how altered receptor function leads to premature cessation of sutural growth. The position of the coronal suture and the effects of coronal synostosis on prenatal growth and shape of the human skull are illustrated in Fig. 1.

Growth of the skull vault is synchronized with growth of the brain, suggesting that during normal development, signals from the brain are involved in controlling the balance between proliferation and differentiation of the osteogenic stem cells. Very little is known about these brain-derived signals, which could include mechanical pressure from the growing brain; however, experimental evidence indicates that the meningeal tissue underlying the sutures is required for maintaining sutural patency and skull growth (Opperman et al 1993, Kim et al 1998). Signals from the brain and meningeal tissues clearly act through signal transduction systems within the skeletogenic membrane itself, which is the subject of our study. Understanding the molecular and functional nature of the events regulating normal sutural development and maintenance is clearly a prerequisite for understanding how the osteogenic proliferation–differentiation balance is biased towards differentiation in the presence of certain abnormal forms of FGFR and TWIST proteins.

Experimental strategy: the mouse model

The aim of the studies to be described here was to analyse how the components of the FGFR signalling system are functionally related to proliferation and differentiation within the osteogenic cell population in the region of the coronal suture. The skull vault is formed from the same bones in all mammals, and in this study we have used mouse fetuses as a model for understanding human sutural development. The investigations fall into two parts: (1) a developmental anatomy approach, in which expression patterns of mouse *Fgfr* and *Twist* genes and FGF ligand are defined, and related to cell function in terms of cell proliferation and osteogenic differentiation; and (2) an experimental approach, in which exogenous FGF ligand is added to the fetal coronal suture to mimic the effects of activating FGFR mutations. Although a number of FGFs are present in the skeletogenic membrane, we found FGF2 to be particularly abundant and have therefore concentrated on this ligand in most of our studies. FGF2 is a strong activator of the IgIIIc isoforms of FGFR1, FGFR2 and FGFR3 (Ornitz et al 1996), which are commonly affected in human craniosynostosis syndromes (see Wilkie et al 2000, this volume). Cell proliferation was detected by means of bromodeoxyuridine uptake or immunohistochemical detection of proliferating cell nuclear antigen; the differentiation markers used were *Osteopontin* and *Osteonectin* RNA transcripts and histochemical detection of alkaline phosphatase activity (see Iseki et al 1997, 1999 and Johnson et al 2000 for details of the methods and probes used).

Skeletogenesis-related gene expression is first detectable in the skeletogenic membrane of mouse fetuses on embryonic day (E) 14, and the frontal and parietal components of the coronal suture show their characteristic overlap by E16 (Figs 2 and 3). The results to be described here will concentrate on the E16 stage, by which

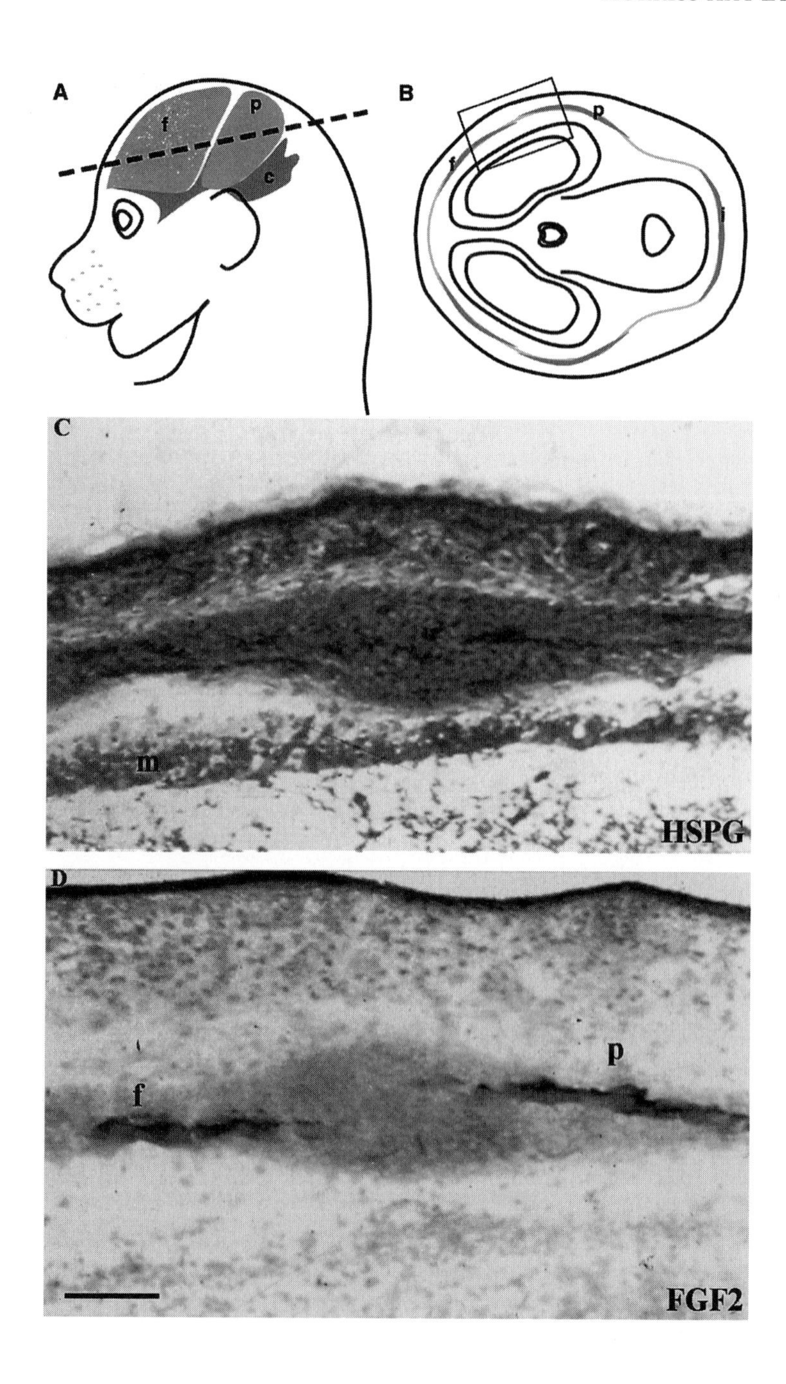
A
f
p
c
B
p
f
C
m
HSPG
D
p
f
FGF2

time a relationship between gene expression and functional tissue organization has been established that will endure throughout the period of sutural growth.

Extracellular and cell membrane components of the FGFR signalling pathway

FGFR signalling requires the binding of two molecules of FGF to two receptors, resulting in homodimerization, *trans*-autophosphorylation of the tyrosine kinase domains, and signal transduction to the nucleus. The subsequent effects of the signal on gene transcription result in altered cell behaviour, involving either a change in the level of proliferation, presumably through an effect on cell cycle-related gene expression, or an effect on differentiation-related gene expression. The binding of FGF ligand to its receptor also requires the cooperation of a non-specific receptor, heparin or a heparin-like molecule such as heparan sulfate proteoglycan (HSPG) (Yayon et al 1991). The activated dimeric receptor complex consists of a sandwich of 2×FGFR–heparin–FGF (Huhtala et al 1999). We have used immunohistochemistry of FGF2 ligand and an HSPG commonly found in developing tissues (perlecan), to discover the availability of these molecules for binding to FGFRs in the region of the coronal suture.

The immunohistochemical localization of FGF2 and perlecan in the E16 coronal suture are illustrated in Fig. 2B, C. Perlecan is present throughout the skeletogenic membrane, both within the suture and in the region of differentiated bone. This molecule is therefore a good candidate for involvement in FGF binding to FGFR throughout the growing skull vault, although it is likely that other HSPGs also play important roles, since disruption of the gene encoding perlecan has more severe effects on long bones than on the skull vault (Arikawa-Hirasawa et al 1999). Since the signals from the brain to the skeletogenic membrane may also be FGF ligands, it is interesting to observe that perlecan is also present in the meningeal layers between the brain and the skeletogenic membrane.

In contrast to perlecan, the distribution of FGF2 is much more specific: following secretion from the differentiating osteoblasts, it is present at high levels attached to the osteoid (unmineralized bone matrix) of the differentiated region of

FIG. 2. The mouse coronal suture. (A, B) Plane and organization of the sections shown in subsequent figures; the boxed area in (B) indicates the region illustrated. (C) Immunoreactivity of HSPG (perlecan) in the skin, skeletogenic membrane (frontal, parietal and coronal suture region) and meningeal tissue. (D) Immunoreactivity of secreted FGF2 in the region of the coronal suture: high levels are attached to osteoid and low levels diffuse into the extracellular matrix of the sutural mesenchyme and around the cells above and below the osteoid plates. c, cartilage; f, frontal bone domain; m, meningeal tissue; p, parietal bone domain. Scale bar = 100 μm.

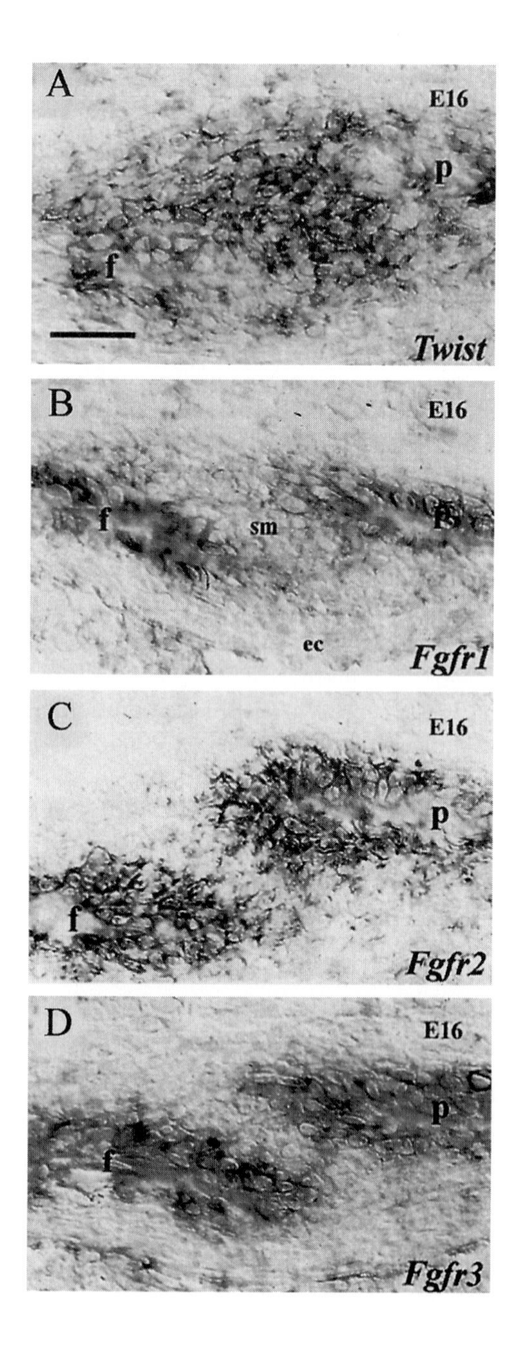
A
E16
p
f
Twist
B
E16
f
sm
p
ec
Fgfr1
C
E16
p
f
Fgfr2
D
E16
p
f
Fgfr3

the frontal and parietal bones, but at low levels in the extracellular matrix around the mesenchymal cells of the suture. This distribution suggests that there is a gradient (or possibly a stepwise change in concentration) of FGF from the region of differentiation to the region of proliferation, both within the suture, where proliferation leads to increase in size of the calvarial bones, and on the outer (skin side) and inner (brain side) surfaces of the bony plates, where proliferation allows increase in thickness of the bones. It is interesting to note that an FGF gradient has been reported to play a key developmental role in quite a different context, that of anteroposterior patterning (the determination of head, trunk and tail structures) in amphibian embryos, in which high levels of FGF–FGFR signalling are required for the specification of tail structures and low levels for head structures (Christen & Slack 1999).

Expression of *Fgfr* and *Twist* genes in the coronal suture

We have used *in situ* hybridization to reveal the tissue distribution of all four mouse *Fgfr* genes and *Twist* in the developing coronal suture (the *Fgfr* probes used do not distinguish between the IIIb and IIIc isoforms of *Fgfr* 1, -2 and -3). The pattern at E16 is illustrated in Fig. 3 (*Fgfr4*, which is expressed only in muscle, is not shown).

Fgfr1 is expressed specifically in cells close to the osteoid; *Fgfr2* expression does not overlap with that of *Fgfr1*, and is seen in two distinct frontal and parietal sutural cell populations, as well as in scattered cells on the outer and inner surfaces of the osteoid plates. The expression domain of *Fgfr3* appears to be intermediate between those of the other two receptor genes, but it should be noted that detection of *Fgfr3* RNA transcripts requires a much longer colour reaction time, suggesting a lower level of expression. A higher level of *Fgfr3* expression was detected in the cartilaginous precursors of the endochondral bones of the skull, including the exoccipital cartilage and the greater wing of the sphenoid bone, which is at this stage a thin cartilaginous plate underlying the lower part of the coronal suture (illustrated in Iseki et al 1999).

The parietal and frontal domains of *Fgfr* gene expression in the coronal suture are separated by an area of midsutural mesenchyme in which no *Fgfr* gene expression can be detected. This separation of two distinct *Fgfr* domains is

FIG. 3. *In situ* hybridization of embryonic day 16 mouse calvarial transverse sections showing expression of *Twist* and *Fgfr* genes, as indicated, in the region of the coronal suture. Only *Twist* is expressed in the midsutural mesenchyme (A), overlapping with the domain of *Fgfr2* expression (C), which is mutually exclusive with that of *Fgfr1* (B). The *Fgfr3* domain (D) is intermediate between *Fgfr1* and *Fgfr2*, but the transcript level is very low, requiring a long colour development time (see Iseki et al 1999, for details). ec, endocranium; f, frontal bone domain; sm, mid-sutural mesenchyme; p, parietal bone domain. Scale bar = 50 μm.

observed from E14, the first stage at which gene expression patterns indicate the site of the future suture, and is maintained throughout the prenatal period (to E18), during which the edges of the two bones show an increasing degree of overlap (Johnson et al 2000). Of the genes studied, only *Twist* was found to be expressed in the midsutural mesenchyme; its domain is broader than this *Fgfr*-free tissue, overlapping with both frontal and parietal *Fgfr2* domains (Fig. 3).

Functional correlates of the gene expression patterns

Differentiation is a multistage process involving the sequential expression of a number of different genes. Comparison of three indicators of bone differentiation indicates that *Osteonectin* expression and alkaline phosphatase activity are present in cells close to the region of proliferation, in preosteoblasts and osteoblasts at an early stage of differentiation. They are co-expressed with *Fgfr1*. In contrast, *Osteopontin* is only expressed in more mature osteoblasts (cells in contact with osteoid), indicating that it is a marker for later stages of differentiation. *Fgfr1* is down-regulated in mature osteoblasts, which continue to express *Osteopontin*. These observations suggest that signalling through FGFR1 is associated with the onset of differentiation, and that its role is complete before the osteoblasts are fully differentiated. This result is analogous to observations in developing muscle, in which down-regulation of *Fgfr1* is essential for completion of the differentiation process (Itoh et al 1996).

Proliferating osteoprogenitor cells (detected by bromodeoxyuridine uptake or immunoreactivity for proliferating cell nuclear antigen) are present at the sutural edges of the bones and on the outer and inner surfaces; *Fgfr2* is expressed in these cells. The midsutural mesenchyme, in which only *Twist* is expressed, shows neither differentiation-related gene expression nor cell proliferation markers. Anatomically it appears to be a kind of buffer zone between the frontal and parietal proliferating populations, in continuity at its periphery with proliferating cells that coexpress *Twist* and *Fgfr2*. These results are summarized in Fig. 4.

Experimental application of FGF2 to the developing coronal suture

The observation that FGF2 levels are higher in the differentiating region than in the region of proliferating osteogenic stem cells suggests that proliferation is only maintained in cells in which the level of FGFR signalling is relatively low. To test this hypothesis we stimulated FGFR signalling by inserting FGF2-soaked heparin acrylic beads onto the coronal suture of E15 fetuses and allowed development to continue for up to 48 hours. PBS-soaked beads were used for control fetuses. The operative technique is described in detail in Iseki et al (1997). In some experiments, FGF2 was conjugated with digoxygenin in order to monitor its rate and distance of diffusion from the beads (Iseki et al 1999).

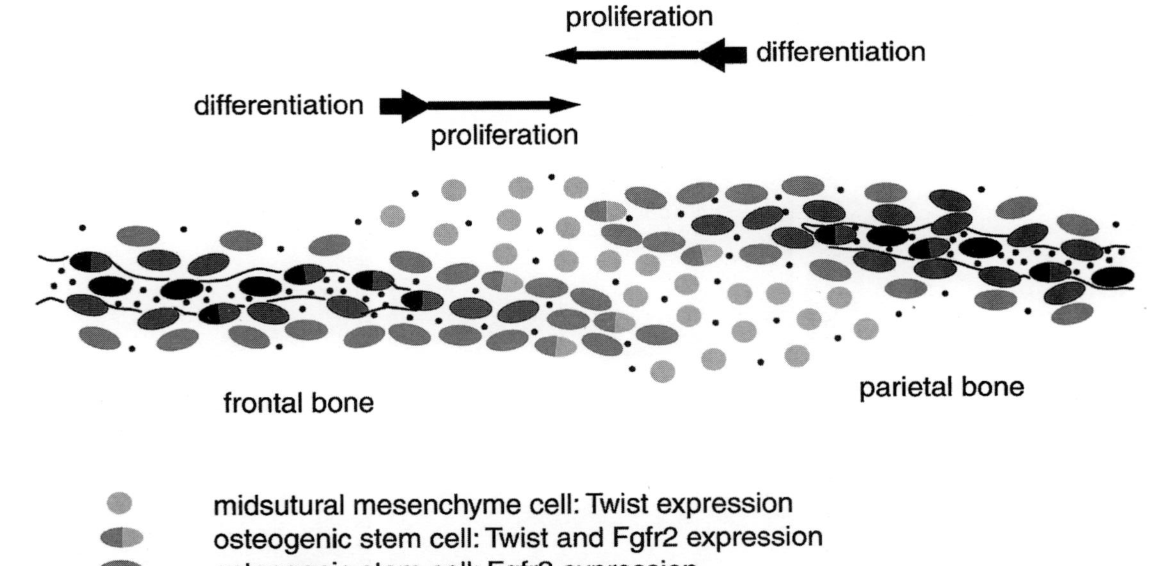

FIG. 4. Diagrammatic summary of the region of the coronal suture with respect to cell proliferation (osteogenic stem cells), differentiation-related gene expression (*Osteopontin*), expression patterns of *Fgfr* genes and *Twist*, and differential FGF2 levels.

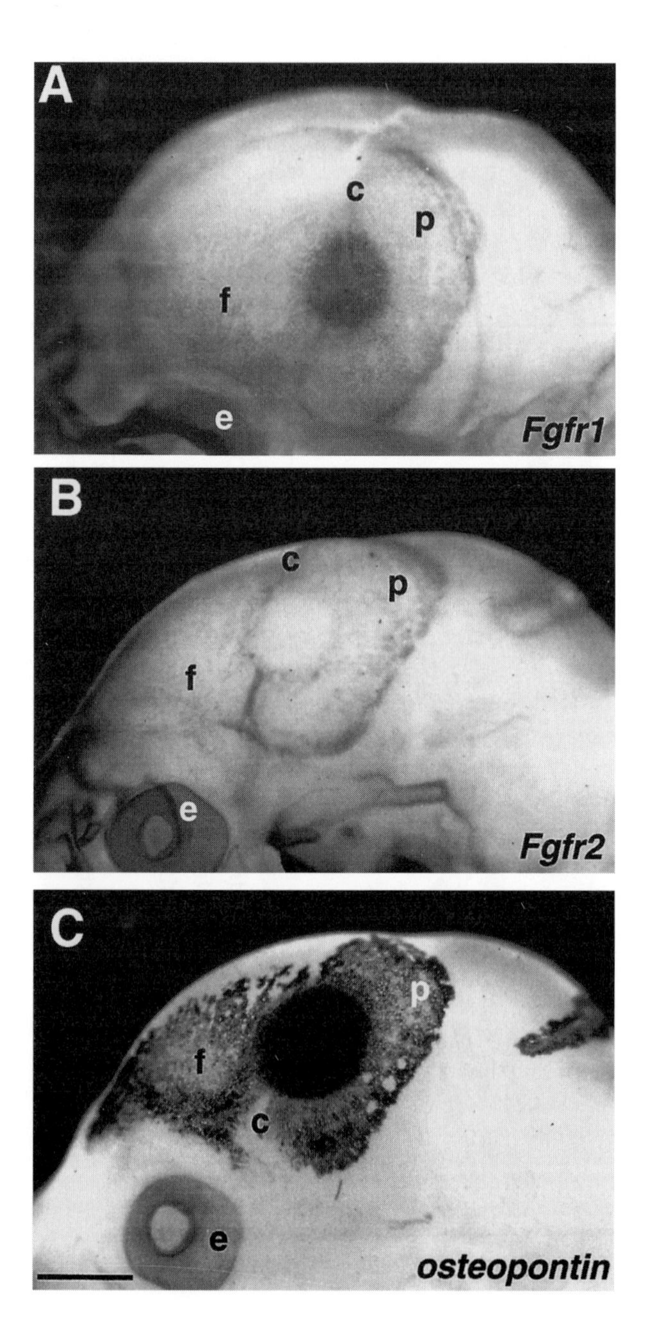
A
c
p
f
e
Fgfr1
B
c
p
f
e
Fgfr2
C
p
f
c
e
osteopontin

FGF2 reached the skeletogenic membrane within four hours of bead insertion. By this stage changes in gene expression and in cell proliferative activity were already apparent. Two hours later (six hours after bead insertion), the following changes were clear in all specimens, in the region around the beads: expression of *Fgfr1* and *Osteopontin* was up-regulated; expression of *Fgfr2* and *Fgfr3* was down-regulated. Cells in the skeletogenic membrane close to the FGF beads showed poor immunoreactivity for proliferating cell nuclear antigen (PCNA), indicating that the osteogenic stem cells had left the cell cycle.

The same effects were observed after 24 hours (Fig. 5). By 48 hours some reversal of these changes had taken place; by this stage, FGF was no longer diffusing from the beads, which had in any case been lifted away from the skeletogenic membrane by the considerable increase in bulk of the subcutaneous mesenchyme.

Summary and conclusions

These results suggest that maintenance of proliferating osteogenic stem cells at the margins of the membrane bones that form the coronal suture (and on the inner and outer surfaces of these bones) depends on FGF levels being relatively low; higher levels of FGF are associated with osteogenic differentiation. The mitogenic signal is mediated by FGFR2 (and FGFR3). Higher levels of FGF2-stimulated FGFR2 signalling lead to down-regulation of *Fgfr2* (and *Fgfr3*) gene expression and up-regulation of *Fgfr1*. Signalling through FGFR1 does not have a mitogenic outcome in this context, but leads to the expression of osteogenic differentiation genes. Once differentiation is well established, *Fgfr1* is down-regulated.

In the normal suture, this mechanism, involving differential levels of FGF from high in the differentiated region to low in the suture, ensures that sutural stem cell populations are maintained at the periphery of the growing bones. However, when receptor activation is increased, either experimentally (by the addition of exogenous FGF), or pathologically (by altered receptor function), *FGFR2* is prematurely down-regulated and proliferation ceases. These conclusions are supported by the observation that FGF2 levels are increased within sutures at the time of fusion (Mehrara et al 1998). Also, in the synostosed coronal sutures of patients with Crouzon syndrome, which is due to activating mutation of *FGFR2* (Reardon et al 1994), FGFR2 immunoreactivity is reduced (Bresnick & Schendel 1995).

FIG. 5. The effect of exogenous FGF2 on gene expression in the region of the coronal suture: whole-specimen *in situ* hybridization on embryonic day (E) 16 mouse heads 24 hours after subcutaneous insertion of FGF2-soaked beads onto the coronal suture on E15. (A) Up-regulation of *Fgfr1* expression; (B) down-regulation of *Fgfr2*; (C) up-regulation of the bone differentiation marker, *Osteopontin*, in the region of the beads (see Iseki et al 1999, for details). Scale bar = 1 mm.

The role of FGFR3 signalling in development and maintenance of the coronal suture differs from that of FGFR2 in being involved in growth of both membrane and endochondral components of the skull: *Fgfr3* is the only *Fgfr* gene expressed in cranial cartilage (Iseki et al 1999). However, the phenotypes of human *FGFR3* mutations suggest that the skeletal growth response to FGFR3 signalling in endochondral bones of the skull and long bones are different. The achondroplasia phenotype (Gly380Arg mutation of *FGFR3*; Rousseau et al 1994, Shiang et al 1994) is characterized by decreased proliferation of epiphyseal chondrocytes, causing deficient growth of the long bones, but in the skull, growth is excessive, leading to macrocephaly. In thanatophoric dysplasia, a more severe phenotype resulting from mutation at different sites of the *FGFR3* gene, the disparity between deficiency of long bone development and disproportionately increased head growth is even greater (see this volume: Wilkie et al 2000, Ornitz 2000, for details).

The hypothesis that *Twist* expression is essential for maintenance of osteogenic cell proliferation is supported by the observation that human *TWIST* mutations associated with craniosynostosis exert their effect via haploinsufficiency, in contrast to the gain-of-function mechanism of action of *FGFR* mutations. *Twist* expression is down-regulated during differentiation of cultured mouse calvarial osteoprogenitor cells, suggesting either that it promotes proliferation or that it inhibits differentiation (Murray et al 1992); it is known to inhibit differentiation during myogenesis (Hebrok et al 1994). If *Twist* is upstream of FGFR signalling in vertebrates, the overlap of the *Twist* and *Fgfr2* expression domains in the coronal suture suggests that loss of function of *Twist* could lead to decreased osteogenic cell proliferation through failure to maintain adequate levels of *Fgfr2* expression.

We suggest (1) that in the coronal suture, the rate of progression from proliferation to differentiation is maintained by interactions between TWIST-regulated transcription, FGFR2/FGFR3 mitogenic signalling and FGFR1 differentiation-related signalling; (2) that differential levels of FGF2 (and possibly other FGF ligands) play a key role in these interactions through regional regulation of expression of the receptor genes.

Acknowledgements

We are grateful to F. Perrin-Schmitt, J. Heath and D. Ornitz for donation of probes. This study was funded by a Wellcome Clinical Training Fellowship award to D. Johnson, and an Action Research grant to G. M. Morriss-Kay.

References

Arikawa-Hirasawa E, Watanabe H, Takami H, Hassell JR, Yamada Y 1999 Perlecan is essential for cartilage and cephalic development. Nat Genet 23:354–358

Beiman M, Shilo BZ, Volk T 1996 Heartless, a *Drosophila* FGF receptor homolog, is essential for cell migration and establishment of several mesodermal lineages. Genes Dev 10:2993–3002

Bellus GA, Gaudenz K, Zackai EH et al 1996 Identical mutations in three different fibroblast growth factor receptor genes in autosomal dominant craniosynostosis syndromes. Nat Genet 14:174–176

Bresnick S, Schendel S 1995 Crouzon's disease correlates with low fibroblastic growth factor receptor activity in stenosed cranial sutures. J Craniofac Surg 6:245–248

Christen B, Slack JM 1999 Spatial response to fibroblast growth factor signalling in *Xenopus* embryos. Development 126:119–125

Couly GF, Coltey PM, Le Douarin NM 1993 The triple origin of skull in higher vertebrates: a study in quail–chick chimeras. Development 117:409–429

el Ghouzzi V, Le Merrer M, Perrin-Schmitt F et al 1997 Mutations of the *TWIST* gene in the Saethre–Chotzen syndrome. Nat Genet 15:42–46

Hebrok M, Wertz K, Fuchtbauer EM 1994 M-twist is an inhibitor of muscle differentiation. Dev Biol 165:537–544

Howard TD, Paznekas WA, Green ED et al 1997 Mutations in *TWIST*, a basic helix-loop-helix transcription factor, in Saethre–Chotzen syndrome. Nat Genet 15:36–41

Huhtala MT, Pentikinen OT, Johnson MS 1999 A dimeric ternary complex of FGFR1, heparin and FGF-1 leads to an 'electrostatic sandwich' model for heparin binding. Structure Fold Des 7:699–709 (erratum: 1999 Structure Fold Des 15:R198)

Iseki S, Wilkie AOM, Heath JK, Ishimaru T, Eto K, Morriss-Kay GM 1997 *Fgfr2* and *osteopontin* domains in the developing skull vault are mutually exclusive and can be altered by locally applied FGF2. Development 124:3375–3384

Iseki S, Wilkie AOM, Morriss-Kay GM 1999 *Fgfr1* and *Fgfr2* have distinct differentiation- and proliferation-related roles in the developing mouse skull vault. Development 126:5611–5620

Itoh N, Mima T, Mikawa T 1996 Loss of fibroblast growth factor receptors is necessary for terminal differentiation of embryonic limb muscle. Development 122:291–300

Johnson D, Horsley SW, Moloney DM et al 1998 A comprehensive screen for *TWIST* mutations in patients with craniosynostosis identifies a new microdeletion syndrome of chromosome band 7p21.1. Am J Hum Genet 63:1282–1293

Johnson D, Iseki S, Wilkie AOM, Morriss-Kay GM 2000 Expression patterns of *Twist* and *Fgfr1*, *-2* and *-3* in the developing mouse coronal suture suggest a key role for *Twist* in suture initiation and biogenesis. Mech Dev 91:341–345

Kim HJ, Rice DP, Kettunen PJ, Thesleff I 1998 FGF-, BMP- and Shh-mediated signalling pathways in the regulation of cranial suture morphogenesis and calvarial bone development. Development 125:1241–1251

Mehrara BJ, Mackool RJ, McCarthy JG, Gittes GK, Longaker MT 1998 Immunolocalization of basic fibroblast growth factor and fibroblast growth factor receptor-1 and receptor-2 in rat cranial sutures. Plast Reconstr Surg 102:1805–1817

Muenke M, Gripp KW, McDonald-McGinn DM et al 1997 A unique point mutation in the fibroblast growth factor receptor 3 gene (*FGFR3*) defines a new craniosynostosis syndrome. Am J Hum Genet 60:555–564

Murray SS, Glackin CA, Winters KA, Gazit D, Kahn AJ, Murray EJ 1992 Expression of helix-loop-helix regulatory genes during differentiation of mouse osteoblastic cells. J Bone Miner Res 7:1131–1138

Noden DM 1988 Interactions and fates of avian craniofacial mesenchyme. Development (suppl) 103:121–140

Opperman LA, Sweeney TM, Redmon J, Persing JA, Ogle RC 1993 Tissue interactions with underlying dura mater inhibit osseous obliteration of developing cranial sutures. Dev Dyn 198:312–322

Ornitz DM 2001 Regulation of chondrocyte growth and differentiation by fibroblast growth factor receptor 3. In: The molecular basis of skeletogenesis. Wiley, Chichester (Novartis Found Symp 232) p 63–80

Ornitz DM, Xu J, Colvin JS et al 1996 Receptor specificity of the fibroblast growth factor family. J Biol Chem 271:15292–15297

Paznekas WA, Cunningham ML, Howard TD et al 1998 Genetic heterogeneity of Saethre–Chotzen syndrome, due to TWIST and FGFR mutations. Am J Hum Genet 62:1370–1380

Reardon W, Winter RM, Rutland P, Pulleyn LJ, Jones BM, Malcolm S 1994 Mutations in the fibroblast growth factor receptor 2 gene cause Crouzon syndrome. Nat Genet 8:98–103

Rousseau F, Bonaventure J, Legeai-Mallet L et al 1994 Mutations in the gene encoding fibroblast growth factor receptor-3 in achondroplasia. Nature 371:252–254

Shiang R, Thompson LM, Zhu YZ et al 1994 Mutations in the transmembrane domain of *FGFR3* cause the most common form of dwarfism, achondroplasia. Cell 78:335–342

Wall SA 1997 Diagnostic features of the major non-syndromic craniosynostosis and the common deformational conditions which may be confused with them. Curr Paediatr 7:8–17

Wilkie AOM, Oldridge M, Tang Z, Maxson RE Jr 2001 Craniosynostosis and related limb anomalies. In: The molecular basis of skeletogenesis. Wiley, Chichester (Novartis Found Symp 232) p 122–143

Yayon A, Klagsbrun M, Esko JD, Leder P, Ornitz DM 1991 Cell surface, heparin-like molecules are required for binding of basic fibroblast growth factor to its high affinity receptor. Cell 64:841–848

DISCUSSION

Bard: I was interested in one aspect of your pictures of the early expression of the heparan sulphate proteoglycan (HSPG) perlecan, and of FGF2 (Fig. 2). Around the forming bone, the perlecan and FGF expression seems coincident while the subjacent meningeal layer has perlecan but not FGF associated with it. I therefore wonder whether the HSPG around the early bone layer traps and immobilizes the FGF, and, if so, whether this affects any FGF gradient?

Morriss-Kay: The perlecan staining is very widespread and just shows that an HSPG is generally available for FGF binding—it doesn't provide any information on gradients. Perlecan is an extracellular matrix HSPG; the HSPG species that are likely to be involved in formation of the FGFR/HSPG/FGF complex are the transmembrane forms such as syndecan, which we haven't yet investigated. However, perlecan could be binding any FGF that moves away from its site of synthesis, and could therefore sharpen the difference between FGF levels in the differentiated region compared with those in the sutural mesenchyme where the osteogenic stem cells reside. FGF2 was not detected in the meningeal layers; other FGFs may be present there, but by E16 the distance is rather great for FGF diffusion from the brain to have a direct effect on events in the skeletogenic membrane. In the bead experiments, the FGF released from the bead didn't diffuse deeper than the skeletogenic membrane.

Beresford: This is going to sound a bit pedantic, but can you really say that based on the localization of the labelled FGF? How sensitive is your detection method, and how much FGF do you need to see an effect? It could be below the level of detection, but still be perfectly capable of eliciting a biological response?

Morriss-Kay: It could be, but the digoxygenin labelling detection technique is quite sensitive. We are looking at them with the same technique all the way through. We have some sense of there being a dynamic of diffusion from the beads. Of course I can't say that none goes through the suture, but the main point I want to make is about the FGFR1 and FGFR2 effects, because we seem to understand these a bit better than FGFR3.

Kronenberg: I'm a little confused about the FGF2R hypothesis. The impression I got was that your idea is that when you added the FGF2, FGFR2 was down-regulated. To project to cransisynostosis, the idea might be that the activating mutations will have a similar down-regulating effect.

Morriss-Kay: I should have explained more clearly what I think is happening in the normal sequence of events from proliferation to differentiation. As cells differentiate they lay down osteoid, which binds FGF2. Thus the extension of high levels of FGF2 is all the time progressing closer into the suture as these cells undergo differentiation and start to secrete FGF2. As FGF2 levels rise in the region of sutural stem cells that are expressing FGFR2, signalling through that receptor is increased, which feeds back to down-regulate the expression of that receptor and up-regulate expression of FGFR1. The two are mutually exclusive in their expression. Clearly, a down-regulation of receptor 2 is happening before upregulation of receptor 1. This process, which is part of the normal pattern of differentiation, appears to be speeded-up in the activating mutations. We mimic this with the FGF2 beads.

Russell: Can you say a little more about what happens to the sutures? Presumably, synostosis goes to the point where they are almost fusing. Do you actually get bone-to-bone fusion?

Morriss-Kay: There is bone-to-bone fusion in synostosis, obviously, but this doesn't happen until very late stages in the normal human skull. In the normal mouse, there is very little bony sutural fusion, presumably because they have a relatively short lifespan. The fact that bony fusion occurs in human cranio-synostosis seems to me one of the most interesting aspects of this abnormality, because the mid-sutural mesenchyme normally forms connective tissue that persists for years after growth has ceased (consider that most of us in this room have not got all of our cranial sutures fused, because there is still connective tissue within the suture). Something is inhibiting bony fusion of the sutures until an advanced age in the normal human skull. In craniosynostosis there is clearly an effect on tissue that normally remains fibrous for many years, so it is more than just the speeding up of a normal physiological process. It may be that, at least for the coronal suture, the expression of *TWIST* in the mid-sutural mesenchyme is very important for maintaining that separation of the bones within the suture.

Russell: Presumably, in post-fetal life other processes such as remodelling will be occurring. So the main question to Brian Hall is still development, I guess.

Kingsley: I am still having trouble understanding the Twist expression domains. If I understand what you said, in *Drosophila* Twist is upstream of the FGFR. In the developmental anatomy of the skull, Twist is the only one expressed in the mid-suture mesenchyme in a pattern that was non-overlapping.

Morriss-Kay: It overlaps with *Fgfr2* but not *Fgfr1*.

Kingsley: But there would still be a central region where Twist would be expressed and FGFR2 would not be expressed. Then, in the bead experiments, FGFR2 went down to undetectable levels while FGFR1 went up. FGFR1 is normally the receptor found in the differentiated osteocytes, and yet in that same situation Twist, which is normally the marker of the mid-suture mesenchyme, is also upregulated despite the fact that the beads are essentially causing a mimic of a craniosynostosis fusion between the bone plates. I'm surprised the Twist expression is up; I'm having trouble seeing what you think the regulatory loops are, given that many of the regions that normally express twist are regions that don't express either of the FGFRs.

Morriss-Kay: That's an important point: obviously, that result of the Twist expression effect on the FGF2 beads was quite puzzling. That is why I said that my interpretation was somewhat teleogical. It could be some kind of natural defence mechanism in which, if differentiation or the change over to differentiation is proceeding too fast, then *Twist* up-regulation would help slow it down. There are lots of checks and balances in these systems. The transforming growth factor (TGF)βs were mentioned earlier: there are TGFs expressed in sutures, and also bone morphogenetic proteins (BMPs), and the sonic hedgehog (SHH) signalling system is involved at a slightly later stage. So there are clearly lots of interacting signalling systems here, and I think one needs to hold on to that when we find difficulty in actually interpreting a single observation.

Wilkins: Also basic helix-loop-helix (bHLH) proteins have lots of partners, and the partner they interact with can change them from activators to repressors. It is therefore possible that Twist is actually doing something different when it's up-regulated, if it has a different partner.

Kronenberg: If in craniosynostosis the goal is to down-regulate FGFR2, wouldn't it be easier to have inactivating mutations? Why wouldn't that be a cause of the disease?

Wilkie: There, the problem we face is the fact that FGF receptors are ubiquitously expressed. The homozygous knockout of FGFR2 is early embryonic lethal (Arman et al 1998).

Tickle: Gillian, you raised some general ideas about craniosynostosis, and you suggested that perhaps the brain produces signals that control the size of the skull. We have been talking predominantly about the local interactions in the sutures, and you mentioned that for these children, the problem is that the sutures fuse and then the brain keeps on growing. Do you think there is an uncoupling of some of these interactions in these patients? How do you see your work on the local events in the sutures interfacing with this idea of signals from the brain?

Morriss-Kay: Some work has been done on signals from the meningeal layers. I hoped that Lynn Opperman was going to be able to come to this meeting, and I'm sorry she isn't here, because her group has done some very good work on the effects of separating the skeletogenic layer from the dura mater, followed by calvarial organ culture (Opperman et al 1999). Clearly, signals that lie within the dura affect the timing of sutural fusion. However, I can't find anything in the literature that gives us clues as to what kinds of signals may be emanating from the brain itself.

Poole: One of the signals could be pressure. Mechanical forces have such fundamental effects on cellular differentiation in the skeletal system. The fact that the sutures don't really close is probably a safety mechanism that provides an opportunity for adjustment and compensation. It is a finely balanced system. It would be very interesting to do some *in vitro* pressure studies just as we do with cartilage and bone to explore how this kind of effect might operate.

Mundlos: This experiment has frequently been done by nature. If you have a microcephalus, you have early fusion of the sutures. If the brain doesn't grow the sutures close within the first few years of life, whereas in hydrocephalus they never close.

Wilkins: But there could be chemical events correlated with those growth events as well: I wouldn't say that those observations have proved that pressure is the mechanism, although they are consistent with this idea.

Mundlos: I agree, there is an obvious correlation.

Bard: There was a nice paper by Chicurel and colleagues in *Nature* a couple of years ago, showing how pressure affects the actin cytoskeleton (Chicurel et al 1998).

Hall: What molecules is the dura producing?

Morriss-Kay: TGFs, BMPs and Msx2 have been detected (Opperman et al 1997, Rice et al 1999).

Ornitz: There is a nice set of experiments by Longaker (Levine et al 1998, Most et al 1998) where he was examining regulation of the frontal suture. He cut out the suture and rotated it 180° and showed differences between the proximal and distal regions of the suture. This was correlated with TGF and FGF expression in the underlying suture.

Bard: Given that there is so much proteoglycan in the meningeal layer, do you really think that you can get signals out of it? I would have thought that everything would get trapped in all that proteoglycan.

Pizette: But the binding of the FGF to proteoglycans is a dynamic process. There are proteases present that cleave FGF together with pieces of proteoglycans, releasing this complex from the extracellular matrix. In this case FGF could diffuse again.

Ornitz: In addition, there are heparinases which could release soluble fragments of FGF complexed with heparin.

Meikle: Many of the patients that have craniosynostosis also have maxillary hypoplasia. Is there any evidence that maxillary sutures are affected in any way by these mutations, or is it confined to the calvaria?

Wilkie: It's certainly not just confined to the calvaria. The whole facial skeleton is distorted in Apert syndrome (Cohen & Kreiborg 1996, see Fig. 1A of Wilkie et al 2000, this volume). There are many other systemic effects of FGFR2 mutations in Apert syndrome. There is an increased frequency of cardiac, gastrointestinal, genitourinary and central nervous system abnormalities (Cohen & Kreiborg 1990, 1993). It is also not understood why Apert mutations cause craniosynostosis and syndactyly in 100% of cases, whereas the other malformations are more variably present. What's the difference in pathogenic processes that makes some abnormalities absolute features of the condition, whereas in other cases it just biases the likelihood of the malformation?

Kingsley: One of the things that surprises me when I think about the data presented in the last two papers, is that very similar mutations in FGFR1 and FGFR2 give a craniosynostosis phenotype, despite the fact that when we've done the detailed developmental anatomy of these two receptors, they're expressed in non-overlapping patterns. FGFR1 is primarily associated with differentiating cells; FGFR2 is primarily associated with the cells at the edge that are also BrdU-positive in the double-labelling experiments. The expression patterns look like they're identifying separate cell types, but if you make an amino acid mutation in either there is essentially an identical effect.

Morriss-Kay: Our observations suggest that an activating mutation in FGFR2 would lead to its own premature down-regulation in osteogenic stem cells. There would thus be a turnover from proliferation to differentiation too quickly. Conversely, an activating mutation in FGFR1 would enhance differentiation. Fabienne Perrin-Schmitt has been studying FGFR1 knockout heterozygotes, which have an abnormality of sutural development.

Kingsley: I was struck in the whole-skeleton skull slides that both of you showed, that there appears to be another boundary within the flat bones of the skull. This appears to be what was labelled as the parietal and the frontal eminence. It is not a suture boundary, but it is an obvious morphological boundary that shows up in the skull.

Morriss-Kay: It isn't a boundary; it is a focal point from which the spicules of bone radiate out. It is to do with thc rcshaping of the skull. It is possibly the area which represents where the bones first started forming, but it also must be made more prominent like that by the remodelling process.

Kingsley: It looks like a visual lightening or darkening across the bone.

Morriss-Kay: By the time you see this feature in a skull at birth, an enormous amount of re-modelling has taken place since the initial laying down of the first bone in the fetus.

References

Arman E, Haffner-Krausz R, Chen Y, Heath JK, Lonai P 1998 Targeted disruption of fibroblast growth factor (FGF) receptor 2 suggests a role for FGF signaling in pregastrulation mammalian development. Proc Natl Acad Sci USA 95:5082–5087

Chicurel ME, Singer RH, Meyer CJ, Ingber DE 1998 Integrin binding and mechanical tension induce movement of mRNA and ribosomes to focal adhesions. Nature 392:730–733

Cohen MM Jr, Kreiborg S 1990 The central nervous system in the Apert syndrome. Am J Med Genet 35:36–45

Cohen MM Jr, Kreiborg S 1993 Visceral anomalies in the Apert syndrome. Am J Med Genet 45:758–760

Cohen MM Jr, Kreiborg S 1996 A clinical study of the craniofacial features in Apert syndrome. Int J Oral Maxillofac Surg 25:45–53

Levine JP, Bradley JP, Roth DA, McCarthy JG, Longaker MT 1998 Studies in cranial suture biology: regional dura mater determines overlying suture biology. Plast Reconstr Surg 101:1441–1447

Most D, Levine JP, Chang J et al 1998 Studies in cranial suture biology: up-regulation of transforming growth factor-beta1 and basic fibroblast growth factor mRNA correlates with posterior frontal cranial suture fusion in the rat. Plast Reconstr Surg 101:1431–1440

Opperman LA, Nolen AA, Ogle RC 1997 TGF-1, TGF-2, and TGF-3 exhibit distinct patterns of expression during cranial suture formation and obliteration *in vivo* and *in vitro*. J Bone Miner Res 12:301–310

Opperman LA, Chhabra A, Cho RW, Ogle RC 1999 Cranial suture obliteration is induced by removal of transforming growth factor (TGF)-3 activity and prevented by removal of TGF-2 activity from fetal rat calvaria *in vitro*. J Craniofac Genet Dev Biol 19:164–173

Rice DP, Kim HJ, Thesleff I 1999 Apoptosis in murine calvarial bone and suture development. Eur J Oral Sci 107:265–275

Wilkie AOM, Oldridge M, Tang Z, Maxson RE Jr 2001 Craniosynostosis and related limb anomalies. In: The molecular basis of skeletogenesis. Wiley, Chichester (Novartis Found Symp 232) p 122–143

Craniosynostosis and related limb anomalies

Andrew O. M. Wilkie, Michael Oldridge, Zequn Tang* and Robert E Maxson Jr*

*Institute of Molecular Medicine, John Radcliffe Hospital, Oxford OX3 9DS, UK, and *University of Southern California/Norris Hospital, USC Medical School, 1441 Eastlake Avenue, Los Angeles, CA 93012, USA*

Abstract. Many genetically determined craniosynostosis syndromes feature limb anomalies, implying that pathways of cranial suture and limb morphogenesis share some identical components. Identification of heterozygous mutations in *FGFR1*, *FGFR2*, *FGFR3*, *TWIST* and *MSX2* in craniosynostosis has focused particular attention on these genes. Here we explore two themes: use of clinical/molecular analysis to provide new clues to pathophysiology and the contrasting effects of loss- and gain-of-function mutations. Apert syndrome is a severe craniosynostosis/syndactyly disorder usually caused by specific substitutions (Ser252Trp or Pro253Arg) in FGFR2. The relative severity of cranial and limb malformations varies in opposite directions for the two mutations, suggesting that these phenotypes arise by different mechanisms. Clinical and biochemical evidence supports a model in which alternative splice forms of FGFR2 mediate these distinct effects. Pro → Arg substitutions equivalent the Pro253Arg/FGFR2 mutation occur in both FGFR1 and FGFR3, and are also associated with craniosynostosis. This suggests a common pathological mechanism, whereby enhanced affinity for a limited repertoire of tissue-specific ligand(s) excessively prolongs signalling in the cranial suture. The first MSX2 mutation in craniosynostosis was described in 1993 but this remains the only example. We have recently identified three MSX2 mutations associated with a different cranial phenotype, parietal foramina. DNA binding studies show that the craniosynostosis and parietal foramina arise from gain and loss of function, respectively.

2001 The molecular basis of skeletogenesis. Wiley, Chichester (Novartis Foundation Symposium 232) p 122–143

The bones of the cranial vault are unusual in that they develop directly from the membrane overlying the skull. In the human, ossification begins around the ninth week of development and apposition of the paired frontal and parietal bones occurs at about 15 weeks. Continued growth of the skull is made possible by the cranial sutures, narrow undifferentiated seams of mesenchyme that separate the cranial bones. This growth needs to be accurately tailored to growth of the underlying brain.

Until 1993, there had been little exploitation of modern molecular genetic techniques to understand how the skull develops. However, identification of mutations in humans with craniosynostosis — the premature fusion of the cranial sutures — served both to focus interest on this problem and simultaneously, to identify some of the key molecules involved. Several craniosynostosis syndromes are associated with diagnostic patterns of limb abnormality, indicating that development of the skull and limbs shares some critical genes, and perhaps developmental pathways in common. An accompanying article (Morriss-Kay et al 2001, this volume) illustrates what has been learnt about the role of these genes in cranial suture development. Here we briefly summarize the history of the human genetic findings and then, by focusing on three examples, illustrate how clinical molecular genetics can raise new developmental questions for further study in biochemical or animal model systems.

Mutations in human craniosynostosis

Heterozygous mutations in five genes implicated in skull development have been identified in craniosynostosis: in all cases, a positional candidate approach was used. These genes encode the homeodomain protein MSX2, the fibroblast growth factor receptors FGFR1, FGFR2 and FGFR3, and the helix–loop–helix transcription factor TWIST (reviewed by Wilkie 1997).

The first mutation to be described was in MSX2, a Pro148His substitution occurring at the seventh position of the 60 amino acid homeodomain, in a single family with 'Boston-type' craniosynostosis (Jabs et al 1993). Paradoxically, no further MSX2 mutations have been described in craniosynostosis (but see section below). However, the real explosion in interest started in 1994 when mutations were identified in the fibroblast growth factor receptors (FGFRs), a family of four transmembrane receptor tyrosine kinase proteins. Initially FGFR3 mutations were described in achondroplasia, followed shortly after by FGFR1 and FGFR2 mutations in Pfeiffer and Crouzon syndromes respectively. It emerged that allelic missense mutations of FGFR2 are associated with a variety of craniosynostosis phenotypes, one of the most severe of which is Apert syndrome (Fig. 1), characterized by additional bony syndactyly of the hands and feet (Wilkie et al 1995a). Mutations of FGFR3 have been identified in disorders characterized by either bone dysplasia or by craniosynostosis (Table 1).

Most FGFR mutations are specific missense changes and many are highly recurrent, exhibit an elevated mutation rate and are exclusively paternal in origin. Biochemical studies show that these FGFR mutations act by a variety of distinct gain-of-function mechanisms: many, but not all, bypass the normal requirement for fibroblast growth factor (FGF)-mediated FGFR dimerization (Table 2). Evidence, mostly from the study of bone dysplasias, suggests that this has

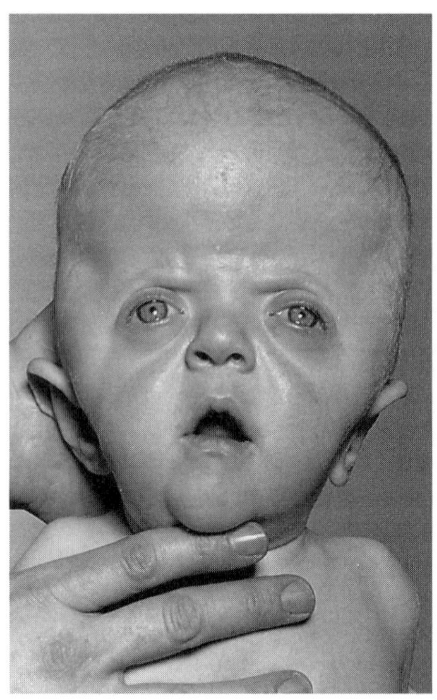

(a)

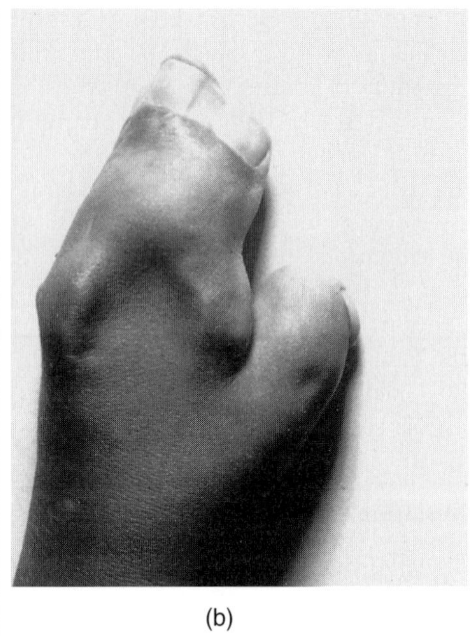

(b)

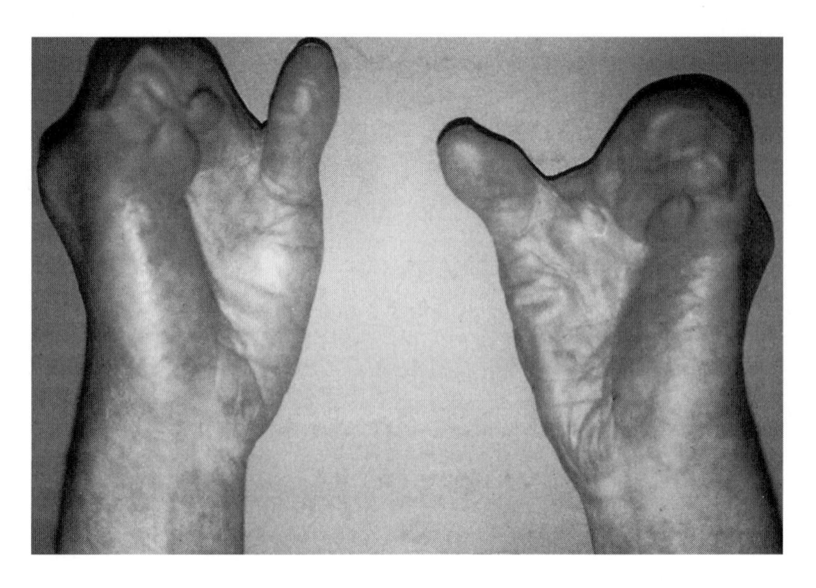

(c)

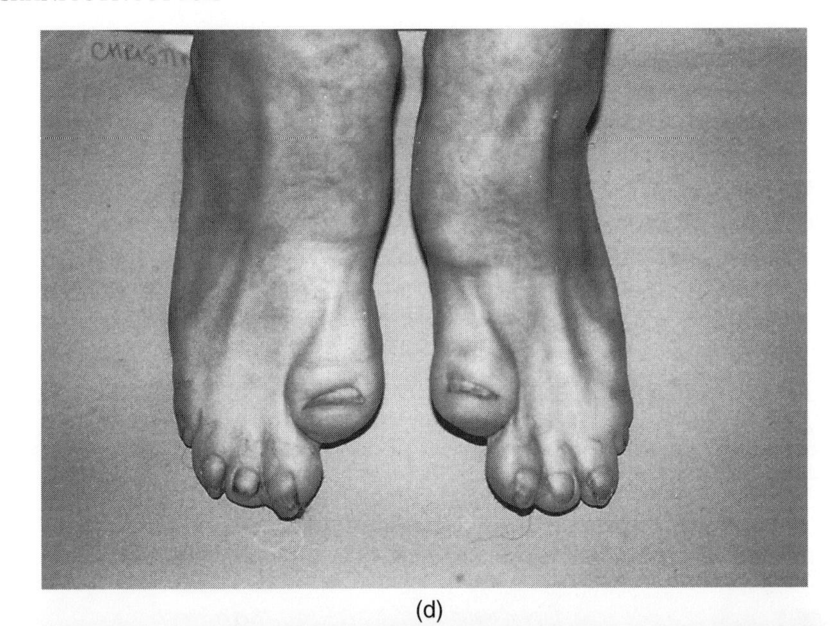

(d)

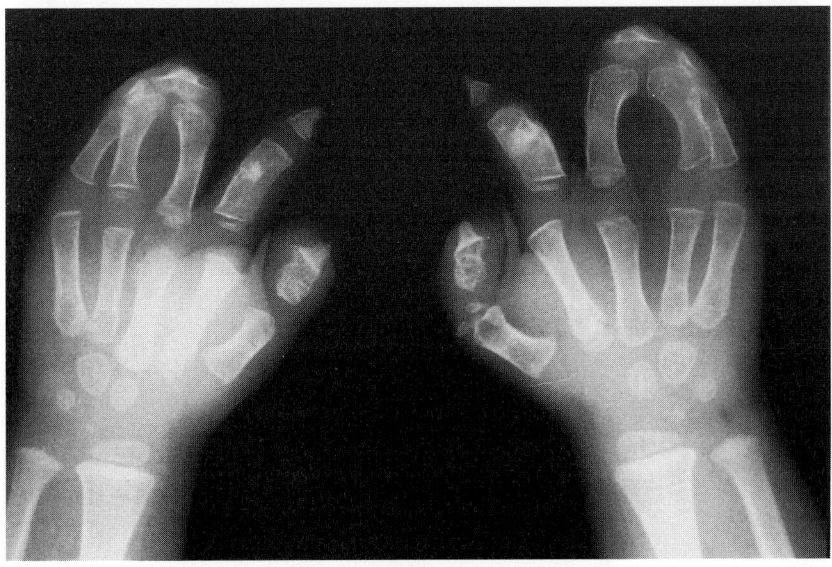

(e)

FIG. 1. Clinical features of Apert syndrome. (a) Pre-operative craniofacial appearance. Note broad high forehead, prominent eyes and underdeveloped midface caused by craniosynostosis. (b, c) Hands in two unrelated patients. The pattern of syndactyly differs between individuals but is highly symmetrical in a single individual. (d) The feet also manifest syndactyly. (e) Radiographic appearance of the hands. In this individual, both longitudinal and transverse fusions of the phalanges are apparent.

TABLE 1 Genes mutated in craniosynostosis

Gene	*Localization*	*Phenotypes*[a]
MSX2	5q34-q35	Boston craniosynostosis, parietal foramina
FGFR1	8p11.2-p12	Pfeiffer syndrome
FGFR2	10q26	Crouzon, Pfeiffer, Apert, Jackson–Weiss, Beare–Stevenson syndromes
FGFR3	4p16.3	Muenke, crouzonodermoskeletal syndromes, thanatophoric dysplasia; achondroplasia[b], hypochondroplasia[b], SADDAN[b]
TWIST	7p21.1	Saethre–Chotzen syndrome

[a]See Johnson et al (1998) and Passos-Bueno et al (1999) for further references.
[b]These disorders are characterized by dysplasic growth of the long bones rather than by craniosynostosis but are included for completeness.

TABLE 2 Gain of function in FGFR mutations

Mechanism	*Type of mutation*	*Key reference*
Reduced dissociation of ligand	IgII–IgIII linker mutation	Anderson et al (1998)
Covalent cross-linking of Cys	Loss of cysteine	Neilson & Friesel (1995)
	Gain of cysteine	Naski et al (1996)
	Disruption of IgIII domain structure	Robertson et al (1998)
Ectopic expression of IgIIIb spliceform	IgIIIc acceptor splice site mutation	Oldridge et al (1999)
Transmembrane hydrogen bonding	Neutral → charged mutation	Webster & Donoghue (1996)
Autoactivation of kinase domain	Substitution of inhibitory lysine	Webster et al (1996)

See Webster & Donoghue (1997) and Wilkie (1997) for further references.

diverse consequences for the cell (Table 3): the most consistent finding is the importance of STAT1 activation. Many mysteries about these FGFR mutations remain, not least of which are why they are not lethal prenatally (given the essential role of FGFR1 and FGFR2 in embryogenesis) and why there is no obvious increase in neoplasia associated with syndromes of FGFR mutation. A more complete description and interpretation of the molecular genetic findings

TABLE 3 Cellular consequences of FGFR mutations

	Reference
Mutant stabilization and nuclear translocation	Delezoide et al (1997)
STAT1 and p21WAF1/CIP1 activation	Su et al (1997), Sahni et al (1999)
Reduced calcium signalling	Nguyen et al (1997)
FGFR down-regulation	Bresnick & Schendel (1998)
STAT1 activation and increased apoptosis	Legeai-Mallet et al (1998)
STAT1 and ink4 activation	Li et al (1999)

in FGFR mutations may be found in several reviews (Wilkie et al 1995b, Webster & Donoghue 1997).

More recently, TWIST mutations were described in Saethre–Chotzen syndrome. These are more diverse in nature (both intragenic mutations and whole gene deletions) but more uniform in phenotype, suggesting that they are caused by TWIST haploinsufficiency (reviewed by Rose & Malcolm 1997, Johnson et al 1998).

Apert syndrome suggests distinct mechanisms of abnormal FGFR signalling in craniosynostosis and syndactyly

Apert syndrome is caused in ~99% of cases by one or other of two specific mutations (Ser252Trp and Pro253Arg) in adjacent amino acids of FGFR2 (Wilkie et al 1995a). These are located in the conserved linker between the second and third immunoglobulin (Ig)-like domains (Fig. 2), which orientates the two domains and also contacts bound FGF (Plotnikov et al 1999). What explains this very narrow spectrum of mutation, and why do these result in such specific morphological abnormalities? Three observations from clinical molecular genetics, combined with current knowledge of the biology of FGF signalling, lead to a testable hypothesis.

(1) The relative severity of cleft palate and syndactyly is reversed between the two mutations. The genotype–phenotype study of Slaney et al (1996) showed that cleft palate (which is likely to reflect severity of craniofacial malformation generally) was more frequently present with the Ser252Trp mutation. In contrast, syndactyly of both the hands and the feet was more severe for the Pro253Arg mutation. These trends have been confirmed in two independent studies (Lajeunie et al 1999, von Gernet et al 2000). These observations suggest that craniofacial and limb malformations in Apert syndrome arise by distinct molecular mechanisms.

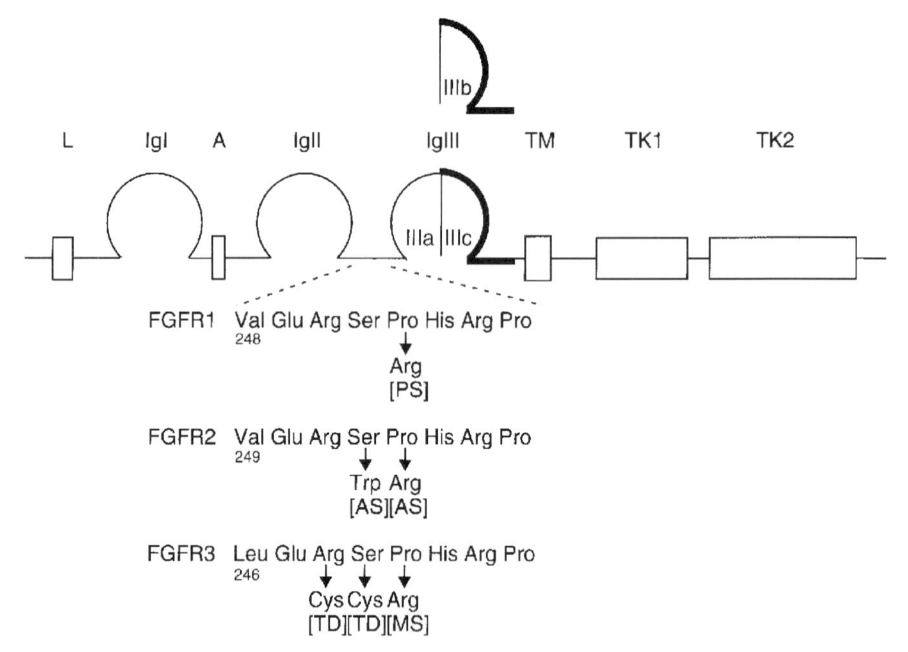

FIG. 2. Schematic structure of FGFR protein showing the leader region (L), immunoglobulin-like domains (IgI–IgIII), acid box (A), transmembrane segment (TM), and split tyrosine kinase domains (TK1 and TK2). Note the alternatively spliced second half of the IgIII domain (IgIIIb or IgIIIc), shown as bold lines. The amino acid sequence of the conserved linker segment between IgII and IgIII is shown for FGFR1, -2 and -3 together with recurrent mutations causing Pfeiffer (PS), Apert (AS) and Muenke (MS) syndromes, and thanatophoric dysplasia type 1 (TD).

(2) The Apert phenotype only occurs with specific amino acid substitutions of the serine–proline dipeptide. We have identified three other mutations of the Ser-Pro dipeptide (Oldridge et al 1997). Two of these (Ser252Leu, and the double substitution Ser252Phe, Pro253Ser) are chemically distinct from the Apert mutations, and neither results in an Apert phenotype. The third, Ser252Phe, is chemically similar to Ser252Trp and causes Apert syndrome. This indicates that morphological specificity depends on the nature, as well as the position of the mutations.

(3) Two patients with Apert syndrome have mutations that are entirely different in position and nature. Perhaps the greatest surprise was the identification of two Apert patients with an entirely different molecular basis for their disorder: *de novo* Alu element insertions within, or upstream of, the alternatively spliced *FGFR2* exon IgIIIc (Oldridge et al 1999). In one case we demonstrated ectopic expression of the alternative IgIIIb spliceform of FGFR2 in a fibroblast cell line from the patient.

The last observation suggests an explanation for the conclusion based on observation (1), that the mechanisms of the craniofacial and limb malformations are distinct. A particular feature of the biology of FGFR1, -2 and -3 is that mutually exclusive alternative splicing between a pair of exons results in two different IgIII domains (termed IgIIIb and IgIIIc; Fig. 2) with different ligand binding properties (Ornitz et al 1996). In the case of FGFR2, the IgIIIb isoform shows a rather narrow binding specificity, predominantly FGF1, -3, -7 and -10, the last three of which do not bind significantly to the IgIIIc isoform. Ectopic IgIIIb expression would extend a cell's binding characteristics to include these FGFs. FGF10 is one of the earliest expressed mesenchymal markers of the future limb field and in the mouse, homozygous loss of function of *Fgf10* causes a limbless phenotype (Min et al 1998, Sekine et al 1999). This led us to propose that the syndactyly of Apert syndrome is IgIIIb-mediated (Oldridge et al 1999). By contrast, craniosynostosis must be largely IgIIIc mediated, because many Crouzon patients have mutations in the IgIIIc exon, leaving the IgIIIb isoform unaffected.

On the basis of observation (1), our model predicts that the Apert Pro253Arg mutation should exhibit greater enhancement of binding to one or more of the IgIIIb-specific ligands relative to Ser252Trp. This would contrast with binding studies of the IgIIIc isoform (Anderson et al 1998), which consistently demonstrated a greater effect (slower ligand dissociation) for the Ser252Trp mutation, in line with the higher frequency of cleft palate associated with this mutation. We are currently performing analogous experiments on the IgIIIb isoform to test this model.

Which FGF ligands are critical for signalling in the cranial suture?

The two canonical Apert mutations occur in a 16 amino acid linker segment, conserved in sequence between all four FGFRs. Strikingly, mutations exactly equivalent to the Pro253Arg mutation occur in both FGFR1 and FGFR3. These are uniquely and specifically associated with mild Pfeiffer (Pro252Arg in FGFR1) and Muenke (Pro250Arg in FGFR3) craniosynostosis syndromes respectively. In the case of FGFR3, a further genotype–phenotype correlation is that mutations of the adjacent amino acids in the linker (Arg248Cys and Ser249Cys) cause an entirely distinct disorder, type 1 thanatophoric dysplasia (Fig. 2). These differences in phenotype point to differences in the pathophysiological mechanism of the FGFR3 mutations. Whereas the Arg248Cys and Ser249Cys mutations create unpaired cysteines that predispose to ligand independent, constitutive activation via covalent cross-linking of receptor monomers (Table 2 and Naski et al 1996), the mechanism of the Pro250Arg mutation in FGFR3 has not been reported. By analogy with the Pro253Arg mutation in FGFR2, it is likely that this mutation acts by a *ligand dependent* mechanism (see previous section). In the study by

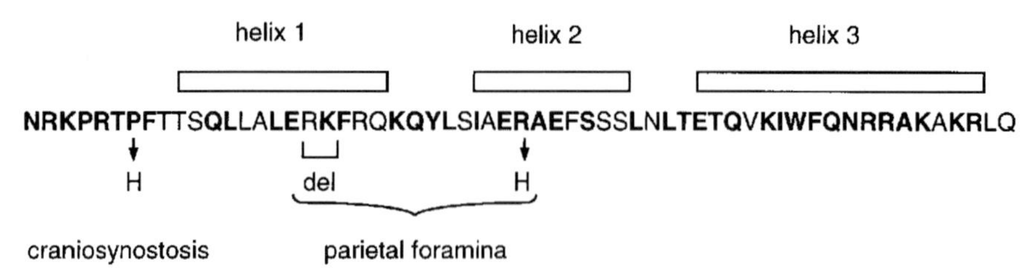

FIG. 3. Structure and amino acid sequence of the MSX2 homeobox. Amino acids that are conserved across all known msh-class homeoboxes are shown in bold. The positions of mutations mentioned in the text are indicated. The major DNA contacts are made by helix 3 and the flexible N-terminal arm, whereas helix 1 and 2 stabilize the domain folding.

Anderson et al (1998), FGF2 exhibited the greatest enhancement of binding affinity amongst those ligands tested (FGF1, -2, -4 and -6). However it seems unlikely that FGF2 is normally limiting for signalling in the cranial suture because *Fgf2* knockout mice have normal skulls (Zhou et al 1998). An intriguing possibility is that there exists another FGF ligand, relatively specific to the cranial sutures, the binding affinity for which is enhanced by corresponding linker Pro → Arg mutations in the three FGFRs. The restriction of phenotypic expression of these FGFR linker mutations largely to the skull would then be explained by ligand limitation. To date, however, no FGF has been shown to have the requisite combination of properties.

Contrasting phenotypes of gain- and loss-of-function mutations of *MSX2*

As described above, a specific MSX2 mutation (Pro148His in the homeodomain) was the first molecular mutation in craniosynostosis. Homeodomains are highly conserved DNA binding motifs (Fig. 3), and it was subsequently demonstrated that this mutation enhanced binding, compared with wild-type, to an optimal target DNA sequence (Ma et al 1996). This remains the solitary MSX2 mutation causing craniosynostosis.

Rare patients with three wild-type copies of *TWIST* (owing to a chromosome duplication) have wide gaps between the skull bones, a phenotype termed cranium bifidum (Stankiewicz et al 1998). This suggests that craniosynostosis and cranium bifidum may represent opposite ends of a developmental spectrum of cranial suture anomalies, corresponding respectively to increased and decreased differentiation. By analogy we speculated that parietal foramina, a mild variant of cranium bifidum, might in some cases be caused by reduced signalling by MSX2. We have recently confirmed this idea by demonstrating three independent heterozygous MSX2 mutations associated with congenital parietal foramina (Wilkie et al 2000). The mutations identified are a complete deletion of MSX2 and two mutations of the homeodomain, an in-frame deletion of two amino acids (Arg159 and Lys160) and an Arg172His substitution. The Lys160 and Arg172 residues occur at positions 19 and 31, respectively of the homeodomain protein, which are identical (Fig. 3) across all known msh-class sequences (including those of several invertebrate phyla). Comparison of the structure of the closely related protein MSX1 to the homeodomain of engrailed, indicates that the residues play important roles in stabilising the protein or contacting the phosphate backbone of the bound DNA target (Li et al 1997).

For the MSX2 homeodomain mutants, we demonstrated reduction in binding to an optimal DNA target sequence of 97% (ArgLys159-160del) and 85% (Arg172His) respectively (Wilkie et al 2000). We conclude that the phenotype of MSX2 haploinsufficiency in humans is parietal foramina. Most mutations of the homeodomain are likely to reduce, rather than increase, the binding of the domain to a target DNA sequence. Hence, parietal foramina are expected to be a more frequent phenotype caused by MSX2 mutation than craniosynostosis. Of note, the anatomical distribution of the skull defects in humans mirrors two lacunae in the calvarial expression of an *Msx2* promoter–*LacZ* transgene in the mouse (Liu et al 1999).

This work confirms the hypothesis that cranium bifidum/parietal foramina and craniosynostosis may be viewed as phenotypes at opposite ends of a spectrum of abnormal calvarial development. We and others (Wuyts et al 1999) have demonstrated genetic heterogeneity in the aetiology of parietal foramina, which suggests new avenues to identify disease genes in craniosynostosis through the isolation of disease loci for parietal foramina.

Acknowledgements

A.O.M.W. thanks Yvonne Jones, John Heath and Gillian Morriss-Kay for discussions and the Wellcome Trust for financial support (Senior Research Fellowship in Clinical Science). He is honoured to record that he is the third member of his family to have contributed to the Ciba/Novartis series of symposia. His mother, June Hill, participated in one and his father Douglas in two of these meetings (Hill 1961, Wilkie 1975, 1981). This chapter is dedicated to their memory.

References

Anderson J, Burns HD, Enriquez-Harris P, Wilkie AOM, Heath JK 1998 Apert syndrome mutations in fibroblast growth factor receptor 2 exhibit increased affinity for FGF ligand. Hum Mol Genet 7:1475–1483

Bresnick S, Schendel S 1998 Apert's syndrome correlates with low fibroblast growth factor receptor activity in stenosed cranial sutures. J Craniofac Surg 9:92–95

Delezoide A-L, Lasselin-Benoist C, Legeai-Mallet L et al 1997 Abnormal FGFR3 expression in cartilage of thanatophoric dysplasia fetuses. Hum Mol Genet 6:1899–1906

Hill JR 1961 The physics and physiology of the development of homoeothermy. In: Somatic stability in the newly born. JA Churchill, London (Ciba Found Symp 64) p 156–169

Jabs EW, Müller U, Li X et al 1993 A mutation in the homeodomain of the human MSX2 gene in a family affected with autosomal dominant craniosynostosis. Cell 75:443–450

Johnson D, Horsley SW, Moloney DM et al 1998 A comprehensive screen for TWIST mutations in patients with craniosynostosis identifies a new microdeletion syndrome of chromosome band 7p21.1. Am J Hum Genet 63:1282–1293

Lajeunie E, Cameron R, El Ghouzzi V et al 1999 Clinical variability in patients with Apert's syndrome. J Neurosurg 90:443–447

Legeai-Mallet L, Benoist-Lasselin C, Delezoide AL, Munnich A, Bonaventure J 1998 Fibroblast growth factor receptor 3 mutations promote apoptosis but do not alter chondrocyte proliferation in thanatophoric dysplasia. J Biol Chem 273:13007–13014 (erratum: 1998 J Biol Chem 273:19358)

Li C, Chen L, Iwata T, Kitagawa M, Fu XY, Deng CX 1999 A Lys644Glu substitution in fibroblast growth factor receptor 3 (FGFR3) causes dwarfism in mice by activation of STATs and ink4 cell cycle inhibitors. Hum Mol Genet 8:35–44

Li H, Tejero R, Monleon D et al 1997 Homology modeling using simulated annealing of restrained molecular dynamics and conformational search calculations with CONGEN: application in predicting the three-dimensional structure of murine homeodomain Msx-1. Protein Sci 6:956–970

Liu YH, Tang Z, Kundu RK et al 1999 Msx2 gene dosage influences the number of proliferative osteogenic cells in growth centers of the developing murine skull: a possible mechanism for MSX2-mediated craniosynostosis in humans. Dev Biol 205:260–274

Ma L, Golden S, Wu L, Maxson R 1996 The molecular basis of Boston-type craniosynostosis: the Pro148→His mutation in the N-terminal arm of the MSX2 homeodomain stabilizes DNA binding without altering nucleotide sequence preferences. Hum Mol Genet 5:1915–1920

Min H, Danilenko DM, Scully SA et al 1998 Fgf-10 is required for both limb and lung development and exhibits striking functional similarity to Drosophila branchless. Genes Dev 12:3156–3161

Morriss-Kay GM, Iseki S, Johnson D 2001 Genetic control of the cell proliferation–differentiation balance in the developing skull vault: roles of fibroblast growth factor receptor signalling pathways. In: The molecular basis of skeletogenesis. Wiley, Chichester (Novartis Found Symp 232) p 102–121

Naski MC, Wang Q, Xu J, Ornitz DM 1996 Graded activation of fibroblast growth factor receptor 3 by mutations causing achondroplasia and thanatophoric dysplasia. Nat Genet 13:233–237

Neilson KM, Friesel RE 1995 Constitutive activation of fibroblast growth factor receptor-2 by a point mutation associated with Crouzon syndrome. J Biol Chem 270:26037–26040

Nguyen HB, Estacion M, Gargus JJ 1997 Mutations causing achondroplasia and thanatophoric dysplasia alter bFGF-induced calcium signals in human diploid fibroblasts. Hum Mol Genet 6:681–688

Oldridge M, Lunt PW, Zackai EH et al 1997 Genotype–phenotype correlation for nucleotide substitutions in the IgII–IgIII linker of FGFR2. Hum Mol Genet 6:137–143

Oldridge M, Zackai EH, McDonald-McGinn DM et al 1999 De novo Alu element insertions in FGFR2 identify a distinct pathological basis for Apert syndrome. Am J Hum Genet 64: 446–461

Ornitz DM, Xu J, Colvin JS et al 1996 Receptor specificity of the fibroblast growth factor family. J Biol Chem 271:15292–15297

Passos-Bueno MR, Wilcox WR, Jabs EW, Sertié AL, Alonso LG, Kitoh H 1999 Clinical spectrum of fibroblast growth factor receptor mutations. Hum Mutat 14:115–125

Plotnikov AN, Schlessinger J, Hubbard SR, Mohammadi M 1999 Structural basis for FGF receptor dimerization and activation. Cell 98:641–650

Robertson SC, Meyer AN, Hart KC, Galvin BD, Webster MK, Donoghue DJ 1998 Activating mutations in the extracellular domain of the fibroblast growth factor receptor 2 function by disruption of the disulfide bond in the third immunoglobulin-like domain. Proc Natl Acad Sci USA 95:4567–4572

Rose CSP, Malcolm S 1997 A TWIST in development. Trends Genet 13:384–387

Sahni M, Ambrosetti DC, Mansukhani A, Gertner R, Levy D, Basilico C 1999 FGF signaling inhibits chondrocyte proliferation and regulates bone development through the STAT-1 pathway. Genes Dev 13:1361–1366

Sekine K, Ohuchi H, Fujiwara M et al 1999 Fgf10 is essential for limb and lung formation. Nat Genet 21:138–141

Slaney SF, Oldridge M, Hurst JA et al 1996 Differential effects of FGFR2 mutations on syndactyly and cleft palate in Apert syndrome. Am J Hum Genet 58:923–932

Stankiewicz P, Baldermann C, Thiele E et al 1998 The TWIST gene is triplicated in trisomy 7p syndrome. Eur J Hum Genet 6:61

Su W-CS, Kitagawa M, Xue N et al 1997 Activation of Stat1 by mutant fibroblast growth-factor receptor in thanatophoric dysplasia type II dwarfism. Nature 386:288–292

von Gernet S, Golla A, Ehrenfels Y, Schuffenhauer S, Fairley JD 2000 Genotype-phenotype analysis in Apert syndrome suggests opposite effects of the two recurrent mutations on syndactyly and outcome of craniofacial surgery. Clin Genet 57:137–139

Webster MK, Donoghue DJ 1996 Constitutive activation of fibroblast growth factor receptor 3 by the transmembrance domain point mutation found in achondroplasia. EMBO J 15:520–527

Webster MK, Donoghue DJ 1997 FGFR activation in skeletal disorders: too much of a good thing. Trends Genet 13:178–182

Webster MK, D'Avis PY, Robertson SC, Donoghue DJ 1996 Profound ligand-independent kinase activation of fibroblast growth factor receptor 3 by the activation loop mutation responsible for a lethal skeletal dysplasia, thanatophoric dysplasia type II. Mol Cell Biol 16:4081–4087

Wilkie AOM 1997 Craniosynostosis: genes and mechanisms. Hum Mol Genet 6:1647–1656

Wilkie AOM, Slaney SF, Oldridge M et al 1995a Apert syndrome results from localized mutations of FGFR2 and is allelic with Crouzon syndrome. Nat Genet 9:165–172

Wilkie AOM, Morriss-Kay GM, Jones EY, Heath JK 1995b Functions of fibroblast growth factors and their receptors. Curr Biol 5:500–507

Wilkie AOM, Tang Z, Elanko N et al 2000 Functional haploinsufficiency of the human homeobox gene MSX2 causes defects in cranial ossification. Nat Genet 24:387–390

Wilkie DR 1975 Muscle as a thermodynamic machine. In: Energy transformation in biological systems. Elsevier/Excerpta Medica/North-Holland, Amsterdam (Ciba Found Symp 31) p 327–339

Wilkie DR 1981 Shortage of chemical fuel as a cause of fatigue: studies by nuclear magnetic resonance and bicycle ergometry. In: Human muscle fatigue: physiological mechanisms. Pitman, London (Ciba Found Symp 82) p 102–119

Wuyts W, Di Gennaro G, Bianco F et al 1999 Molecular and clinical examination of an Italian DEFECT 11 family. Eur J Hum Genet 7:579–584

Zhou M, Sutliff RL, Paul RJ et al 1998 Fibroblast growth factor 2 control of vascular tone. Nat Med 4:201–207

DISCUSSION

Ornitz: With the Alu insertion where you are switching splice forms, one would expect that expressing a IgIIIb splice form of FGFR2 in mesenchyme would be catastrophic to the developing embryo, in that it may activate an autocrine loop within mesenchymal tissue, leading to activation of the receptor. Why don't you see that?

Wilkie: We would expect all these FGFR mutations to be catastrophic. We know that FGFR1 and FGFR2 are essential early in embryogenesis and the knockout mutations are lethal (Deng et al 1994, Yamaguchi et al 1994, Arman et al 1998). We don't understand why any of these activating mutations are compatible with life at all. I suspect the answer to your question is another side of that same coin. Clearly, what we have demonstrated is ectopic IgIIIb isoform expression in a fibroblast cell line from a child. This is completely different from saying that we know that the same thing is happening in the mesenchyme in very early limb buds. There is no way we can demonstrate that in the human, so at the moment this is only a hypothesis. But I think it's a reasonable working hypothesis.

Newman: A recent paper has shown that fibroblasts from Apert's patients overexpress transforming growth factor (TGF)β (Locci et al 1999). How do you see that in this context?

Wilkie: The osteoblasts from these patients show alterations in the expression of many different molecules, not just TGFβ, but also extracellular matrix and osteogenesis markers, like type I collagen, glycosaminoglycans and alkaline phosphatase (Lomri et al 1998, Fragale et al 1999, Locci et al 1999). Some of the findings (for alkaline phosphatase, for example) have been inconsistent, and it is hard to dissect out what is primary, and what are secondary effects of alterations in the growth characteristics of the cell. At the moment, what has happened is that people have picked out a few candidates and looked at their expression. They have had good reasons to choose those candidates, but we don't know the context to understand the significance of the results in relation to overall changes of expression. This is an obvious situation where microarray analysis, where you can analyse in parallel a large number of different transcripts, is going to be helpful to give an idea of the real pattern (Iyer et al 1999).

Beresford: Can I push you a little bit on the ligands in this setting? I got completely lost when you were talking about this strange finding that the plasmon resonance suggests that it is the binding to FGF2 that is altered. This immediately creates problems, because we don't really know how that gets out of the cell. But then you went on to say that in any case FGF2 knockouts are normal. Why did you make that point? If the phenotype is hypothetically due to tighter binding of or reluctance to release a bound ligand, which we suspect might be FGF2, why would the fact that the FGF2 knockout doesn't have a phenotype be informative?

Wilkie: Presumably you need an FGF in the suture to get signalling. If FGF2 was the only important FGF in the suture, you would expect there to be a phenotype with the knockout. Obviously, there could be redundancy—in fact, this is very likely. One explanation of the knockout result could therefore be that there is another FGF that can take on the role of FGF2 in that situation. I was really trying to make the point that there are various ligands expressed in cranial sutures, but we don't know what they're doing and which are the important ones. The pattern of enhanced FGF binding that we see with these mutations should enable us to make some predictions about what properties those ligands should have. For example, FGF4 wasn't found to have any greater affinity for the Apert mutations than for wild-type (Anderson et al 1998). It is therefore hard to make an argument that FGF4 is mediating the craniosynostosis, which is consistent with the low expression level of FGF4 in the sagittal suture (Kim et al 1998).

Ornitz: We have made the equivalent Apert syndrome mutation in the context of FGFR1. This has an increased affinity for FGF2 as well. This makes sense, because the mutation is in the linker domain.

As far as the FGF2 knockout mouse is concerned, it does have several phenotypes. There is not necessarily a suture phenotype, although I am not sure how closely this was looked at, but there are effects in migration of neurons in the cortex, and a vascular smooth muscle phenotype. An intriguing observation has been made by Kim et al (1998), showing that FGF9 is also expressed at high levels in the suture. In your assays, have you looked at FGF9 in the surface plasmon resonance assay?

Wilkie: I don't know how this fits in with your own studies on ligand binding, but in fact, FGF9 wasn't found to bind significantly to the constructs.

Ornitz: It should bind.

Wilkie: Have you looked at FGF9 binding to your proline to arginine mutation in FGFR1?

Ornitz: Not yet. I can add that we have also made *Fgf9* knockout mice. These mice die at birth, but there is no obvious suture defect.

Kingsley: In the crystal structure that you showed, you pointed to an arginine: is that the arginine residue involved in thanatophoric dysplasia?

Wilkie: Yes. The mutation is Arg248Cys in FGFR3 (Tavormina et al 1995).

Kingsley: The mutations that are affecting affinities or off-rates for the other ligands would actually be the proline residue at position 250 in FGFR3. Can you look at the crystal structure and see why the proline to arginine substitution at that position might affect affinity? Then, on the ligand side, if you look across the various FGF ligand family members, can you also predict which ones are likely to be affected?

Wilkie: I learnt an interesting thing about the black art of crystallography from this. Although the paper is published in *Cell* (Plotnikov et al 1999), apparently authors are allowed to hold on to the co-ordinates for nine months after publication. Thus the coordinates are not in the public domain. All that our crystallographer Yvonne Jones can go on is the stereo diagram. There are apparently now fancy programs that can be used to reconstruct a 3D structure from stereo diagrams! Basically, what you're asking is a very apposite question, but without the coordinates it is not possible to answer it. It is too subtle a thing to look at unless you know exactly where those residues are.

Ornitz: I would like to make one comment about the crystal structure. The crystal that was made only contains a fragment of the FGF receptor: it is missing some critical sequence between Ig domains 1 and 2 at the 5′ end of the N-terminus of the receptor. Although Ig domain 1 probably has very little to do with ligand binding, we have shown in a number of different experiments that the region between Ig domains 1 and 2 probably constitutes a second ligand binding site, in particular for FGF2, and may be a primary binding site for other ligands such FGF9 in different receptors. I think the crystal structure is therefore by no means the whole story, and they haven't really addressed how dimerization occurs. The other issue is that the receptor dimer that they show in the crystal structure is based on an FGF and FGFR in an asymmetric unit, and not an actual crystallization dimer of the receptors. Some of the interactions that they are showing may not be what's really happening. One final point is that the crystal was made in the absence of heparin, which we know is also a critical factor for binding and receptor activation.

Wilkie: They seem most confident about the ligand–receptor interaction.

Wilkins: It would be nice to study some of these things developmentally. Is anybody engineering mice with a conditional ectopic expression of these mutant forms?

Wilkie: Certainly. The bone dysplasias have led the way: there are now quite a few mouse models of these (Xu et al 1999, Li et al 1999, Wang et al 1999, Chen et al 1999, Segev et al 2000). The craniosynostosis models have been coming more slowly, and I don't think any have yet been published. The proline to arginine mutation in FGFR1 has now been made (Zhou et al 2000) and there are others coming on line.

Newman: I would like to press you a bit on the TGFβ story. It seems to me that no matter how indirect the receptor's effect is on TGFβ, if osteoblasts and precartilage mesenchymal cells are producing more TGFβ and consequently producing more fibronectin and so on, then that would be explanatory of many of the syndromes that you see. It seems to me that it doesn't really matter exactly what mechanism the receptor is using to do it, if it's causing an increase in molecules that promote mesenchymal condensation, that would be your explanation.

Wilkie: I can't add any further comment.

Kronenberg: I like your hypothesis about the syndactyly and the altered splicing. This makes this mutation qualitatively different from the others, and not just a quantitative part of a continuum. In this context, why do some of these mutations result in craniosynostosis and others result in achondroplasia, but never both? Or are they ever both?

Wilkie: Thanatophoric dysplasia combines both craniosynostosis and short-limbed bone dysplasia. It tends to be classified as a short-limbed bone dysplasia, because craniosynostosis is a clinical problem that usually presents in childhood rather than at birth, and thanatophoric dysplasia is neonatally lethal. In this disorder the two phenotypes can occur together. But I would make the general comment that there are some extremely subtle genotype–phenotype correlations that we still don't understand. A question I wanted to ask David Ornitz yesterday, which I will raise now, is that one of the FGFR3 mutations (Ala391Glu) causes a combination of both a Crouzon-like syndrome with craniosynostosis, and a skin disorder called acanthosis nigricans (Meyers et al 1995): do you know of any work that has elucidated the mechanism of that particular mutation?

Ornitz: I am not aware of any biochemical work on that mutation. FGFR3 is expressed in skin, in epidermis and possibly dermis. It is also expressed in a chondrocyte layer underneath the cranial sutures. At least from patterns of expression there is some precedent for a phenotype in these tissues. It's possible that some of these mutations may affect ligand binding specificity, perhaps in more subtle ways than specificity induced by alternative splicing, but they also could affect heterodimerization between FGF receptors, which may give completely novel phenotypes.

Wilkie: That mutation is a particularly good candidate for affecting FGFR heterodimerization (Shi et al 1993).

Mundlos: Can you tell us a little more what MSX is doing? Is it controlling bone growth?

Wilkie: I haven't done any of this work myself: it is based on work in the mouse, from Rob Maxson's lab in collaboration with Richard Maas. He suggests that MSX is involved in controlling both proliferation and differentiation. He proposes that a reduced dose of MSX2 reduces proliferation, and so there is a smaller number of cells that are able to commit to differentiation (Liu et al 1999, Dodig et al 1999). The way I would interpret our own findings, is that quite unequivocally the reduced MSX2 dosage has reduced the amount of differentiation. The actual size of the skull is normal — there's no effect on the head circumference — so it's impossible to comment on the effect on proliferation based on the human phenotype.

Mundlos: So it is an ongoing process.

Wilkie: Yes, it's ongoing in the sense that these parietal foramina persist throughout life but slowly lessen in size.

Newman: I have a hypothesis that I would like to present that relates to the syndactyly seen in Apert syndrome. This is based on results from chick embryology. We became interested in the FGF receptors because we were looking at a culture system where we would get foci of precartilage condensation forming in culture. This would expand and eventually cover the whole culture unless we added ectoderm or FGFs, in which case the condensations and nodules that formed from them would remain confined. We had earlier shown that TGFβ was responsible for producing fibronectin in these nodules (Leonard et al 1991, Downie & Newman 1995). Because TGFβ is positively auto-regulatory, we would get a natural expansion. But with ectoderm, which is a source of various FGFs, we would get perinodular inhibition. If you look at the development of the limb, in the pre-condensed mesenchyme the main FGF receptor present is FGFR1. In precartilage condensations, the main receptor present is FGFR2. When cartilage is differentiated, FGFR3 is present (Szebenyi et al 1995). Therefore the relevant FGFR in precartilage condensations is FGFR2. Now if a source of FGFs causes a confinement of the condensations, then activating FGFR2 may be causing the release of the putative lateral inhibitor for a Turing-type reaction–diffusion system. To reiterate, this is a system in which there is positive auto-regulation of an activator, which also induces the release of a lateral inhibitor of its own activity. If that inhibitory effect was abrogated by a mutation in FGFR2, then without the inhibition you would get expansion of condensations and syndactyly rather than individual elements. Let's see if this relates to these actual mutations. Since we are dealing with mesenchymal cells, you would expect to have the IgIIIc type of FGFR2. This is generally what's seen here. But if you have a mis-splicing so that you have the IgIIIb form, then you would have the wrong kind of signalling: you would have a receptor that wouldn't be appropriate here, and whatever signal transduction that might lead to these the inhibitory effects could be abrogated under these conditions. The other mutations that preserve the IgIIIc form are described as gain-of-function mutations, because they bind more strongly to FGF, but my understanding is that it's only FGF2 that binds more strongly to the mutated IgIIIc forms.

Wilkie: This is the point about the two classical Apert mutations showing opposite effects in the severity of craniofacial malformations versus the syndactyly. The experiments that were done on the IgIIIc form are in keeping with the severity of the craniofacial problems, but they're actually the reverse of what you see for the syndactyly. This suggests that this is not the mechanism of the syndactyly.

Newman: The IgIIIc is the normal one: when you get the IgIIIb form that arises from the Alu insertions, do those patients exhibit syndactyly?

Wilkie: Yes, by definition, because that's why they have got Apert syndrome (Oldridge et al 1999). But what I was trying to get at is the question of why you get syndactyly in classical Apert syndrome. What we need to do is to close the loop by repeating the experiments on the IgIIIb form. What I would hope to find is that there is a ligand like FGF10 which binds with enhanced affinity to the IgIIIb form, but this time the enhancement in affinity is greater for the Pro253Arg mutation than it is for the Ser252Trp mutation. This would then explain the opposite effects on severity between the two mutations.

Ornitz: Why would you expect to have an IgIIIb form of the receptor in the classical Apert mutation? It shouldn't affect splicing.

Wilkie: The site of the mutations (in the linker between IgII and IgIIIa, see Fig. 2) is such that they are present both in the normal IgIIIb isoform and the IgIIIc isoform.

Kronenberg: Do you know you don't affect splicing? Adjacent sequences can sometimes affect the binding of proteins that regulate splicing.

Wilkie: We have excluded this (Oldridge et al 1999).

Newman: In any case, there are two possible ways of abrogating the normal signal that you would get from the FGFR2. One would be mis-splicing, so that you would have the IgIIIb form in the mesenchymal tissue rather than the IgIIIc form. Another way of doing it is to have enhanced affinity for a ligand that is not the normal ligand for this process. For example, the endoderm is putting out a whole variety of FGFs, including FGF2 and FGF4. If the receptor exhibited enhanced affinity for FGF2, for example, and that was not the ligand responsible for mediating this inhibitory effect, then you might have competition with the normal ligand. The crux of this hypothesis is that it's the expression in the condensations of the correct form of FGFR2, with the appropriate ligand, that is responsible for mediating an inhibitory effect on chondrogenesis. If that's abrogated, then you get expansion of the condensations and syndactyly.

Kingsley: What's the phenotype of a simple FGFR2 knockout?

Wilkie: This is an early embryonic lethal, owing to a defect shortly after implantation (Arman et al 1998). An independent construct with leaky expression is lethal at day 10–11, with absent limbs (Xu et al 1998).

Newman: In our culture system, we've done antisense oligos against FGFR2, and have got expanded condensations. It's essentially a knockout in the culture. When we remove FGFR2, instead of getting individual condensations we get the whole mass of the culture turning into a sheet of cartilage.

Bard: And if you take out the antisense oligos, what happens then?

Newman: We are electroporating these constructs in, so we can't easily take them away.

Kronenberg: Why do you see this effect in the culture? Is it because of proliferation?

Newman: We don't know. Normally, these cells are not really proliferating very much at all. TGFβ is causing matrix production, and fibronectin and cell surface adhesive proteins are causing the precartilage cells to condense. TGFβ will spread out from those sites because of its positive auto-regulation and diffusion, and tend to cause an expansion of these centres. If you have a lateral inhibitor of TGFβ the expansion will be restricted, and you will get spaced-out elements. This is why I found the report of Locci et al (1999), that Apert syndrome fibroblasts and osteoblasts have enhanced TGFβ so interesting. This is exactly what is predicted by this model: the cells are not responsive to whatever is causing TGFβ to limit itself.

Kronenberg: So in your oligo experiment, where you remove FGFR2, why do all the cells become chondrocytes?

Newman: There is no FGFR2 at these centres, so when the condensations meet with ectodermal products, such as FGFs, no inhibitor is moving out from those centres, and we therefore get an expansion of the condensations.

Mundlos: From a clinical perspective this would mean that individuals with syndactyly should have very large bones.

Newman: You would see hard tissue fusion.

Mundlos: Whereas in most syndactyly cases there are normal bones with different fusions, either bony fusions or soft tissue fusions. My understanding was that this is more commonly attributed to the effects of BMPs in apoptosis, rather than these anlage being enlarged. In this case we would expect a hand like in the noggin mutations, which is all bone and cartilage.

Newman: Some of the X-rays that I've seen in published reports have very mis-shapen bones.

Wilkie: In Apert syndrome, I wouldn't say the bones were increased in size, in proportion to the soft tissues. As regards the shape of the bone, you get both longitudinal fusions of the proximal and middle phalanges, and transverse fusions involving the distal phalanges of the middle three digits, so they're fused into one single bone (Fig. 1E). These fusions tend to reflect the morphology of the autopod as a whole. I would see them in terms of the actual mesenchymal condensations being primarily abnormal, rather than as a secondary process of failure of apoptosis.

Ornitz: But there is no change in the length of the bones when they are fused?

Wilkie: If anything they're a little bit shorter than normal.

Kingsley: These were the hand phenotypes: in the clinical picture you showed it looked like the foot phenotypes are more confined to soft tissues.

Wilkie: In the feet there are certainly no transverse fusions. As you know it's often quite difficult to see the foot phalanges on X-ray, so I wouldn't want to comment on longitudinal fusions. In fact, the hallux of the foot is often quite strikingly abnormal with pre-axial polydactlyly involving the first metatarsal bone (Cohen & Kreiborg 1995).

Ornitz: Would you like to comment on why the bone fusions are restricted to the digits and not to more proximal bones?

Wilkie: This is not strictly true: there are other fusions that can occur. One that is quite frequent with FGFR2 mutations — in fact, I find it a useful clinical clue as to whether to look for mutations—is elbow fusion. Humero–radial or humero–ulnar synostosis are both described (Cohen 1993, Cohen & Kreiborg 1993). You can also get longitudinal vertebral fusions, particularly of the cervical vertebrae (Thompson et al 1996).

Morriss-Kay: It is also potentially informative that the digital fusions are progressively severe from proximal to distal, which could suggest that whatever the mechanism is, the longer the cells in the progress zone are exposed to it, the more they are prone to the loss of interdigital tissue. In that sense, what the digits tell you is that you would expect there to be the mildest effect on the most proximal elements of the limb.

Regarding the IgIIIb and IgIIIc isoform expression patterns, Sachiko Iseki in my lab has carried out isoform-specific *in situ* hybridization for *Fgfr2* on mouse limb buds. Although in general the IgIIIb isoform is expressed in epithelia and IgIIIc isoform in mesenchyme, when you look at the development of the hand plate, it's actually not as clear-cut as that. Around the condensations and in the interdigital mesenchyme there is strong expression of the IgIIIc splice variant, but there is also some expression of IgIIIb. It could be that there is an interaction between the two splice variants, or that, when co-expressed, they're both working towards the same end. It is important not to have hypotheses concerning isoform-specific functions on the assumption that their expression patterns show a simple epithelial–mesenchymal reciprocity.

Newman: But even if there's a balance of the two isoforms, if there was a mutation that shifted that balance, it could have the hypothesized effect.

Morriss-Kay: Certainly, one would expect that to have an effect. This is what was recently suggested on the basis of altered isoform expression in Apert syndrome (Oldridge et al 1999).

References

Anderson J, Burns HD, Enriquez-Harris P, Wilkie AOM, Heath JK 1998 Apert syndrome mutations in fibroblast growth factor receptor 2 exhibit increased affinity for FGF ligand. Hum Mol Genet 7:1475–1483

Arman E, Haffner-Krausz R, Chen Y, Heath JK, Lonai P 1998 Targeted disruption of fibroblast growth factor (FGF) receptor 2 suggests a role for FGF signaling in pregastrulation mammalian development. Proc Natl Acad Sci USA 95:5082–5087

Chen L, Adar R, Yang X et al 1999 Gly369Cys mutation in mouse FGFR3 causes achondroplasia by affecting both chondrogenesis and osteogenesis. J Clin Invest 104:1517–1525

Cohen MM Jr 1993 Pfeiffer syndrome update, clinical subtypes, and guidelines for differential diagnosis. Am J Med Genet 45:300–307

Cohen MM Jr, Kreiborg S 1993 Skeletal abnormalities in the Apert syndrome. Am J Med Genet 47:624–632

Cohen MM Jr, Kreiborg S 1995 Hands and feet in the Apert syndrome. Am J Med Genet 57: 82–96

Deng C-X, Wynshaw-Boris A, Shen MM, Daugherty C, Ornitz DM, Leder P 1994 Murine FGFR-1 is required for early postimplantation growth and axial organization. Genes Dev 8:3045–3057

Dodig M, Tadic T, Kronenberg MS et al 1999 Ectopic *Msx2* overexpression inhibits and *Msx2* antisense stimulates calvarial osteoblast differentiation. Dev Biol 209:298–307

Downie SA, Newman SA 1995 Different roles for fibronectin in the generation of fore and hind limb precartilage condensations. Dev Biol 172:519–530

Fragale A, Tartaglia M, Bernardini S et al 1999 Decreased proliferation and altered differentiation in osteoblasts from genetically and clinically distinct craniosynostotic disorders. Am J Pathol 154:1465–1477

Iyer VR, Eisen MB, Ross DT et al 1999 The transcriptional program in the response of human fibroblasts to serum. Science 283:83–87

Kim H-J, Rice DPC, Kettunen PJ, Thesleff I 1998 FGF-, BMP- and Shh-mediated signalling pathways in the regulation of cranial suture morphogenesis and calvarial bone development. Development 125:1241–1251

Leonard CM, Fuld HM, Frenz DA, Downie SA, Massagué J, Newman SA 1991 Role of transforming growth factor-β in chondrogenic pattern formation in the embryonic limb: stimulation of mesenchymal condensation and fibronectin gene expression by exogenous TGF-β and evidence for endogenous TGF-β-like activity. Dev Biol 145:99–109

Li C, Chen L, Iwata T, Kitagawa M, Fu X-Y, Deng C-X 1999 A Lys644Glu substitution in fibroblast growth factor receptor 3 (FGFR3) causes dwarfism in mice by activation of STATs and ink4 cell cycle inhibitors. Hum Mol Genet 8:35–44

Liu Y-H, Tang Z, Kundu RK et al 1999 *Msx2* gene dosage influences the number of proliferative osteogenic cells in growth centers of the developing murine skull: a possible mechanism for MSX2-mediated craniosynostosis in humans. Dev Biol 205:260–274

Locci P, Baroni T, Pezzetti F et al 1999 Differential *in vitro* phenotype pattern, transforming growth factor-β_1 activity and mRNA expression of transforming growth factor-β_1 in Apert osteoblasts. Cell Tissue Res 297:475–483

Lomri A, Lemonnier J, Hott M et al 1998 Increased calvaria cell differentiation and bone matrix formation induced by fibroblast growth factor receptor 2 mutations in Apert syndrome. J Clin Invest 101:1310–1317

Meyers GA, Orlow SJ, Munro IR, Przylepa KA, Jabs EW 1995 Fibroblast growth factor receptor 3 (FGFR3) transmembrane mutation in Crouzon syndrome with acanthosis nigricans. Nat Genet 11:462–464

Oldridge M, Zackai EH, McDonald-McGinn DM et al 1999 *De novo* Alu element insertions in *FGFR2* identify a distinct pathological basis for Apert syndrome. Am J Hum Genet 64: 446–461

Plotnikov AN, Schlessinger J, Hubbard SR, Mohammadi M 1999 Structural basis for FGF receptor dimerization and activation. Cell 98:641–650

Segev O, Chumakov I, Nevo Z et al 2000 Restrained chondrocyte proliferation and maturation with abnormal growth plate vascularization and ossification in human FGFR-3^{G380R} transgenic mice. Hum Mol Genet 9:249–258

Shi E, Kan M, Xu J, Wang F, Hou J, McKeehan WL 1993 Control of fibroblast growth factor receptor kinase signal transduction by heterodimerization of combinatorial splice variants. Mol Cell Biol 13:3907–3918

Szebenyi G, Savage MP, Olwin BB, Fallon JF 1995 Changes in the expression of fibroblast growth factor receptors mark distinct stages of chondrogenesis *in vitro* and during chick limb skeletal patterning. Dev Dyn 204:446–456

Tavormina PL, Shiang R, Thompson LM et al 1995 Thanatophoric dysplasia (types I and II) caused by distinct mutations in fibroblast growth factor receptor 3. Nat Genet 9: 321–328

Thompson DNP, Slaney SF, Hall CM, Shaw D, Jones BM, Hayward RD 1996 Congenital cervical spinal fusion: a study in Apert syndrome. Pediatr Neurosurg 25:20–27
Wang Y, Spatz MK, Kannan K et al 1999 A mouse model for achondroplasia produced by targeting fibroblast growth factor receptor 3. Proc Natl Acad Sci USA 96:4455–4460
Xu X, Weinstein M, Li C et al 1998 Fibroblast growth factor receptor 2 (FGFR2)-mediated reciprocal regulation loop between FGF8 and FGF10 is essential for limb induction. Development 125:753–765
Xu X, Weinstein M, Li C, Deng CX 1999 Fibroblast growth factor receptors (FGFRs) and their roles in limb development. Cell Tissue Res 296:33–43
Yamaguchi TP, Harpal K, Henkemeyer M, Rossant J 1994 *fgfr-1 is* required for embryonic growth and mesodermal patterning during mouse gastrulation. Genes Dev 8:3032–3044
Zhou YX, Xu X, Chen L, Li C, Brodie SG, Deng CX 2000 A Pro250Arg substitution in mouse FgFr1 causes increased expression of CbFa1 and premature fusion of calvarial sutures. Hum Mol Genet 9:2001–2008

The parathyroid hormone-related protein and Indian hedgehog feedback loop in the growth plate

Henry M. Kronenberg and Ung-il Chung

Endocrine Unit, Massachusetts General Hospital and Harvard Medical School, Boston, MA 02114 USA

Abstract. Normal development of the growth plate requires coordinated proliferation and differentiation of chondrocytes and osteoblasts. In previous work, we have shown that Indian hedgehog (IHH), produced by prehypertrophic and hypertrophic chondrocytes, stimulates production of parathyroid hormone-related protein (PTHrP) by perichondrial and early chondrocytic cells. PTHrP then maintains chondrocytes in a proliferative, less differentiated state. Because this less differentiated state delays the production of IHH, IHH and PTHrP may participate in a negative feedback loop that synchronizes and determines the pace of differentiation of chondrocytes in the growth plate. To establish the roles of physiological levels of PTHrP and IHH, we have now injected PTH/PTHrP receptor (−/−) embryonic stem (ES) cells into normal blastocysts to generate mice with chimeric growth plates. The PTH/PTHrP receptor cells leave the proliferative cycle and differentiate prematurely in the middle of the normal proliferative columns. The columns of wild-type cells are longer than normal and the adjacent bone collar is also longer than normal. Patterns of gene expression and the use of chimeras using PTH/PTHrP receptor (−/−); IHH (−/−) ES cells suggest that modified patterns of IHH and PTHrP synthesis explain these abnormalities. Thus, IHH is a master regulator of both chondrocyte and osteoblast differentiation.

2001 The molecular basis of skeletogenesis. Wiley, Chichester (Novartis Foundation Symposium 232) p 144–157

Endochondral bone formation

Most bones form through the process of endochondral bone formation. Mesenchymal cells first move close together (condense), and then turn on the chondrogenic cell programme. The chondrocytes proliferate and secrete a characteristic matrix, yielding a mould in the general shape of the future bone that will replace it. Cells in the centre of the mould stop proliferating and synthesize proteins characteristic of hypertrophic chondrocytes. The hypertrophic chondrocytes secrete a distinct matrix, which becomes calcified, and the hypertrophic chondrocytes undergo apoptosis. Perichondrial cells adjacent to the hypertrophic chondrocytes become osteoblasts and lay down a bone collar that eventually becomes cortical bone. Blood vessels invade the region of dying hypertrophic chondrocytes, in association with the appearance of marrow elements, including osteoclasts that remove the cartilage matrix, and of osteoblasts that lay down a bone matrix that forms the primary spongiosa. Remaining chondrocytes nearer the ends of the bone continue to proliferate, forming prominent orderly columns in the long bones of the limb. The regions of proliferating and hypertrophying chondrocytes are called the growth plates and provide the major engine for further bone growth.

Thus, bone formation requires the proper coordination of proliferation, differentiation, and movement of multiple cell types. Such coordination requires a number of local signalling pathways involving paracrine factors and cell–cell interactions. The importance of these factors, which include bone morphogenetic proteins, transforming growth factor β, fibroblast growth factors, Indian hedgehog (IHH), insulin-like growth factors (IGFs), and parathyroid hormone-related protein (PTHrP), has been demonstrated through the study of transgenic and gene knockout mice and human genetic disorders.

PTHrP regulates endochondral bone formation

The importance of PTHrP in bone development is illustrated by the phenotype of mice missing the PTHrP gene (Karaplis et al 1994). All bones formed by endochondral bone formation are abnormal; they exhibit striking foreshortening of the columns of proliferating chondrocytes. The mice die at birth, probably because the small ribcages preclude normal ventilation. Further, calcified cartilage and bone appear sooner than normal in a variety of sites. Most strikingly, in the anterior portions of ribs, which normally have only proliferative chondrocytes at the time of birth, the chondrocytes become hypertrophic and a bone collar forms adjacent to these chondrocytes. The pattern of cellular differentiation suggests that, normally, PTHrP keeps chondrocytes in the proliferative mode and prevents further differentiation into hypertrophic chondrocytes.

Mice missing the PTH/PTHrP receptor exhibit a similar phenotype in their growth plates (Lanske et al 1996); this similarity suggests that PTHrP acts on the PTH/PTHrP receptor to keep chondrocytes in the proliferative mode. PTHrP mRNA is made by perichondrial cells and by chondrocytes near the ends of bone (periarticular chondrocytes) in fetal life; the PTH/PTHrP receptor mRNA is made by prehypertrophic chondrocytes and, to a much lesser extent, by all but the earliest proliferating chondrocytes (Lee et al 1993, 1995). These expression patterns suggest that PTHrP, synthesized near the ends of bone, acts on the PTH/PTHrP receptor on chondrocytes to maintain them in the proliferative pool.

Studies of transgenic mice overproducing PTHrP in chondrocytes are consistent with this model (Weir et al 1996). These mice have few hypertrophic chondrocytes at birth and exhibit a delay in chondrocyte differentiation. Similarly, mice expressing a constitutively active PTH/PTHrP receptor (Schipani et al 1997) show delay in the movement of chondrocytes from the proliferative to the hypertrophic chondrocyte compartment and show delay in death of hypertrophic chondrocytes. The constitutively active receptor sequence was derived from that of some patients with Jansen chondrodystrophy, a human disorder of short stature and hypercalcaemia (Schipani et al 1995). When the Jansen transgene is introduced into growth plates of *PTHrP*$^{-/-}$ mice by appropriate matings, the overactive PTH/PTHrP receptor rescues the growth plate abnormality and allows the mice to live for several months after birth (Schipani et al 1997). An analogous rescue of the *PTHrP*$^{-/-}$ mice occurs when the transgene directing overproduction of PTHrP in chondrocytes is introduced into these mice (Philbrick et al 1998). Thus, a variety of genetic manipulations support the postulated roles of PTHrP and the PTH/PTHrP receptor in growth plate development.

The mice over-expressing PTHrP or expressing a constitutively active PTH/PTHrP receptor demonstrate the importance of proper regulation of this ligand–receptor system. Too much or too little signalling leads to abnormal bones. Further, mechanisms for regulating activity of this signalling pathway clearly exist. Mice missing the *PTH/PTHrP receptor* gene, for example, have increased *PTHrP* mRNA in cells near the top of the growth plate (U. I. Chung & H. M. Kronenberg, unpublished results). Several studies have suggested that IHH is one important regulator of the PTHrP signalling pathway.

IHH and endochondral bone formation

When IHH was overexpressed in fetal chicken limbs, the conversion of proliferative to hypertrophic chondrocytes was delayed and expression of PTHrP was increased at the ends of the bones (Vortkamp et al 1996). Similarly, when an active fragment of sonic hedgehog (known to act like IHH in bone) was added to fetal mouse limb explants, chondrocyte differentiation was delayed and PTHrP expression was increased. In contrast, when the sonic hedgehog fragment was added to explants of limbs from either *PTHrP*$^{-/-}$ or *PTH/PTHrP receptor*$^{-/-}$ mice, no change in chondrocyte differentiation was seen. These results suggest that IHH acts to keep chondrocytes in the proliferative compartment by stimulating the synthesis of PTHrP. IHH is synthesized by prehypertrophic and early hypertrophic chondrocytes (Bitgood & McMahon 1995). The range of IHH action is limited, however, and it is not yet clear whether IHH directly stimulates

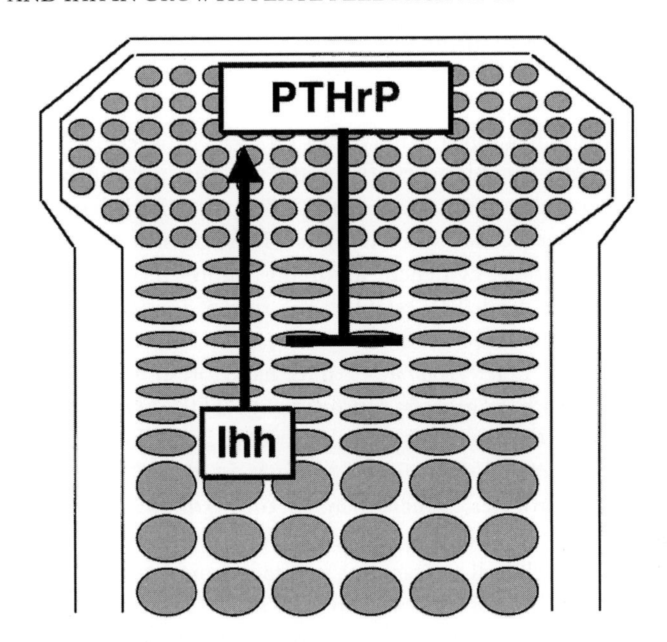

FIG. 1. Feedback loop involving PTHrP and IHH. PTHrP, secreted from perichondrial cells and chondrocytes near the end of the growth plate, delays the differentiation of chondrocytes that synthesize Ihh. Ihh, made by chondrocytes that have left the proliferative pool and have begun further differentiation, stimulates the synthesis of PTHrP.

PTHrP synthesis or, instead, indirectly does so by activating a cascade of signalling molecules.

The interactions of PTHrP and IHH suggest that they participate in a negative feedback loop in the growth plate (Fig. 1). IHH is synthesized by cells just as they are leaving the proliferative compartment and turning on the hypertrophic cell programme. This IHH increases the synthesis of PTHrP, which then acts to delay the movement of chondrocytes from the proliferative to the hypertrophic compartments. Thus, PTHrP action delays the appearance of cells that synthesize IHH. The negative feedback loop involving PTHrP and IHH might be a way for the growth plate to sense the length of the proliferative columns and make local adjustments. This could assure coordination of the proliferation/differentiation of cohorts of cells within one growth plate, in the face of stochastic variations and multiple inputs from local and systemic factors. It is of interest, in this regard, that the $PTHrP^{-/-}$ mice not only have short proliferative columns, but also have admixture of proliferative and hypertrophic cells within the same growth plate. In the absence of the postulated negative feedback loop, the coordination of chondrocyte differentiation is diminished.

Studies of chimeric mice support the importance of IHH–PTHrP interactions

Despite this suggestion from the phenotype of *PTHrP*$^{-/-}$ mice that the negative feedback loop actually functions to fine-tune the pace of differentiation of chondrocytes, the postulated importance of this loop thus far is largely based on experiments in which large amounts of IHH or PTHrP are introduced into bones cultured *in vitro*. Further evidence involving more modest experimental perturbations *in vivo* are needed to establish the role of the feedback loop. The model also leaves unclear which chondrocytes respond directly to PTHrP. The *PTHrP*$^{-/-}$ and *PTH/PTHrP receptor*$^{-/-}$ mice have foreshortened proliferative columns, but the cells with the greatest expression of *PTH/PTHrP receptor* mRNA are prehypertrophic chondrocytes, which appear unchanged in the mutant mice. One possible explanation of these findings might be that PTHrP acts on the prehypertrophic chondrocytes, which then, in turn, stimulate proliferation of adjacent proliferating chondrocytes by secreting a factor such as IGF1. To explore more fully the cell–cell interactions in intact growth plates, we generated embryonic stem (ES) cells missing both copies of the *PTH/PTHrP receptor* gene. Through appropriate matings, these cells were genetically marked with a gene expressing β-galactosidase in all cells descended from the marked ES cell. The mutant ES cells were then introduced into blastocysts, the blastocysts were inserted into the uteri of pseudo-pregnant foster mothers, and the bones of resultant chimeric embryos were examined at various times (Chung et al 1998).

When control, wild-type ES cells marked with β-galactosidase were injected into blastocysts, the resultant bones were normal, with cohorts of blue cells (after staining for β-galactosidase) forming the expected columns of proliferating chondrocytes. In contrast, the mutant PTH/PTHrP receptor chondrocytes differentiated prematurely and became hypertrophic, even though they were surrounded by the much more numerous normal chondrocytes in proliferative columns. This result demonstrates the cell autonomy of the PTH/PTHrP receptor chondrocyte phenotype and eliminates the model in which PTHrP acts only on prehypertrophic cells to stimulate proliferation of chondrocytes through the actions of another ligand.

Strikingly, the columns of genetically normal chondrocytes did not appear normal at all. Instead, the normal chondrocyte columns were much longer than expected. The extra length of columns was roughly in proportion to the contribution of *PTH/PTHrP receptor*$^{-/-}$ chondrocytes in the chimeric growth plates. More mutant cells were associated with longer columns of normal cells. One possible model that could explain these results is shown in Fig. 2. The mutant prehypertrophic cells are expected to synthesize IHH much closer to the ends of bones than normal, because of the premature differentiation of the

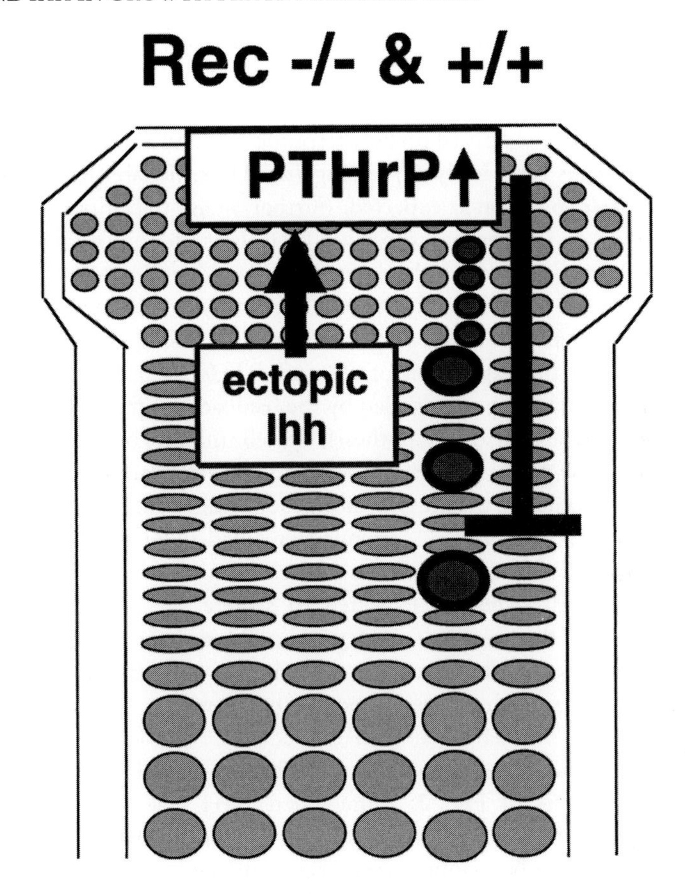

FIG. 2. Abnormalities in chimeric growth plates. Dark cells represent cells derived from *PTH/PTHrP receptor*$^{-/-}$ ES cells. These cells leave the proliferative pool early and synthesize IHH much closer to the end of the bone than normal cells would. This ectopically produced IHH then increases the synthesis of PTHrP at the end of the bone; this acts on the normal chondrocytes to delay their differentiation.

PTH/PTHrP receptor$^{-/-}$ cells. This IHH might then initiate a cascade that leads to an increase in PTHrP synthesis at the ends of the bone. This increase in PTHrP expression would be expected to delay the differentiation of the normal chondrocytes but not affect the differentiation of the *PTH/PTHrP receptor*$^{-/-}$ chondrocytes.

To test this hypothesis, we analysed *Ihh* mRNA by *in situ* hybridization. As expected, *Ihh* was synthesized both in its normal location in the normal prehypertrophic layer of cells and in ectopic, mutant prehypertrophic cells near the top of the growth plate. *Patched* (*Ptc*) is a gene that encodes the IHH receptor and is itself stimulated by IHH action. The expression of *Ptc* is a good index of IHH action. *In situ* hybridization for *Ptc* mRNA demonstrated expression in its normal location adjacent to the normal prehypertrophic cells and in chondrocytes adjacent to the mutant cells ectopically expressing *Ihh*. Thus, the abnormally located *Ihh* stimulates gene expression as expected. Further, *in situ* hybridization for *PTHrP* mRNA showed an increase in expression by cells at the top of the growth plate, just as was found in the growth plates of *PTH/PTHrP receptor*$^{-/-}$ mice. The abnormal location of *Ihh* and *Ptc* expression and the increase in *PTHrP* expression are all predicted by the model in Fig. 2 and support the idea that minor changes in the expression of *Ihh* and *PTHrP* can be sensed by growth plate chondrocytes in ways demanded by the feedback loop hypothesis.

To further substantiate this hypothesis, ES cells missing both the *PTH/PTHrP receptor* and the *Ihh* genes were established and used to generate chimeric mice. In preliminary experiments, the mutant chondrocytes differentiated prematurely, just as the *PTH/PTHrP receptor*$^{-/-}$ chondrocytes had differentiated. This result is expected, because normal cells surrounding the [*Ihh*$^{-/-}$; *PTH/PTHrP receptor*$^{-/-}$] chondrocytes are expected to secrete Ihh and make up for the loss of Ihh from the relatively low number of mutant chondrocytes. Thus, the double knockout chondrocytes resemble the *PTH/PTHrP receptor*$^{-/-}$ cells in chimeric growth plates. Importantly, however, the normal chondrocyte columns are not lengthened in the chimeric limbs containing [*Ihh*$^{-/-}$; *PTH/PTHrP receptor*$^{-/-}$] chondrocytes. This is precisely the result predicted from the hypothesis of Fig. 2, namely that IHH made by the mutant cells triggers the cascade leading to lengthening of the normal columns. Without IHH, the mutant cells cannot trigger the synthesis of PTHrP and, thereby, stimulate the proliferation of adjacent cells. Thus, the negative feedback loop between IHH and PTHrP can synchronize and determine the pace of chondrocyte differentiation in the growth plate.

IHH as a master coordinator of endochondral bone formation

IHH has actions in endochondral bone formation that do not depend on the stimulation of PTHrP synthesis. This is most clearly demonstrated by the phenotype of *Ihh*$^{-/-}$ mice (St-Jacques et al 1999). These mice have very short limbs because of a dramatic decrease in chondrocyte proliferation. Further, in the limbs, these mice are missing osteoblasts both in the bone collar and in the primary spongiosa. Neither of these abnormalities is found in mice missing either PTHrP or the PTH/PTHrP receptor. The absence of osteoblasts in the bone collar and primary spongiosa is consistent with prior tissue culture work that showed that IHH stimulated the osteoblastic differentiation of mesenchymal cells *in vitro* and that sonic hedgehog induced bone formation when injected subcutaneously *in vivo* (Kinto et al 1997, Nakamura et al 1997). These data help us to interpret another abnormality found in chimeric bones containing both normal and *PTH/PTHrP receptor*$^{-/-}$ cells. In these bones, bone collar forms much closer to the ends of bones than normal, adjacent to the abnormally located hypertrophic *PTH/PTHrP receptor*$^{-/-}$ chondrocytes. In light of the evidence just summarized that Ihh stimulates formation of osteoblasts, perhaps Ihh made by ectopic hypertrophic *PTH/PTHrP receptor*$^{-/-}$ chondrocytes stimulates the formation of the abnormally extended bone collar. In support of this idea, the bone collars of the chimeras containing [*Ihh*$^{-/-}$; *PTH/PTHrP receptor*$^{-/-}$] chondrocytes are normal.

These experiments suggest that IHH acts to coordinate the differentiation of both chondrocytes and osteoblasts during endochondral bone formation. IHH, in actions independent of PTHrP, stimulates proliferation of chondrocytes. By stimulating the synthesis of PTHrP, IHH also keeps cells from advancing out of the proliferative pool. At the same time, IHH signals to mesenchyme adjacent to the chondrocytes to stimulate differentiation of cells into osteoblasts. PTHrP action serves to delay the generation of cells that produce IHH. Thus, PTHrP, and probably many other local factors, serve to regulate the regulator, IHH.

References

Bitgood MJ, McMahon AP 1995 Hedgehog and BMP genes are coexpressed at many diverse sites of cell–cell interaction in the mouse embryo. Dev Biol 172:126–138

Chung UI, Lanske B, Lee K, Li E, Kronenberg H 1998 The parathyroid hormone/parathyroid hormone-related peptide receptor coordinates endochondral bone development by directly controlling chondrocyte differentiation. Proc Natl Acad Sci USA 95:13030–13035

Karaplis AC, Luz A, Glowacki J et al 1994 Lethal skeletal dysplasia from targeted disruption of the parathyroid hormone-related peptide gene. Genes Dev 8:277–289

Kinto N, Iwamoto M, Enomoto-Iwamoto M et al 1997 Fibroblasts expressing Sonic hedgehog induce osteoblast differentiation and ectopic bone formation. FEBS Lett 404:319–323

Lanske B, Karaplis AC, Lee K et al 1996 PTH/PTHrP receptor in early development and Indian hedgehog-regulated bone growth. Science 273:663–666

Lee K, Deeds J, Bond A, Jüppner H, Abou-Samra A-B, Segre G 1993 In situ localization of PTH/PTHrP receptor mRNA in the bone of fetal and young rats. Bone 14:341–345

Lee K, Deeds JD, Segre GV 1995 Expression of parathyroid hormone-related peptide and its receptor messenger ribonucleic acids during fetal development of rats. Endocrinology 136:453–463

Nakamura T, Aikawa T, Iwamoto-Enomoto M et al 1997 Induction of osteogenic differentiation by hedgehog proteins. Biochem Biophys Res Commun 237:465–469 (erratum: 1998 Biochem Biophys Res Commun 247:910)

Philbrick WM, Dreyer BE, Nakchbandi IA, Karaplis AC 1998 Parathyroid hormone-related protein is required for tooth eruption. Proc Natl Acad Sci USA 95:11846–11851

Schipani E, Kruse K, Jüppner H 1995 A constitutively active mutant PTH-PTHrP receptor in Jansen-type metaphyseal chondrodysplasia. Science 268:98–100

Schipani E, Lanske B, Hunzelman J et al 1997 Targeted expression of constitutively active receptors for parathyroid hormone and parathyroid hormone-related peptide delays endochondral bone formation and rescues mice that lack parathyroid hormone-related peptide. Proc Natl Acad Sci USA 94:13689–13694

St-Jacques B, Hammerschmidt M, McMahon A 1999 Indian hedgehog signaling regulates proliferation and differentiation of chondrocytes and is essential for bone formation. Genes Dev 13:2072–2086

Vortkamp A, Lee K, Lanske B, Segre GV, Kronenberg HM, Tabin CJ 1996 Regulation of rate of cartilage differentiation by Indian hedgehog and PTH-related protein. Science 273:613–621

Weir EC, Philbrick WM, Amling M, Neff LA, Baron R, Broadus AE 1996 Targeted overexpression of parathyroid hormone-related peptide in chondrocytes causes chondrodysplasia and delayed endochondral bone formation. Proc Natl Acad Sci USA 93:10240–10245

DISCUSSION

Poole: When you have premature expression of the hypertrophic phenotype, do you see any evidence of angiogenesis?

Kronenberg: In high level chimeras, we can see vascular invasion. When the chimerism is low level, there is still hypertrophy, but there is no mineralization or vascular invasion. It seems that a critical mass of chondrocytes is needed to get mineralization, and this is the only setting in which we see the vascular invasion. There must be some signal build-up or removal of inhibitors, or something like that. The differentiation state is cell autonomous, but secondary phenomena require groups of cells.

Poole: That's interesting, because in development one tends to see hypertrophy in the forming epiphyses without mineralization. With hypertrophy, vascularization is then seen.

Kronenberg: Along those lines, what this would mean is that one needs enough cells together to create an environment or a strong enough signal or whatever.

Kingsley: One of the things that surprised me was that the chimeras that were made with just the receptor mutant cells show this ectopic bone collar, but they didn't show the internal mineralization of the hypertrophic cartilage. In contrast, when you made the chimeras with the double receptor *Ihh* knockout mutant cells, you didn't get the bony collar, but you seemed to get a much more prominent internal mineralization.

Kronenberg: The mineralization of 'ectopic' hypertrophic chondrocytes is variable. It seems to correlate most with the extent of chimerism. It is difficult to get high-level chimerism in the growth plate with *PTH/PTHrP receptor* knockout cells, because the knockout of the *PTH/PTHrP receptor* limits the proliferation of those cells. The normal cells are not limited at all. In fact, they extend, so as you go down the growth plate what you see, even in high-level chimeras, is that as soon as you get past the knockout hypertrophic cells almost all the cells are normal chondrocytes. You are thus selecting against the mutant cells. But if you make chimeras that have high enough chimerism, then you do see normal mineralization, but not with isolated hypertrophic cells even though they make type X collagen perfectly normally in a way that doesn't require any critical mass of cells. But they won't mineralize the matrix unless there is a critical mass.

Blair: Does that suggest a relationship to the vitamin D response?

Kronenberg: The whole vitamin D story is very mysterious. Two different groups have knocked out the vitamin D receptor. Marie Demay in the Massachusetts General Hospital Endocrine Unit has shown that these mice develop rickets and mineralization abnormalities, but the growth plate is 100% completely normal if you give the mice enough calcium and phosphate, even though they don't have any possible receptor-mediated vitamin D action. Thus, while there may well still be some abnormalities left to be seen in the growth plate of the vitamin D receptor knockout mice, in terms of *in situ* gene expression and histology, even without vitamin D receptor the growth plate functions completely normally if there is enough calcium and phosphate in the diet.

Beresford: Are you talking about the classical vitamin D receptor or the recently described membrane-bound form (Nemere et al 1998)?

Kronenberg: The classical one. Obviously, this is a wonderful model for studying the non-genomic actions of vitamin D.

Russell: Many years ago, we looked at bisphosphonates, such as etidronate, which inhibits mineralization, and found remarkable changes in the growth plate (Schenk et al 1973). Of course, by blocking mineralization you get an accumulation of the hypertrophic cells so that the hypertrophic zone becomes extraordinarily enlarged, and mineralization seems to be the trigger to vascular invasion.

Burger: I would like to raise this concept of cross-talk between these different molecules. This implies that there is transportation of the molecules preferentially through the cartilage matrix, otherwise you would get diffusion of these molecules, that are formed for instance in the interzone area, elsewhere. Do you have any idea how this works?

Kronenberg: I'm not alone in not understanding this! The way that the hedgehog family moves from cell to cell is really quite confusing and is not known. There is a fruit fly mutant *tout velu* which has its mammalian equivalent in the exostosis genes that are involved in heparan sulfate synthesis. These genes are essential for hedgehog to be able to diffuse at all. Exactly how hedgehog gets through matrix instead of getting just handed from one cell to another, nobody knows. Hedgehog diffusion is a tightly regulated process. Cholesterol certainly has a lot to do with it. Chuang & McMahon (1999) have shown that there is a membrane protein (hedgehog interacting protein) that binds hedgehog and limits its diffusion. The receptor for hedgehog itself is present in sufficient amounts, both in fruitfly and in vertebrates, to limit how far hedgehog can travel. The diffusion of ligands has to be thought of as a biologically regulated event. It is not just a physicochemical phenomenon, certainly not with hedgehog, nor with the majority of paracrine factors. I have come into this area from the perspective of an endocrinologist, where hormones can diffuse all over the place, and then they diffuse away — their main regulatory control after their secretion is their rapid diffusion, metabolism and destruction. In paracrine biology, things are quite different: moving three cells instead of five cells changes everything, and there are 10 other levels of regulation of simple diffusion.

Burger: The matrix also changes in the area of chondrocyte hypertrophy.

Kronenberg: So what this is showing is that the hedgehog is moving sideways — patched expression shows that — and is acting on these perichondrial cells, as well as in some presumed cascade eventually leading to increased PTHrP production. The problem is not that different from the mysteries of how Sonic hedgehog gets to the fibroblast growth factors (FGFs) in the limb bud, and we know that some of that is through the inhibition of bone morphogenetic protein (BMP) inhibitors locally (Zúñiga et al 1999). In analogous fashion sonic hedgehog in chicken affects BMP action the lateral plate mesoderm, hundreds of cell layers away, by stimulating the production of the BMP antagonist, caronte, which then diffuses (Yokouchi et al 1999). It seems likely that IHH isn't moving very far here either, and yet it must be at the start of some cascade, perhaps acting through BMP or transforming growth factor (TGF)β.

Tickle: We've been working on an interesting chicken mutant called *Talpid3*, which has a defect in responding to hedgehog signalling, in that the high level *Patched* expression is not induced in response to hedgehog signals in the limb bud. This has led us to various hypotheses. The mutant has limb polydactyly, but in the context of your work it's also interesting that it lacks bone. We are currently looking at the idea that this failure of high-level *Patched* induction is downstream of all hedgehog signalling, and so this could fit quite nicely with the idea that IHH signals bone formation.

Kronenberg: Is it missing all bone?

Hall: I thought that the membrane bones were present, but not the endochondral bone.

Kronenberg: The *Ihh* knockout has vertebrae, skull and scapula, but it doesn't have limb bones.

Karsenty: Vertebrae are endonchondral.

Hall: There's also the interesting mouse mutant *brachypod*, where the fibula delays hypertrophy for several weeks: it doesn't hypertrophy until well after birth, and then osteogenesis begins when hypertrophy is switched on. This would again be an interesting model in which to look at *Ihh* and *PTHrP* expression patterns. I think brachypod is BMP based.

Kingsley: Yes, it is a mutation in a member of the BMP family called *Gdf5*.

I have a question about the expression of *PTHrP* normally, which is periarticular. Of course, articular cartilage never hypertrophies. I was struck by the chimeras that you showed, that it looked like mutant cells that don't have the PTH receptor are excluded from the blue surface of the bone. Again, you can only show some of the pictures, but in the ones that you did show the chimeras that were made with wild-type had a lot of blue contribution to the very surface of the skeletal element, and in the ones that you showed right next to it that were the PTHrP mutant cells, all the blue that was along the surface in the wild-type chimeras was missing. Does this suggest that PTHrP and this receptor signalling pathway plays an important role in maintaining the undifferentiated state of the articular surface on the top of the bone?

Kronenberg: Either that is not so, or we may have missed something very interesting, because I haven't noticed this. But I think what you have just suggested is generally not the case. That is, what seems quite striking to us is that in the first few cell layers closest to the articular surface, the receptor knockout cells and normal cells behave pretty much identically. We like that idea, because this is a region that doesn't have PTH/PTHrP receptors. We, therefore, wouldn't have expected normal and mutant cells to behave differently there. This is a layer that seems to be expressing FGFR1 and as soon as you get past that, you start making PTH/PTHrP receptors, and that's where the columns stop in the mutants.

Kingsley: What is your current thinking on why direct effects of IHH on chondrocytes were missed earlier? In the first papers, the Patched receptor was not detected in chondrocytes and all the effects were postulated to go through the perichondrium and then to PTHrP. There are now expression reasons to think that the receptors are there in chondrocytes, and there's also phenotypic evidence for direct effects, and yet when the first experiments were done adding IHH to the bone cultures of the PTHrP in knockout mice, it appeared that eliminating PTHrP eliminated all responsiveness to IHH?

Kronenberg: We didn't look at BrdU incorporation in the bone explants. What we know for sure is that the growth plates didn't grow: there were not more chondrocytes. I think the difference between the *in vivo* data and the bone explant data is that in one experiment you're going from normal levels in IHH to none (*Ihh* knockout vs. wild-type mice), and in the other you're going from normal levels of IHH to even higher levels (bone explants). I've been assuming that going from normal IHH to even more doesn't do much to proliferation. There is a dose–response curve, and the normal level of IHH must be near the top of the proliferative response. Therefore the phenotype is dominated by the effect of the increase in PTHrP production. The two over-expression paradigms taught the same lesson: the effect in the chicken wing and of adding hedgehog to bone explants both led to the dramatic suppression of hypertrophy, without a dramatic proliferative effect.

Ornitz: One other issue we have to keep in mind, is that either IHH or PTHrP could effect the expression of an unknown FGF ligand, which must be present in the growth plate.

Kronenberg: That is certainly possible. We're studying these two signalling systems—PTHrP and IHH—not because we think they are the only two important ligands controlling how long the growth plate is but because our own interests led us in this direction. Other signalling systems are likely to be very important as well.

Chen: The other reason could be that IHH is necessary but not sufficient to stimulate chondrocyte proliferation. Therefore, if you are just adding IHH it doesn't stimulate proliferation. But if you knock it out proliferation is affected.

Morriss-Kay: I'd like to ask David Ornitz to expand on what he just said, about the possible relationship between FGFR3 function in chondrocyte development and the PTHrP/IHH system that Henry Kronenberg described. These two signalling systems appear to be studied rather separately, as far as I can see from the literature, yet clearly they're both acting on the same developmental process.

Ornitz: In the data that I showed yesterday, activating FGFR3 mutations suppresses IHH signalling. That's one way you can tie into this system.

Morriss-Kay: Can you describe what you think is happening in normal development in terms of the relationship between these two systems?

Ornitz: In normal development, FGFR3 is probably upstream of this feedback loop. I think this feedback loop is likely to be a key regulator of the length of the growth plates. There may be some direct effects of FGFR3 on proliferating chondrocytes, but just as easily it could be indirect effects by modulating the activity of the hedgehog–PTHrP pathway. There is some *in vitro* evidence that FGFR3 can have a direct effect, because when the activated receptor is expressed in chondrocyte cell lines in culture it can slow down the growth of those cells.

Mundlos: We have talked about proliferation, but isn't apoptosis another factor that regulates the length of the growth plate?

Kronenberg: Yes, it certainly does. We know that the hypertrophic cells are vividly TUNEL positive and have turned off *Bcl2* expression. If you express more PTHrP and delay that part of the differentiation programme, you're also delaying apoptosis. In experiments in which Schipani and Jüppner expressed a constitutively active PTH/PTHrP receptor in a transgenic mouse driven by the collagen II promoter, it is possible to have PTHrP-like action in proliferative chondrocytes and in hypertrophic chondrocytes. In that setting, the hypertrophic chondrocytes don't die and they fill up the bone marrow space. Apoptosis is part of the chondrocyte differentiation programme, and if you slow down the programme, you stop apoptosis.

Burger: Do you get interference with the excavation of the primitive marrow cavity, and influx of osteoclasts in that situation?

Kronenberg: We wanted that to be the case in the *PTHrP* and *PTH/PTHrP receptor* knockout mice because PTHrP is an important stimulator of osteoclastogenesis. When we've looked in our mutants, however, we haven't been able to see any evidence of decreased osteoclast development or action. It seems that at least during this modelling stage the osteoclasts needed for bone development are not importantly regulated by PTHrP.

M. Cohn: Goff and Tabin showed a few years ago that retroviral overexpression of *Hoxd11* and *Hoxd13* in hindlimb buds had an effect on leg bone growth. They found that bone length was altered because of an effect on cells in the growth plate rather than in the early limb bud. How do *Hox* genes interface with the IHH/PTHrP loop?

Kronenberg: I don't know this paper, but it sounds like something that we could study pretty easily.

References

Chuang PT, McMahon AP 1999 Vertebrate hedgehog signalling modulated by induction of a hedgehog-binding protein. Nature 397:617–621

Nemere I, Schwartz Z, Pedrozo H, Sylvia VL, Dean DD, Boyan BD 1998 Identification of a membrane receptor for 1,25-dihydroxyvitamin D_3 which mediates rapid activation of protein kinase C. J Bone Miner Res 13:1353–1359

Schenk R, Merz WA, Muhlbauer R, Russell RGG, Fleisch H 1973 Effect of two diphosphonates on bone and cartilage growth and resorption in the tibial epiphysis and metaphysis of rats. Calcified Tissue Res 11:196–214

Yokouchi Y, Vogan KJ, Pearse RV II, Tabin CJ 1999 Antagonistic signaling by *Caronte*, a novel *Cerberus*-related gene, establishes left–right asymmetric gene expression. Cell 98:573–583

Zúñiga A, Haramis AP, McMahon AP, Zeller R 1999 Signal relay by BMP antagonism controls the SHH/FGF4 feedback loop in vertebrate limb buds. Nature 401:598–602

Cartilage matrix resorption in skeletogenesis

William Wu[2], Fackson Mwale, Elena Tchetina, Toshi Kojima[3], Tadashi Yasuda[4] and A. Robin Poole[1]

Joint Diseases Laboratory, Shriners Hospitals for Children and Division of Surgical Research, Department of Surgery, McGill University, 1529 Cedar Avenue, Montreal, Quebec, Canada H3G 1A6

Abstract. Chondrocytes assemble an extracellular matrix in which the relative composition of type IX versus type II collagen and aggrecan changes during assembly. On maturation and differentiation into hypertrophic cells type IX collagen first loses the NC4 globular domain of the $\alpha1$(IX) chain that protrudes from the collagen fibril. Subsequently, collagenase 3 (matrix metalloproteinase 13; MMP13) is up-regulated as type X collagen is expressed leading to extensive cleavage and removal of type II collagen and of the remaining COL2 domain of type IX collagen $\alpha1$(IX) chain. The proteoglycan aggrecan is selectively retained in the extracellular matrix. Inhibition of collagenase leads to arrest of hypertrophy as well as gene expression of MMP13. Thus proteolysis and in particular MMP13 are required for chondrocyte differentiation and for matrix resorption in skeletal development.

2001 The molecular basis of skeletogenesis. Wiley, Chichester (Novartis Foundation Symposium 232) p 158–170

Skeletal growth and development involve not only processes of cellular differentiation but also the establishment and controlled resorption of extra-cellular matrix. In this paper we describe our current state of knowledge of cartilage matrix assembly, turnover and resorption as part of the process of endochondral ossification that leads to the calcification of cartilage matrix and ultimately to bone

[1]The chapter was presented at the symposium by A. Robin Poole, to whom correspondence should be addressed.
Present addresses:
[2]Massachusetts General Hospital, Arthritis Unit, 55 Fruit Street, Boston, Massachusetts, USA
[3]Department of Orthopedic Surgery, Faculty of Medicine, Nagoya University, Tsunami-Cho, Showa-Ku, Nagoya, Japan
[4]Department of Orthopaedic Surgery, Faculty of Medicine, Kyoto University, 54 Kawara-cho, Shogoin, Kyoto, Japan

formation. During endochondral ossification chondrocytes proliferate and elaborate an extracellular matrix. The matrix of this physeal cartilage is remodelled as the cells become hypertrophic and calcify their extracellular matrix. Eventually it is partially resorbed by a process involving angiogenesis. The calcified cartilage that remains provides cartilaginous trabeculae on which osteoblasts settle and elaborate woven bone. This in turn is resorbed by chondroclasts/osteoclasts and replaced by a mature trabecular bone (Poole 1991, 1997).

Composition of the extracellular matrix in the physeal cartilage

An extensive network of collagen fibrils provides an 'endoskeleton' about which other matrix molecules are organized (Fig. 1). Type II collagen is the principle component of the fibrils which are thin ($\sim$25 nm in diameter) because this is a transient cartilage, last only several days and must be easily resorbable. This contrasts to the thick fibrils found in mature articular cartilage which are usually more than 100 nm in diameter and must resist resorption and survive for the life of that individual. These fibrils lack the small proteoglycan decorin which is found on fibrils in the adjacent epiphyseal and articular cartilages (M. Alini & A. R. Poole, unpublished work). Type IX collagen is also present on the fibril during matrix assembly in a molar ratio of $\sim$5:1 (type II/IX). The majority of these molecules ($\sim$90%) lack the basic NC4 domain that is normally found on type IX collagen in articular cartilage (Mwale et al 2000). Type XI collagen is also present within the fibril and on its surface (Poole 1997).

Associated with collagen fibrils is hyaluronic acid to which the proteoglycan aggrecan binds at its G1 domain, this binding being stabilized by link protein (Poole 1997; Fig. 1). Many other molecules are present within this matrix, a number of which may also be associated with these collagen fibrils (Poole 1997).

The extracellular matrix is actively synthesized by the cells of the proliferating zone where the ratio of matrix volume per cell volume is at its greatest (Alini et al 1992). As cells become hypertrophic synthesis of type II procollagen is up-regulated (Poole 1991, 1997). The C-propeptide of this molecule (chondro-calcin), initially fibril associated, accumulates in focal sites where matrix mineralization is concentrated (Poole et al 1984, Lee et al 1996).

Type X collagen is synthesized by hypertrophic chondrocytes. This forms a pericellular lattice work (Schmid & Linsenmayer 1990) and is also closely associated with collagen fibrils (Poole & Pidoux 1989). With hypertrophy matrix volume is markedly and rapidly reduced (Alini et al 1992). This is characterized by a net loss of collagen and the selective retention of the proteoglycan aggrecan (Alini et al 1992, Matsui et al 1991). The latter is concentrated in focal sites where the C-propeptide is found (Poole et al 1989, Poole 1997).

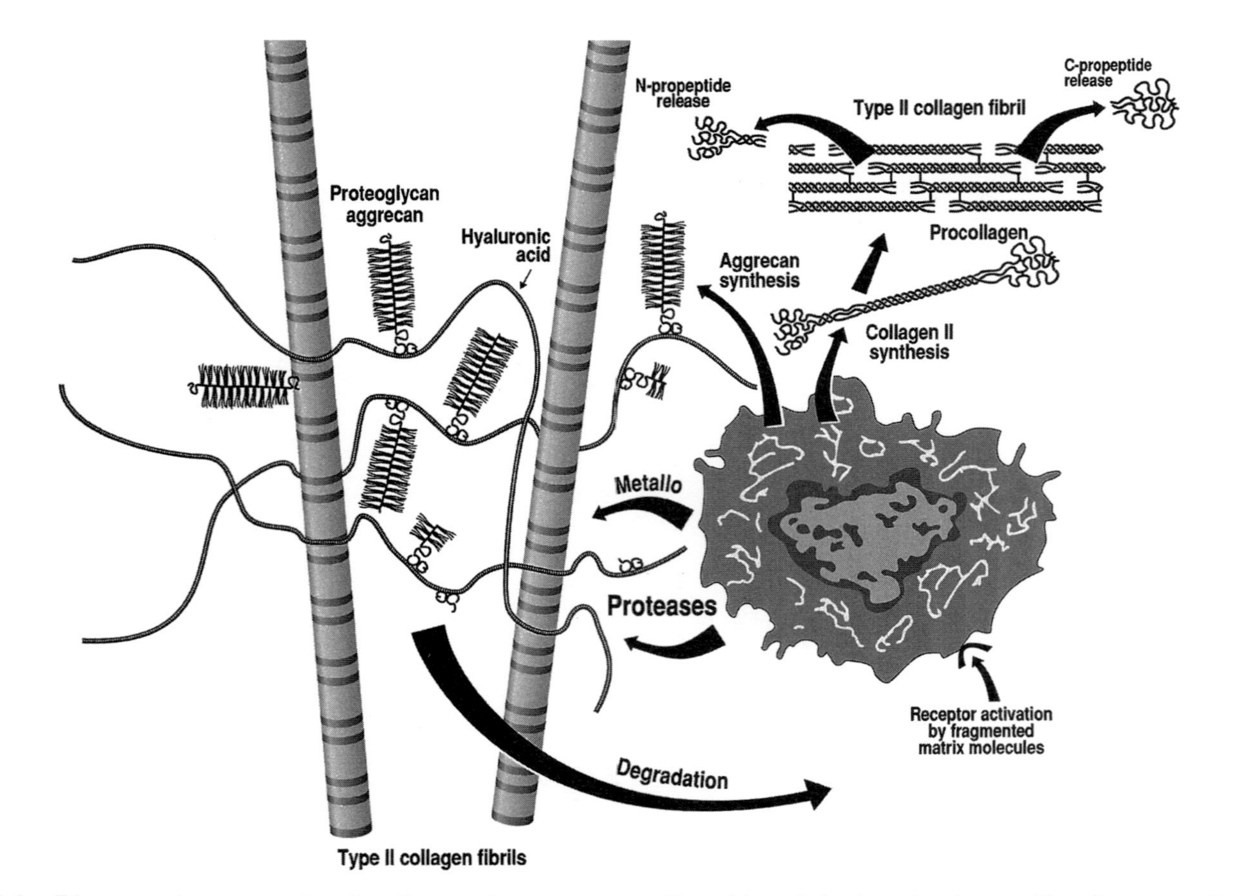

FIG. 1. Diagrammatic representation of cartilage matrix structure, assembly and degradation by a chondrocyte. The substructure of a collagen fibril is shown at the top of the figure to demonstrate its assembly from procollagen II.

The degradation of type II collagen

As the matrix is resorbed by hypertrophic cells, type II collagen is denatured (Alini et al 1992). This can be measured immunologically using an antibody to an intrachain epitope (Dodge & Poole 1989). The principle proteinases that can cleave type II collagen belong to the matrix metalloproteinase (MMP) family and are called collagenases. There are three such enzymes namely collagenase 1 (MMP1 or interstitial collagenase), collagenase 2 (MMP8 or neutrophil collagenase), and collagenase 3 (MMP13). Of these three collagenases, MMP13 is most effective at cleaving type II collagen (see Billinghurst et al 1997). There is another proteinase called membrane type I-MMP (MT1-MMP or MMP14) which has been reported to cleave this collagen and activate MMP13. MMP14 and MMP8 have not been detected in the growth plate. However, a null mutation for MMP14 does not exhibit any lack of collagen II cleavage or matrix resorption in the growth plate (Holmbeck et al 1999, Poole et al 2000) although cleavage of type II collagen in other tissues is inhibited. These collagenases cleave the triple helix at a point approximately three quarters distant from the N-terminus. Following cleavage and denaturation, the α chains are susceptible to cleavage by a wide variety of MMPs. The collagen fibril can also be cleaved in the telopeptide domain (Poole 1997) but as yet there is no published evidence that this occurs *in situ*.

Prior knowledge of collagenases in endochondral ossification

Earlier work has identified excessive collagenolytic activity in the hypertrophic zone of the growth plate (see Poole 1991, 1997). The content of inhibitors of collagenase (tissue inhibitors of metalloproteinases) is also reduced. Net collagen breakdown results. Collagenase has been localized in the extracellular matrix of the hypertrophic zone in the growth plates (see review Poole 1997, Blair et al 1989). Using a newly developed antibody to the collagenase-generated neoepitope in type II collagen (Billinghurst et al 1997), we have shown that collagen cleavage is present throughout the growth plate, although it is enhanced in the hypertrophic zone where net loss of type II collagen is observed (F. Mwale, I. Pidocux, E. Tchetina & A. R. Poole, unpublished results).

Current studies of matrix resorption

Using a system of isolated bovine fetal chondrocytes cultured at high density in serum-free culture (Alini et al 1994, 1996), we have shown that in the presence of thyroid hormones prehypertrophic chondrocytes isolated from fetal bovine fetuses and separated by Percoll gradient centrifugation, elaborate, in culture, an extensive extracellular matrix, undergo hypertropy (synthesize type X collagen) and mineralize their extracellular matrix. This also involves resorption of the extracellular matrix at the time of mineral deposition. We have performed analyses using the reverse transcriptase (RT)-PCR and using Northern blotting on these cultures. They reveal that only MMP13 is present (W. Wu, C. Billinghurst & A. R. Poole, unpublished results). Studies have independently shown the up-regulation of MMP13 by hypertrophic chondrocytes (Gack et al 1995, Johansson et al 1997, Stähle-Bäckdahl et al 1997). Prior to hypertrophy there is more limited collagenase activity, which is inversely related to type II collagen and aggrecan content. This suggests a regulatory role for collagenase in matrix turnover (Mwale et al 2000). The presence of activated enzyme is markedly increased at the time of hypertrophy and coincides with an increase in type II collagen cleavage by collagenase and a rapid loss of this collagen (W. Wu, C. Billinghurst & A. R. Poole, unpublished results). Prior to this increase there is a selective removal of the NC4 domain of the $\alpha1$(IX) chain of type IX collagen which projects from the fibril (Mwale et al 2000). Collagen cleavage occurs at the time that mineral is deposited in the extracellular matrix suggesting that it is required to permit crystal formation and growth.

A synthetic inhibitor of MMP13 at a concentration as low as 1 nM leads to inhibition of collagen degradation in articular cartilages (Billinghurst et al 1997). Unexpectedly in the growth plate cultures this inhibitor arrests not only MMP13 but also inhibits mineralization demonstrating the dependence of mineralization on matrix resorption (W. Wu, C. Billinghurst & A. R. Poole, unpublished results). Moreover, gene expression of MMP13 was also suppressed suggesting that matrix degradation products, such as those derived from type II collagen, may also serve to maintain collagenase expression, synthesis and activity (Fig. 2). In addition, type X collagen expression was suppressed reflecting a more profound linkage of hypertrophy to matrix resorption.

We have tested this hypothesis by adding cyanogen bromide fragments of purified bovine type II collagen to chondrocyte cultures. In articular cartilage these induce up-regulation of type II collagen cleavage by collagenase as well as the expression of this proteinase and of type X collagen. Again, MMP13 expression was inhibited by the synthetic inhibitor (T. Yasuda, T. Kojima & A. R. Poole, unpublished results). So far these experiments have been mainly conducted with articular chondrocytes. But we have evidence to indicate that this may also occur in cultures of prehypertrophic chondrocytes isolated from the growth plate (F. Mwale & A. R. Poole, unpublished results).

In parallel *in situ* analyses of the bovine tibial primary growth plate we have established that a very similar cleavage and loss of type II collagen is associated with hypertrophy and matrix mineralization (F. Mwale, E. Tchetina & A. R. Poole, in preparation).

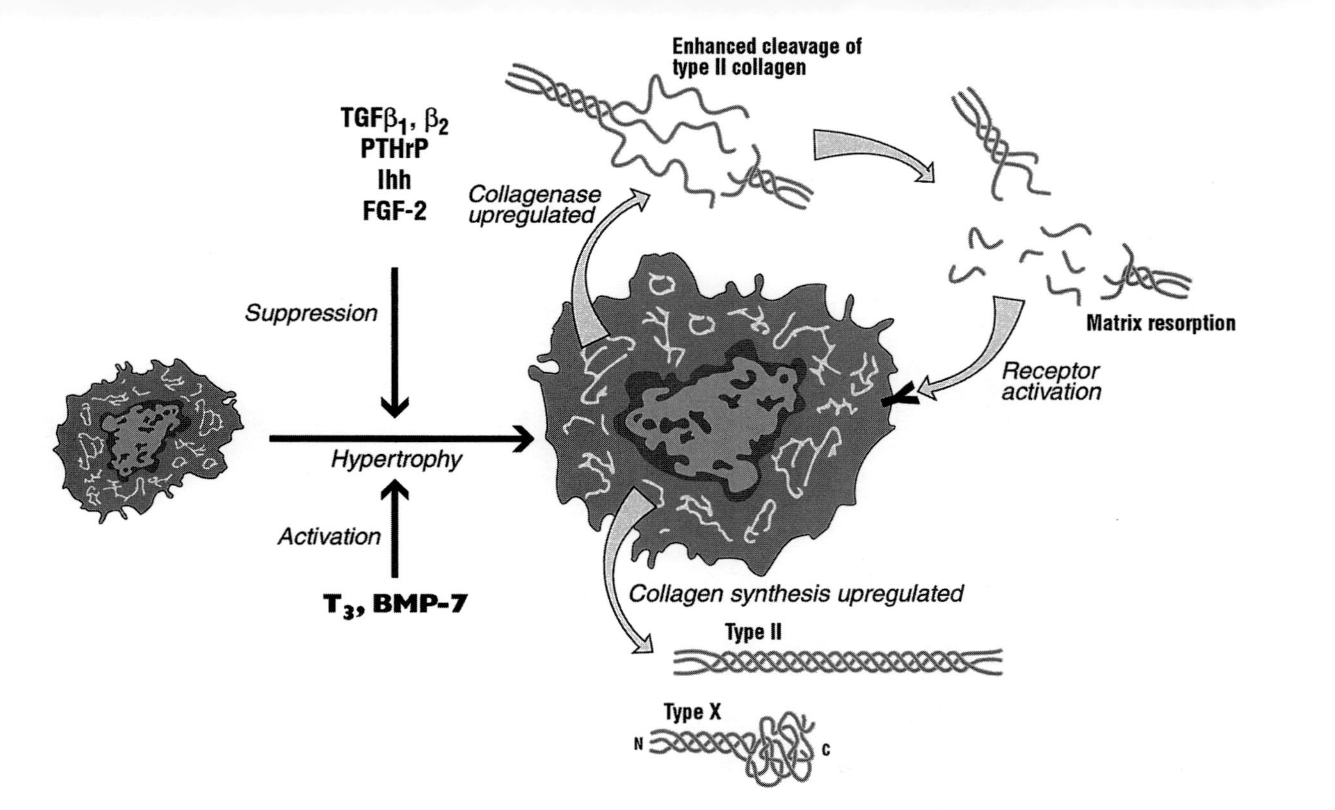

FIG. 2. Diagrammatic representation of chondrocyte hypertrophy. This is suppressed by molecules such as transforming growth factor (TGF)β1 and TGFβ2, parathyroid hormone-related peptide (PTHrP), Indian hedgehog (Ihh) and fibroblast growth factor 2 (FGF2). Triiodothyronine (T_3) and bone morphogenetic protein 7 (BMP7) can each induce hypertrophy from prehypertrophic chondrocytes.

Discussion

These studies provide direct evidence for a role for MMP13 (collagenase 3) and/or MMPs in the resorption of cartilage matrix and its mineralization. Recently, the gelatinase B (MMP9)-deficient mouse was shown to lack angiogenesis and apoptosis in the growth plate (Vu et al 1998). These results suggest that the activity of MMPs such as MMP13 and MMP9 are essential for chondrocyte maturation and differentiation. The experimental evidence also points to the ability of matrix degradation products, in particular of type II collagen, to up-regulate MMP13 expression and activity. Similar observations have previously been made for fibronectin fragments and collagenase and MMP expression and activity in fibroblasts (Werb et al 1989) and chondrocytes (Homandberg et al 1992). Based on the latter studies we believe that this feedback regulation may operate via a receptor-mediated mechanism, in the case of fibronectin involving integrin receptor-mediated signalling.

Recently it has been shown that the MMP13 gene expression can be down-regulated by deletion of the transcription factor Cbfa1 (Inada et al 1999). Gene deletion studies in mice have revealed that not only does this often result in a loss of collagenase expression in a number of growth plates but also an arrest of chondrocyte maturation and hypertrophy. Together these studies point to an important role of MMPs, and MMP13 in particular, in endochondral ossification and hence in growth differentiation and development.

Acknowledgements

These studies have been funded by Shriners Hospitals for Children, the Medical Research Council of Canada, the National Institutes of Health, USA (to ARP) and the Canadian Arthritis Network. Fackson Mwale is a recipient of a Shriners Hospitals Fellowship.

References

Alini M, Matsui Y, Dodge GR, Poole AR 1992 The extracellular matrix of cartilage in the growth plate before and during calcification: changes in composition and degradation of type II collagen. Calcif Tissue Int 50:327–335

Alini M, Carey D, Hirata S, Grynpas MD, Pidoux I, Poole AR 1994 Cellular and matrix changes before and at the time of calcification in the growth plate studied *in vitro*: arrest of type X collagen synthesis and net loss of collagen when calcification is initiated. J Bone Miner Res 9:1077–1087

Alini M, Kofsky Y, Wu W, Pidoux I, Poole AR 1996 In serum-free culture thyroid hormones can induce full expression of chondrocyte hypertrophy leading to matrix calcification. J Bone Miner Res 11:105–113

Billinghurst RC, Dahlberg L, Ionescu M et al 1997 Enhanced cleavage of type II collagen by collagenases in osteoarthritic articular cartilage. J Clin Invest 99:1534–1545

Blair HC, Dean DD, Howell DS, Teitelbaum SL, Jeffrey JJ 989 Hypertrophic chondrocytes produce immunoreactive collagenase *in vivo*. Connect Tissue Res 23:65–73 (erratum: 1989 Connect Tissue Res 23:218)

Dodge GR, Poole AR 1989 Immunohistochemical detection and immunochemical analysis of type II collagen degradation in human normal, rheumatoid and osteoarthritic articular cartilages and in explants of bovine articular cartilage cultured with interleukin 1. J Clin Invest 83:647–661

Gack S, Vallon R, Schmidt J et al 1995 Expression of interstitial collagenase during skeletal development of the mouse is restricted to osteoblast-like cells and hypertrophic chondrocytes. Cell Growth Differ 6:759–767

Holmbeck K, Bianco P, Caterina J et al 1999 MT1-MMP deficient mice develop dwarfism, skeletal dysplasia, arthritis and fibrosis because of inadequate collagen turnover. Cell 99:81–92

Homandberg GA, Meyers R, Xie D-L 1992 Fibronectin fragments cause chondrolysis of bovine articular cartilage slices in culture. J Biol Chem 267:3597–3604

Inada M, Yasui T, Nomura S et al 1999 Maturational disturbance of chondrocytes in Cbfa1-deficient mice. Dev Dyn 214:279–290

Johansson N, Saarialho-Kere U, Airola K et al 1997 Collagenase-3 (MMP-13) is expressed by hypertrophic chondrocytes, periosteal cells, and osteoblasts during human fetal bone development. Dev Dyn 208:387–397

Lee E, Smith CE, Poole AR 1996 Ultrastructural localization of the C-propeptide released from type II procollagen in fetal bovine growth plate cartilage. J Histochem Cytochem 44:433–443

Matsui Y, Alini M, Webber C, Poole AR 1991 Characterization of aggregating proteoglycans from the proliferative, maturing, hypertrophic and calcifying zones of the cartilaginous physis. J Bone Joint Surg (Am) 73:1064–1074

Mwale F, Billinghurst RC, Wu W et al 2000 Selective assembly and remodelling of collagens II and IX associated with regression of the chondrocyte hypertrophic phenotype. Dev Dyn 218:648–662

Poole AR 1991 The growth plate: cellular physiology, cartilage assembly and mineralization. In: Hall BK, Newman S (eds) Cartilage: molecular aspects. CRC Press, Boca Raton, p 179–221

Poole AR 1997 Cartilage in health and disease. In: Koopman W (ed) Arthritis and allied conditions. A textbook of rheumatology, 13th edn. Williams and Wilkins, Baltimore, p 255–308

Poole AR, Pidoux I 1989 Immunoelectron microscopic studies of type X collagen in endochondral ossification. J Cell Biol 109:2547–2554

Poole AR, Pidoux I, Rosenberg L 1982 Role of proteoglycans in endochondral ossification: immunofluorescent localization of link protein and proteoglycan monomer in bovine fetal epiphyseal growth plate. J Cell Biol 92:249–260

Poole AR, Pidoux I, Reiner A, Choi H, Rosenberg LC 1984 Association of an extracellular protein (chondrocalcin) with the calcification of cartilage in endochondral bone formation. J Cell Biol 98:54–65

Poole AR, Matsui Y, Hinek A, Lee ER 1989 Cartilage macromolecules and the calcification of cartilage matrix. Anat Rec 224:167–179

Poole AR, Alini M, Hollander A 1995 Cellular biology of cartilage degradation. In: Henderson B, Pettifer R, Edwards, J (eds) Mechanisms and models in rheumatoid arthritis. Academic Press, New York, p 163–204

Schmid TM, Linsenmayer TF 1990 Immunoelectron microscopy of type X collagen: supramolecular forms within embryonic chick cartilage. Dev Biol 138:53–62

Stähle-Bäckdahl M, Sandstedt B, Bruce K et al 1997 Collagenase-3 (MMP-13) is expressed during human fetal ossification and re-expressed in postnatal bone remodelling and in rheumatoid arthritis. Lab Invest 76:717–728

Vu TH, Shipley JM, Bergers G et al 1998 MMP-9/gelatinase B is a key regulator of growth plate angiogenesis and apoptosis of hypertrophic chondrocytes. Cell 93:411–422

Werb Z, Tremble PM, Behrendtsen O, Crowley E, Damsky CH 1989 Signal transduction through the fibronectin receptor induces collagenase and stromelysin gene expression. J Cell Biol 109:877–889

DISCUSSION

Karsenty: There is some evidence that osteoprotegerin ligand is involved in cartilage degradation, certainly in disease. Have you looked at this?

Poole: We're just starting to look at this. It is an obvious thing to do.

Kronenberg: With your antagonist, you have the opportunity to look at effects not just in primary culture, which is a special circumstance, but also in bone explants or potentially *in vivo*. Are there effects on chondrocyte differentiation and mineralization in those more physiological contexts?

Poole: Yes, there has been quite a lot of *in vivo* work done with this inhibitor. It has profound effects in skeletal development, as one might predict.

Russell: Some of the more broad-spectrum inhibitors are better known in terms of what they do, and I'm sure you could talk about those.

Poole: Yes, they can have quite profound effects on this whole endochondral process, and considerably delay the full expression of hypertrophy, so you end up with a very extended growth plate. Part of that story we would imagine is due to the shutting down of the collagenase activity and matrix resorption. You end up with a condition that looks like a vitamin D deficiency.

Russell: With the recent description of the aggrecanases, which have been long postulated but only recently described, how do you fit them into the picture of the chronicity? There is another whole family of degradation products that could be driving the process based on aggrecan.

Poole: ADAMS (proteases) are very differently regulated because they're membrane enzymes. They are switched on early, if one takes normal cartilage and induces matrix resorption. We would predict that the regulation of these particular proteases is totally different to that of the regulation of collagenase activity. In fact, when we start to probe these systems, as we have in articular cartilage in some detail, and as I've indicated with respect to the type II collagen work here, there is evidence of very marked differences in regulation. Interestingly, for example, with the RGD-containing sequences of fibronectin, you can up-regulate degradation of both type II collagen and aggrecan, presumably involving aggrecanase. But with the collagen fragments, and with one specific peptide we have isolated that embodies all the properties of the whole degraded molecule, there is a selective effect on type II collagen. This is very attractive, because it is of great interest that the substrate of the protease may be able to specifically influence the expression and activity of that protease very selectively. This gives us another level of regulation.

Russell: Just to extend the question, if you were postulating which classes of protease inhibitors might be the most attractive clinical candidates for blocking either osteoarthritis progression or particularly rheumatoid arthritis, would you advise broad spectrum inhibitors, or very focused narrow spectrum inhibitors like the one you described, or specific aggrecanase inhibitors? All those strategies are potentially available.

Poole: I would advocate the much more focused approach. If we are talking about arthritis, I would particularly focus on collagenase 3 (MMP13) inhibitors.

Meikle: When you isolated your chondrocytes from the growth plate, did they constitutively synthesize collagenase or other MMPs, or did you have to stimulate them?

Poole: This is all constitutive; we're not attempting to knowingly stimulate them in any way. The cells naturally mature, elaborate a matrix and turn it over without any stimulation other than the fact that we're putting them into a foreign defined environment in the absence of serum. It is a very natural process, and the process of maturation that we observe *in situ* is exactly what we see in culture albeit slower than *in vivo*.

Meikle: So from a pathological point of view, you could easily invoke cytokines, but what do you think the stimulatory factors are in the growth plate?

Poole: This is something that we're trying to come to grips with. It has been very interesting sitting in on these discussions. We know that fibroblast growth factor (FGF) can have quite interesting effects on the collagenase gene expression in other systems, looking back historically in the literature. This raises some very important issues with respect to the FGF receptor 1 (FGFR1), which may have some bearing on some of these observations. We know that FGFs can be very much involved in these processes. It is a complicated issue, because we also know that transforming growth factor (TGF)β can up-regulate collagenase 3 (MMP13) in certain conditions. Traditionally, we know for example that interleukin (IL)-1 can up-regulate, and in our work on articular cartilage we've demonstrated an IL-1α-dependent loop which has also been demonstrated for fibronectin. Thus the actual signalling through a receptor seems to be mediated at least so far in articular chondrocytes through an IL-1 pathway; if we block the IL-1 generation following addition of collagen or collagen peptide with IL-1RA, we can shut down the up-regulation of gene expression.

Meikle: Do they make TIMP (tissue inhibitors of metalloproteinases)?

Poole: These chondrocytes are certainly making TIMP, but in this type of situation at least, I see TIMP more as a scavenger rather than a true regulator of activity. The regulation of collagenase activity here seems to be more at the level of gene expression than anything else.

Blair: In the presence of your inhibitor at day 22 you were seeing a rather impressive increase in denaturation fragments.

Poole: Yes, when we isolate these chondrocytes and subject them to rather harsh conditions as part of the isolation, we can transiently induce collagenase 1 (MMP1). But immediately after isolation collagenase 1 expression is lost within a day or so, and then you only see collagenase 3 expression. What we might be seeing—and this is something we're working on at the moment—is a restoration via the induction of a salvage pathway of collagenase 1 to overcome this block. We know when we shut down certain proteases, the body has the ability to get round this block and start to cut up the collagen, in this case through other cleavages in other sites. We believe that these cells may be resorting to collagenase 1 expression when the collagenase 3 is blocked. This is being probed at the moment.

Chen: I wonder whether the inhibition of calcification and hypertrophy is a general effect of collagenase inhibitors. The reason I say this is that Tom Schmid's group (Davies et al 1996) has shown that tetracycline, which inhibits collagenase, also down-regulates type X collagen. I also wonder whether there is a more direct approach, such as introducing mutations in the cleavage site of a collagen, to see if the mutation process is affected.

Poole: Yes, we are certainly pursuing this with a much more direct approach. As you correctly say, Tom Schmid has made some interesting observations with respect to doxycycline. Tom also sees down-regulation of collagenase 3 (MMP13) expression in the presence of doxycycline, which inhibits this process of maturation. This is another example where you can perturb the system. When we saw the publications dealing with the *Cbfa1* knockout, we also saw this very interesting affect on maturation. It all fits together nicely.

D. Cohn: Could this paradoxical turnover of type II collagen at the same time that these cells are making lots of type II collagen be there to essentially feed the system, once you've got it initiated to rapidly turn the matrix over, or do you think you're actually remodelling the matrix in some way? What is the fate of the newly synthesized type II collagen? Is it stable in the matrix and are there different interactions in the matrix that gets turned over?

Poole: This is something that we're starting to look at in detail. From the evidence we have, there's quite a striking remodelling of the matrix, although eventually it involves net removal of type II collagen fibrils. But if you actually look morphologically at the collagen fibrils, there is clear evidence of a reduction in number and a reduction in diameter. There is evidence for a significant remodelling. But we believe that this loop is really part and parcel of the maintenance of gene expression in this particular type of system and other systems that we're probing, such as articular cartilage.

Hall: I wondered about this in terms of compartmentalization. We have situations such as the periodontal ligament, where the same cell can be synthesizing and degrading collagen. In the growth plate are you getting compartmentalization of the synthesis and the degradation, or is the same cell affecting both?

Poole: As far as we can make out, it's the same cell doing both. We have come to the same conclusion in articular cartilage, based upon somewhat indirect evidence. We are exploring the importance of this newly-synthesized collagen, because we believe that it could play a pivotal role in this process, rather than simply the resident collagen. This is being explored with respect to selective inhibition of type II collagen synthesis. Is newly synthesized type II collagen a preferred substrate for collagenase?

Kronenberg: Since MMP13 is the secreted protein, there are important distinctions between the cells that make it, and where it is and where it can act, because it can diffuse. When you look at mRNA for MMP13, the only chondrocytes that make it are the one or two sublayers of most differentiated chondrocytes that are about to die. It is appealing, in a way, that those are cells that are sort of finished with their business, and in your model they're making the molecule that perhaps diffuses further up and could therefore affect the differentiation of their precursors. It doesn't take away from the model—it actually makes it more interesting—that the only cells that make it are the most differentiated cells that are practically finished making type X, let alone type II collagen.

Mundlos: The other *in situs* that I remember show that it is basically made by the entire population of hypertrophic chondrocytes.

Kronenberg: Not in our hands. Perhaps those are more sensitive *in situs*. Robin Poole, have you done any *in situs*?

Poole: Yes, we've done some work on this. If we look at the mouse we can see earlier expression with *in situ* hybridization. It is an issue that we are anxious to resolve, because we also see *in situ* evidence of cleavage by collagenase early on, prehypertrophic, although it becomes more pronounced in the hypertrophic region.

Kronenberg: I'm only talking about fetal MMP13 expression, because that's the only thing that we have looked at.

Poole: This is postnatal; we haven't looked at fetal. It could be different.

Mundlos: Is the collagenase 3 (MMP13) necessary for vascular invasion of cartilage?

Poole: There is indirect evidence for this from work on collagenase inhibitors and in the *Cbfa1* knockout.

Chen: Does MMP13 cleave type X or type IX collagen?

Poole: Type X can be cleaved as well. Type IX is controversial: we don't find any evidence for type IX cleavage by MMP13. We have been looking at type IX collagen cleavage and this is a very early event. The initial cleavage involves the removal of the NC4 domain of α1(IX) chain from the fibril, and subsequently the COL2 domain of α1(IX) chain is removed when the type II collagen molecule is cleaved and removed.

Hall: Am I right in thinking that type IX is associated with type II collagen?

Poole: Yes, type IX is a collagen that sits on the surface of the fibril. The molar ratio when the matrix is being assembled in the first few days reaches a minimal value of about five type II, to one of type IX. Then there is this very selective removal of the NC4 domain, which is only present in about 10% of the molecules.

Hall: Is it the degraded type II that is spared or the type IX, which is associated with type II?

Poole: No, the type IX has on its α1 chain a basic NC4 domain, and this protrudes from the fibril. It is removed before the fibril itself is actually attacked. Then the part of the type IX, including the COL2 domain that sits on the fibril, is attacked at the same at the same time as the bulk of the fibril is cleaved by collagenase.

Mundlos: So you postulate that there is a receptor?

Poole: Yes, we're looking at this at the moment. We believe that there is a receptor. We've managed to identify the peptide, which is really quite short, and which has the same properties and the same molar potency that we see for a part of the molecule from which we have been able to isolate the actual activity.

Mundlos: Is that in analogy to the type I collagen?

Poole: This is again something that we're exploring. It is quite possible the same type of mechanism may operate with respect to type I collagen turnover. In that case, probably more with respect to MMP1 activity.

Reference

Davies SR, Cole AA, Schmid TM 1996 Doxycycline inhibits type X collagen synthesis in avian hypertrophic chondrocyte cultures. J Biol Chem 271:25966–25970

Retinoid signalling and skeletal development

T. Michael Underhill*†, Arthur V. Sampaio† and Andrea D. Weston†

*School of Dentistry and †Department of Physiology, Faculty of Medicine & Dentistry, The University of Western Ontario, London, Ontario, Canada N6A 5C1

Abstract. Metabolites of vitamin A, including retinoic acid (RA), comprise a class of molecules known to be important in development and homeostasis. RA functions through a class of nuclear hormone receptors, the RA receptors (RARs), to regulate gene transcription. In the developing mammalian limb, RA affects the differentiation of many cell lineages, including those of the chondrogenic lineage. In excess, RA is a potent teratogen, causing characteristic skeletal defects in a stage- and dose-dependent manner. Genetic analysis has shown that the absence of RARs leads to severe deficiencies in cartilage formation at certain anatomical locations while promoting ectopic cartilage formation at other sites. Expression of either a dominant-negative or a weak constitutively active RAR in the developing limbs of transgenic mice adversely affects chondrogenesis leading to skeletal malformations. Together, these results show that RAR-mediated signalling plays a fundamental role in skeletogenesis. This chapter will focus on the function of RARs in regulating chondroblast differentiation and the contribution of RA signalling to appositional and longitudinal growth of the skeletal primordia.

2001 The molecular basis of skeletogenesis. Wiley, Chichester (Novartis Foundation Symposium 232) p 171–188

Cartilage has many important functions during development and in the mature animal. During development, cartilage provides the foundation of most of the skeleton by providing a template which subsequently mineralizes. In adults, remnants of this embryonic cartilage contribute to the articular surfaces of bones. Numerous studies have focused on the signalling pathways important in patterning of the skeletal elements, but it is still unclear how these patterning cues are translated into a specific cellular differentiation programme that culminates in the formation of a skeletal element. To understand the biological mechanisms regulating this process, it is necessary to identify and functionally characterize factors that play a role in controlling the early stages of skeletogenesis—the elaboration of a chondrogenic template. Numerous model systems have been used to study this problem, with the developing limb proving to be one of the most experimentally accessible.

Overview of vertebrate limb development

Outgrowth of the mammalian limb (reviewed in Johnson & Tabin 1997) is initiated when the lateral plate mesoderm induces the overlying ectoderm to thicken into a ridge containing pseudostratified columnar epithelial cells, termed the apical ectodermal ridge (AER). Limb outgrowth proceeds in a proximodistal (PD) direction, such that the femur or humerus forms prior to more distal elements such as the digits. Sustained outgrowth is maintained by the continued proliferation of distal mesenchymal cells that underlie the AER, in a region termed the progress zone (PZ). The skeletal progenitors of the developing limb emanate from the PZ.

Multiple signals converge to give mesenchymal cells an identity with respect to the three limb axes (Johnson & Tabin 1997). As mesenchymal cells are displaced proximally from the PZ, they are thought to acquire their PD identity, with the length of time spent in the PZ determining their position along the PD axis. Classical transplantation experiments performed in chicks led to the identification of a region in the posterior mesenchyme termed the zone of polarizing activity (ZPA) and the dorsal ectoderm, which are important in anteroposterior and dorsal–ventral patterning, respectively. More recently, several extracellular signalling molecules have been identified that are expressed in these various regions. These include: fibroblast growth factor 4 which is expressed in the AER, and can substitute for loss of the AER; sonic hedgehog, which is present in the ZPA and has ZPA activity; and *Wnt7a*, a gene expressed in the dorsal ectoderm that has been shown by genetic analysis to be important in specifying dorsal–ventral identity. Together these molecules coordinate outgrowth and patterning, and eventually, the positional identity of PZ cells is translated into a cell fate, ending in commitment to a specific terminal differentiation programme. With respect to the formation of the skeletal elements, these various signalling pathways and others converge to specify the spatiotemporal pattern for condensation of mesenchyme.

Embryonic cartilage and development of the limb

Bones within the limb are formed from a cartilage precursor and the cartilage, itself, forms from condensed mesenchyme (reviewed in Hall & Miyake 1992). These condensations represent the earliest stages of limb patterning and are the forebears of the mature limb bones. Following condensation, mesodermal cells within the condensation differentiate into chondrocytes. Differentiation occurs in concert with limb outgrowth, such that proximal mesenchymal cells (i.e. close to the body wall) that are fated to become chondrocytes differentiate prior to more distal cells. The spatiotemporal regulation of mesenchyme differentiation into chondrocytes is a crucial step in endochondral bone formation in that it preserves the pattern of the bone primordia established earlier in limb development and provides a suitable matrix for subsequent ossification.

During limb outgrowth, signals that promote as well as inhibit chondrogenesis are important in regulating skeletal formation (Wolpert 1990). Agents known to promote chondrogenesis *in vivo* and *in vitro* include members of the transforming growth factor (TGF)β superfamily, some bone morphogenetic proteins (BMPs), GDF5, and TGFβ1, -2 and -3. In contrast, a group of vitamin A derivatives collectively referred to as the retinoids are known to inhibit cartilage formation.

BMPs and chondrogenesis

The BMPs regulate many aspects of endochondral bone formation including the commitment and differentiation of mesenchymal cells to the chondrocytic lineage. During limb bud outgrowth, *Bmp2* and *Bmp4* are expressed adjacent to condensing mesenchyme, in the perichondrium and in the interdigital regions. Null mutant embryos do not survive much beyond embryonic day (E) 10.5, and thus have not been informative in sorting out the function of BMP2 and 4 in these regions. Subsequent studies, however, in which dominant-negative or constitutively active BMP type II receptors were used *in vitro* and *in vivo*, indicate that BMP signalling is a requisite step in cartilage formation (Zou et al 1997). These results are complemented by experiments in which BMPs were overexpressed in the developing chick limb, in that BMPs were found to stimulate cartilage formation and modify skeletal patterning (Duprez et al 1996). Under certain conditions, BMPs have also been found to promote apoptosis within the interdigital region (IDR) (Zou & Niswander 1996). Moreover, loss- or gain-of-function studies with Noggin, a secreted inhibitor of BMP2 and 4 with lower affinity for BMP7, have shown that BMP2 and 4 are important in skeletal development and that regulation of BMP signalling is required for delineation of the various skeletal elements (Brunet et al 1998, Capdevila & Johnson 1998). Mice deficient in BMP6 or BMP7 also present with skeletal defects including polydactyly in the hind limbs of *Bmp7*$^{-/-}$ animals (Luo et al 1995). Thus, BMPs, especially 2, 4 and 7 are important in early limb skeletal development, likely in the commitment and differentiation of mesenchymal cells to chondrocytes.

Retinoic acid: an inhibitor of chondrogenesis

One class of molecules important in development and homeostasis is the vitamin A metabolites, including retinoic acid (RA) (reviewed in Underhill et al 1995). In the developing mammalian limb, RA affects the differentiation of many cell lineages, including those of mesenchymal origin. Administration of high doses of RA to mouse embryos *in utero* results in a wide range of birth defects that affect development of the limbs as well as other tissues (Kochhar 1973). The timing of RA treatment and the resultant limb defects appear to coincide with the timing of mesenchyme condensation and differentiation into chondrocytes. Further analysis has shown that RA inhibits chondrogenesis and it is most likely this activity that contributes to the aforementioned limb defects (Underhill & Weston 1998). Together, these results suggest that RA is a potent inhibitor of chondrogenesis.

Retinoic acid signal transduction: CRABPs, RARs and RXRs

Several intracellular RA binding proteins have been identified (Giguère 1994). Cytoplasmic retinoic acid binding proteins (CRABPs) I and II appear to be important in modulating the intracellular RA concentration, whereas the actions of RA are primarily mediated through nuclear receptors for RA, the retinoic acid receptors (RARs) and the retinoid-X-receptors (RXRs). CRABPII has recently been shown to be localized to the nucleus and to directly interact with the RARs, and thus may play a role in delivering ligand to the RARs (Delva et al 1999). The retinoid receptors function as ligand-inducible transcription factors that by sequence homology belong to the steroid hormone nuclear receptor superfamily. Each RAR and RXR subfamily contains three members, α, β and γ. Several isoforms have been identified within each receptor type which have distinct N-termini, are differentially expressed during development, and have different functional properties. Two prominent isoforms for RARα (1 and 2) have been isolated, while 4 and 2 major isoforms have been isolated for RARβ (1, 2, 3 and 4) and RARγ (1 and 2), respectively. The RARs affect gene transcription primarily through two mechanisms, one that involves direct DNA binding of the receptor to RA response elements (RAREs) of DNA and another, more indirect mechanism, whereby the receptors influence gene transcription through cross-talk with other signal transduction pathways (Gttlicher et al 1998).

CRABPs and RARs in chondrogenesis

In the developing murine limb, the RARs are expressed in distinct and sometimes overlapping spatiotemporal patterns. Early during limb skeletogenesis (E9.5–E11.5) RARα and γ are expressed in overlapping regions in the forelimb (Dollé et al 1989). Subsequently, RARγ becomes preferentially localized to precartilage condensations and cartilage, whereas RARα expression becomes restricted to the surrounding mesenchyme and the interdigital region where it is comparatively highly expressed. RARβ is expressed in the proximal mesenchyme during the early stages of limb development and subsequently becomes localized to the interdigital region and in cells adjacent to the phalangeal cartilages. RXRα and β are expressed ubiquitously throughout the developing limb to approximately E16.5. CRABPI and II are initially expressed in the limb mesenchyme and at later stages exhibit a more complex expression pattern. Neither CRABPI nor II mRNAs have been detected in condensing mesenchyme or cartilages during limb development (Dollé et al 1989, Ruberte et al 1992). Prior to overt cytodifferentiation within the developing mouse limb bud, the predominant ligand for RARs all-*trans* retinoic acid is present at slightly higher concentrations in the posterior limb tissue ($\sim$16 nM) than the anterior limb tissue ($\sim$11 nM) (Scott et al 1994).

Surprisingly, null mutants of individual RARs or RXRs exhibit no limb skeletal malformations (reviewed in Underhill & Weston 1998). Approximately 60% of RARα null mutants did, however, have webbed digits, while RARγ null mutants exhibited defects in formation of the tracheal cartilage. The importance of RARα and γ in paraxial development is demonstrated by the range of limb abnormalities seen in RAR$\alpha^{-/-}$ RAR$\gamma^{-/-}$ mice (Lohnes et al 1994). Moreover, these skeletal defects do not appear to have been due to aberrant expression of Sonic hedgehog, fibroblast growth factor 4, BMP2, HOXD9, HOXD13, or MSX1 (Lohnes et al 1994). While no limb malformations were observed in RXRα null animals, the absence of RXRα was associated with reduced severity of RA-induced skeletal defects, providing further support for the requirement of an RXR partner in RAR-mediated signalling (Sucov et al 1995). Skeletal defects were, however, reported in RXR double and triple mutants, which contain a combination of both a mutation in the AF-2 domain of RXRα and a mutant RXRβ allele or mutant RXRβ and RXRγ alleles (Mascrez et al 1998). Interestingly, in compound homozygous null mutants (RAR$\alpha^{-/-}$ RAR$\beta2^{-/-}$ or RAR$\alpha^{-/-}$ RAR$\gamma^{-/-}$) ectopic cartilages are present at several distinct sites, including the heart (at the base of the semilunar cusps), meninges, diaphragm, and peritoneum (Mendelsohn et al 1994). Hence, under certain circumstances the absence of RARα in conjunction with loss of RARβ or γ leads to permissive conditions resulting in ectopic cartilage formation.

Evidence from studies of RA teratogenesis suggests that the presence of ligand-activated RARs and/or continued presence of RARs inhibits chondrogenesis. Genetic analysis has shown that the absence of these receptors leads to severe deficiencies in cartilage formation at certain anatomical locations, while promoting ectopic cartilage formation at other sites. Hence, the RARs appear to have the ability to both promote and inhibit chondrogenesis (Underhill & Weston 1998). To resolve these seemingly disparate activities and to further clarify RAR function in skeletogenesis, we took a transgenic approach, in conjunction with the manipulation of RAR signalling in cultured limb mesenchyme.

RARα-expressing transgenics exhibit limb defects

As mentioned previously, RARα and RARγ are expressed in distinct but overlapping patterns in the developing mouse limb. To better resolve their expression patterns during chondrogenesis we adapted a whole-mount *in situ* hybridization technique for the analysis of gene expression in micromass cultures. From this technique, RARα expression was found to be down-regulated during chondrogenesis, consistent with its pattern of expression *in vivo*. RARα is present in condensed mesoderm but not in chondroblasts, whereas RARγ is abundant in chondroblasts. To evaluate the function of RARα in mammalian limb development, we misexpressed a modified RARα1 (a weakly constitutive active version of RARα1 that contains a fusion to the *E. coli* β-galactosidase gene, denoted herein as tgRARα) in the limbs of transgenic mice (referred to herein as RARα transgenic mice) using a Hoxb-6 promoter fragment to target transgene expression to the developing limb (Cash et al 1997).

Ectopic expression of tgRARα1 in the developing mouse limb was associated with marked preaxial and postaxial limb defects, which included polydactyly, syndactyly, ectrodactyly, fibular deficiencies, and tarsal and carpal fusions (Fig. 1) (Cash et al 1997). The defects displayed in these mice recapitulate many of the congenital limb malformations observed in the fetuses of dams administered high doses of RA. Histological sections prepared from E13.5 limbs of transgenic and non-transgenic litter mates indicate that development of the limb cartilage is delayed in the transgenic animals (Fig. 2). The axial bones of transgenic and control animals were examined using magnetic resonance imaging and found not to be significantly different (Fig. 3). Comparison of the two images clearly shows that the axial structures demarcated by the intense signal are developed to the same extent in the two animals, whereas there is reduction and loss of signal intensity in the tibia and fibula of the transgenic animal in comparison to the control animal. The skeletal structures in these images correlate well with those structures stained with alizarin red S in embryos of the same age.

Chondrogenesis *in vivo* is preceded by the aggregation of chondroprogenitor cells to form condensations. The condensations then enlarge and cells within the aggregates begin to differentiate into chondrocytes. If either the aggregation or differentiation of mesenchymal cells is delayed, then a reduction in the size of the corresponding skeletal element (or even complete loss of that element) would be expected. With the advent of micromass culturing techniques to establish primary limb mesenchymal cultures, it has become possible to examine the differentiation of these cells *in vitro* (Ahrens et al 1977). When plated at high density, dissociated limb

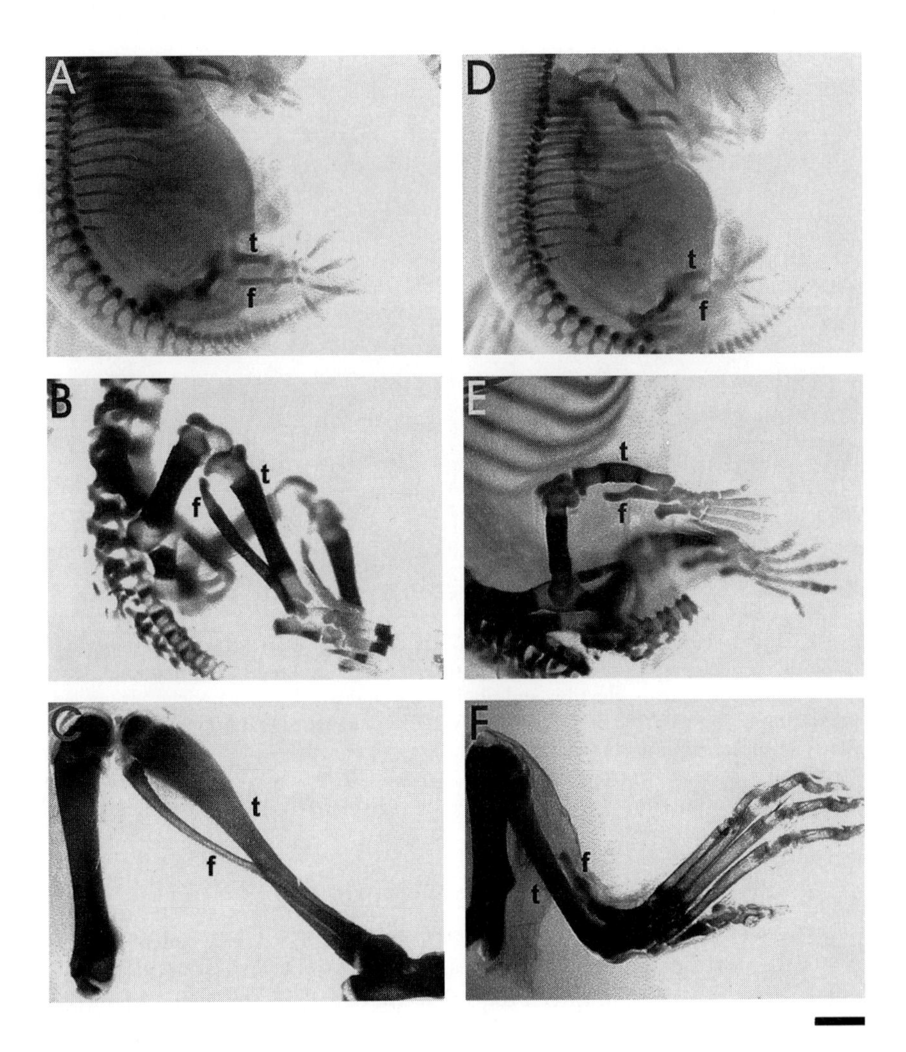

FIG. 1. Expression of tgRARα in the developing limb results in a number of skeletal abnormalities. Alcian blue and alizarin red S-stained preparations of right hind limb skeletons from normal and transgenic fetuses, and 2 week old mice. (A, B, C) lateral views of wild-type fetuses at E13.5 (A), E17.5 (B) and a 2 week old mouse. (D, E, F) lateral view of transgenic fetuses at E13.5 (C), E17.5 (E) and a 2 week old mouse. In comparison to wild-type animals the transgenic animals have a several malformed elements, with the tibia and fibula being the most severely affected. In the transgenic animals the tibia is foreshortened and thickened, while the fibula is reduced in size and truncated proximally. f, fibula and t, tibia. Dark grey staining within skeletal elements represents alizarin red S staining of mineralized tissue, while light grey staining in the same shows alcian blue staining of cartilage. Bar = 1 mm.

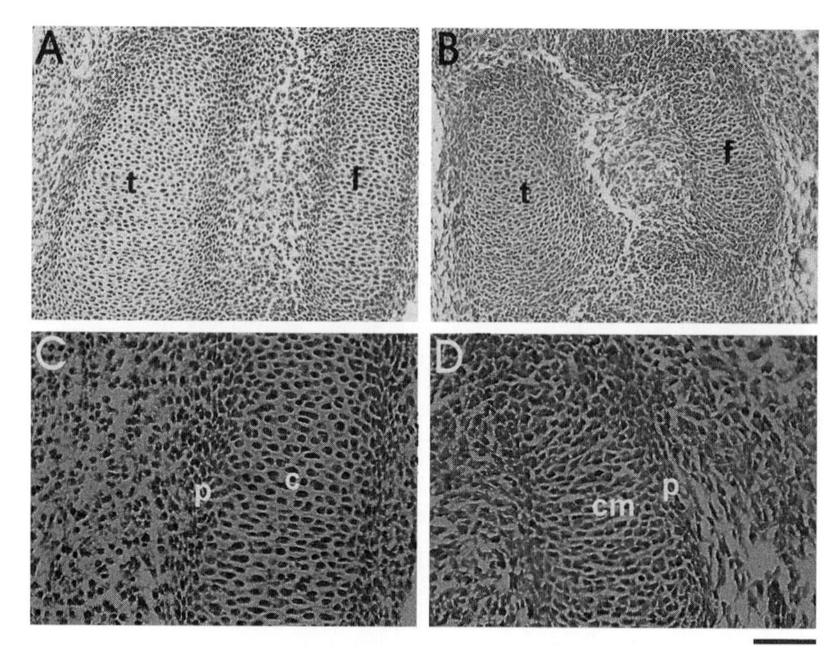

FIG. 2. Chondrogenesis of the tibia and fibula is delayed in RARα transgenic mice. (A–D) Histological sections prepared from E13.5 control (A and C) and transgenic (B and D) fetuses. (A, C) In wild-type animals, the tibia and fibula are well defined by the perichondrium, and a higher magnification view (C) of the fibula shows an abundance of maturing chondrocytes. (B, D) In transgenic fetuses, the tibia and fibula are not as well organized, and a higher magnification view of the fibula shows that it contains condensed mesenchymal cells. f, fibula; p, perichondrium; t, tibia; c, chondrocyte; cm, condensed mesenchyme. Bar=0.1 mm in the upper panel and 0.05 mm in the lower panel.

mesenchymal cells (obtained from E10–E12 forelimbs or E10.5–E12.5 hind limbs) closely follow the progression of limb mesenchyme *in vivo*. These cultures give rise to a number of differentiated cell types, including chondroblasts which form nodules of cartilage within 2 to 3 days that are detectable by alcian blue (binds sulfated proteoglycans in cartilage). Using these primary cultures, the temporal and spatial sequence of events within the chondrogenic programme can be rigorously examined.

Micromass cultures were used to examine the action of the transgene on chondrogenesis. Micromass cultures prepared from limb mesenchyme of transgenic embryos give rise to a similar number of condensations as those from non-transgenic cultures within the first two days of culturing. After 2 days, the condensations in wild-type cultures develop into alcian blue-stained cartilage nodules, whereas those of the transgenic cultures fail to develop into cartilage nodules, with many fewer and weaker-stained cartilage nodules being apparent

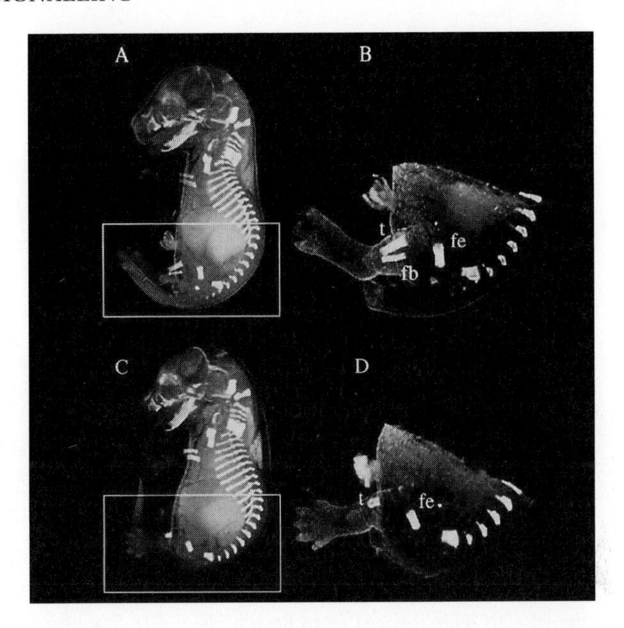

FIG. 3. Magnetic resonance microscopy of RARα transgenic and control animals. Volume rendered magnetic resonance microscopy scans of 16.5 d.p.c. size-matched control and transgenic litter mates highlighting mineralized bone. (A) Control embryo, note the extent of mineralization in the tibia (t) and fibula (fb). (B) Hind quarter was removed from embryo shown in A and scanned to provide greater resolution. (C) Transgenic embryo, note the absence of a mineralized fibula and the reduced degree of mineralization in the tibia as compared to A. (D) Hind quarter from embryo shown in C at greater resolution. t, tibia; fe, femur.

(Fig. 4) (Cash et al 1997). Analysis of gene expression within the cultures indicate that the transgenic cultures contain numerous precartilaginous nodules as demonstrated by expression of *Ncad, Gli1* and *ColI*, but do not exhibit the intense localized expression of *ColII* observed in wild-type cultures (Weston et al 2000). These results suggest that expression of the transgene does not interfere with condensation or chondroprogenitor commitment, but does affect the transition from a chondroprogenitor to a chondroblast.

Additional evidence supporting the functional importance of RAR-signalling in chondroblast differentiation was provided by experiments in which RAR-signalling was selectively manipulated. Cartilage formation can be stimulated in response to an RARα-selective antagonist (Weston et al 2000). Moreover, down-regulation of RARα expression is coincident with chondroblast differentiation in mouse limb mesenchyme, further suggesting an involvement of RARα in this process. In the absence of RARα and RARγ, a single RARα2 allele is sufficient to rescue most of the limb skeletal defects present in these compound null mutants (Lohnes et al 1994). In addition, diminution of RAR expression by addition of

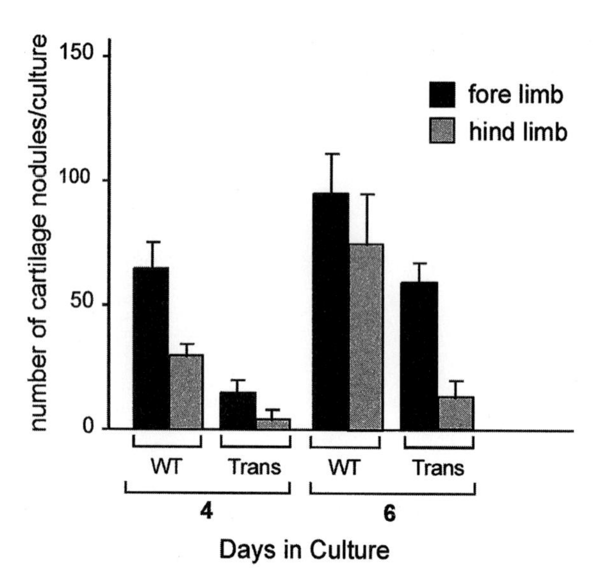

FIG. 4. Expression of the RARα transgene interferes with cartilage formation in micromass cultures. Mesenchymal cells were isolated from limb buds of E11.5 wild-type and transgenic embryos and plated under micromass conditions. Cultures were fixed and stained with alcian blue after 4 or 6 days, and the number of nodules counted. In general, there are fewer cartilage nodules in the transgenic cultures as compared to the corresponding wild-type culture.

antisense oligonucleotides to micromass cultures promotes cartilage nodule formation (Jiang et al 1995). These results suggest that antagonism of RAR-mediated signalling promotes chondroblast differentiation and that continued RAR-mediated signalling disrupts cartilage formation by interfering with chondroblast differentiation. An increase in the expression of an RA-metabolizing enzyme P450RA (also referred to as CYP26) during formation of precartilaginous condensations is also consistent with this model (de Roos et al 1999). Taken together, these results suggest that attenuation of RAR-mediated signalling is important in cartilage formation.

Given that inhibition of RAR-mediated signalling seems critical to chondroblast differentiation, one would expect that alteration of RAR signalling in chondroprogenitors may contribute to precocious chondroblast differentiation. This has been observed in cultured limb mesenchyme *in vitro* and may, in part, explain the phenotypes observed in some of the compound RAR knockouts. Premature differentiation of chondroprogenitors would be expected to lead to smaller hypoplastic cartilages and skeletal deficiencies as observed in these animals. Similarly, loss of RAR-signalling in the ventricular myocardium has been suggested to stimulate precocious differentiation of ventricular myocytes and to contribute to a hypoplastic ventriculum in RARα null animals (Kastner et al 1997). In many cell types, the onset of retinoid-mediated signalling invokes cellular differentiation, whereas, it appears that it is a loss of an RAR-mediated signal that triggers chondroblast differentiation which may, in part, explain the appearance of ectopic cartilages in RAR null mutants.

Recent studies indicate that at later developmental stages RA may be important in regulating chondrocyte maturation. The perichondrium synthesizes RA and expresses retinaldehyde dehydrogenase 2, an enzyme involved in RA synthesis (Niederreither et al 1997, Koyama et al 1999). Furthermore, in transgenic mice containing *lacZ* reporter gene coupled to an RARE, transgene expression is activated in the perichondrium (von Schroeder & Heersche 1998). In this manner, local sources of RA may be important in preventing premature differentiation of prechondrogenic perichondrial cells and contribute to maturation of adjacent chondrocytes.

RAR and BMP signalling in chondrogenesis

As mentioned earlier, the BMPs play a prominent role in skeletogenesis and are important at several stages during the chondrogenic process. Inhibition of BMP signalling leads to a number of skeletal defects, some of which overlap with the defects observed in the aforementioned RARα transgenic animals. Furthermore, several reports have shown that RAR signalling can modify and/or directly regulate the expression of BMPs (Heller et al 1999). Northern blot analyses and whole-mount *in situ* hybridization studies, however, reveal no dramatic differences in *Bmp2* or *4* expression between wild-type and transgenic mice. To directly assess the contribution of BMP signalling to the transgenic phenotype, we examined the ability of BMPs to rescue cartilage formation in transgenic cultures. Treatment of transgenic cultures with either BMP2 or 4 stimulates cartilage nodule formation, but fails to induce differentiation of transgene-expressing cells into chondroblasts. Instead, both molecules promote condensation of transgene-expressing cells. Treatment of wild-type micromass cultures with Noggin (an inhibitor of BMP2 and 4) suppresses cartilage formation, but the addition of an RARα antagonist to these Noggin-treated cultures can restore nodule formation (Weston et al 2000). Taken together, these results suggest that RA signalling functions downstream of a BMP-mediated signal to regulate chondroblast differentiation.

RAR and BMP signalling in interdigital apoptosis

In addition to regulating chondrogenesis, RA and BMP signalling have also been shown to be important in regulating apoptosis of limb mesenchyme. Single and compound null RAR mutants exhibit webbed digits and, in cultured limbs, apoptosis in the interdigital region is stimulated by exogenous RA (Lussier et al

1993, Dupé et al 1999). Thus, the retinoid signalling pathway appears to play a role in interdigital apoptosis. During limb outgrowth, the BMPs act on regions in which RARα and RARγ are expressed, namely in the condensing mesenchyme, the perichondrium and the interdigital region (IDR). All of these regions have the potential to form cartilage. Alteration of the BMP or RA signalling pathways compromises cartilage formation during skeletal development. Interference with BMP or RA signalling, however, leads to cessation of apoptosis in the IDR, and in some instances, to ectopic cartilage formation (Zou & Niswander 1996, Dupé et al 1999, Rodriquez-Leon et al 1999). Hence, these two signalling pathways play an important role in both the development of the skeletal elements and in the separation of the digits. A recent report suggests that BMP7 is involved in IDR apoptosis and that retinoids function upstream of a BMP signal to regulate apoptosis, possibly by regulating *Bmp7* expression (Rodriquez-Leon et al 1999). Dupé et al (1999) showed that *Bmp7* is down-regulated in RAR null mutants, but that this is not sufficient to account for the presence of webbed digits in these animals, as heterozygous compound RAR mutants with normal *Bmp7* expression still present with webbed digits. Further experiments need to be carried out to reconcile these opposing observations which, in part, may be due to intrinsic differences in limb development between the chick and the mouse. In general, these results suggest that BMP signalling in combination with a retinoid-mediated signal stimulate apoptosis in interdigital mesenchyme, whereas BMP signalling coupled with a reduction in RAR-mediated signalling favours cartilage formation.

Summary and prospects

Diverse experimental approaches have been used to demonstrate that RA signalling is important in many aspects of skeletal development (summarized in Fig. 5). RAR activity appears to be important in the regulation of

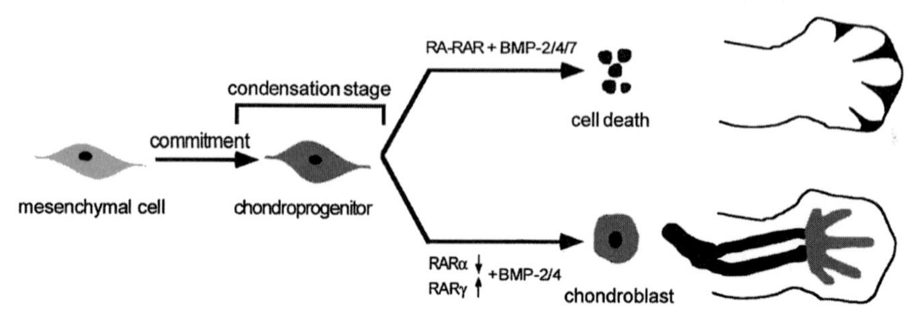

FIG. 5. Hypothetical roles for BMPs and RARs in limb development. Expression of BMP2/4 coupled with ligand activated-RAR α, β, γ contributes to apoptosis of the IDR limb mesenchyme, whereas attenuation of RAR activity leads to chondroprogenitor differentiation and cartilage formation.

chondroprogenitor differentiation. In this manner, RAR-mediated signalling may specify the size of progenitor cell populations, and/or influence cell fate decisions by modulating the competency of cells to respond to inductive signals, such as BMPs. Therefore, the status of cellular RAR activity is an important determinant in the spatiotemporal regulation of cell differentiation in the developing limb and, as such, contributes to both the size and shape of the skeletal primordia. Future studies aimed at defining the regulatory networks operating upstream and downstream of RAR-mediated signalling to regulate chondroblast differentiation will be useful to further elucidate the role of RARs in chondrogenesis. For example, questions concerning availability of RA, mechanisms regulating receptor expression, molecules that modulate RAR signalling (i.e. co-activators, co-repressors, cross-talk with other signal transduction pathways) and RAR-target genes, all remain unanswered. Such knowledge will, however, not only further our understanding of fundamental processes important in chondrogenesis, but will also help to develop rational approaches for manipulating RAR activity *in vivo* to stimulate cartilage formation for treatment of disorders of cartilage such as arthritis.

Acknowledgements

We apologize to those whose work was not referenced for the sake of brevity. We thank Drs S. J. Dixon and S. M. Sims for helpful comments on the manuscript. Work from the authors' laboratory reviewed here was supported by the Medical Research Council of Canada and the Canadian Arthritis Network. A.V.S. and A.D.W. were supported by an Ontario Graduate Scholarship in Science and Technology, and an MRC predoctoral scholarship, respectively. The magnetic resonance images were collected at the Duke University Centre for *In vivo* Microscopy by Dr B. R. Smith while T.M.U. was a post doctoral fellow in Dr E. Linney's laboratory.

References

Ahrens PB, Solursh M, Reiter RS 1977 Stage related capacity for limb chondrogenesis in cell culture. Dev Biol 60:69–82

Brunet LJ, McMahon JA, McMahon AP, Harland RM 1998 Noggin, cartilage morphogenesis, and joint formation in the mammalian skeleton. Science 280:1455–1457

Capdevila J, Johnson RL 1998 Endogenous and ectopic expression of *noggin* suggests a conserved mechanism for regulation of BMP function during limb and somite patterning. Dev Biol 197:205–217

Cash DE, Bock CB, Schughart K, Linney E, Underhill TM 1997 Retinoic acid receptor α function in vertebrate limb skeletogenesis: a modulator of chondrogenesis. J Cell Biol 136:445–457

de Roos K, Sonneveld E, Compaan B, ten Berge D, Durston AJ, van der Saag PT 1999 Expression of retinoic acid 4-hydroxylast (CYP26) during mouse and *Xenopus laevis* embryogenesis. Mech Dev 82:205–211

Delva L, Bastie JN, Rochette-Egly C et al 1999 Physical and functional interactions between cellular retinoic acid binding protein II and the retinoic acid-dependent nuclear complex. Mol Cell Biol 19:7158–7167

Dollé P, Ruberte E, Kastner P et al 1989 Differential expression of genes encoding α, β and γ retinoic acid receptors and CRABP in the developing limbs of the mouse. Nature 342: 702–705

Dupé V, Ghyselinck NB, Thomazy V et al 1999 Essential roles of retinoic acid signaling in interdigital apoptosis and control of BMP-7 expression in mouse autopods. Dev Biol 208:30–43

Duprez D, Bell EJ, Richardson MK et al 1996 Overexpression of BMP-2 and BMP-4 alters the size and shape of developing skeletal elements in the chick limb. Mech Dev 57:145–157

Giguère V 1994 Retinoic acid receptors and cellular retinoid binding proteins: complex interplay in retinoid signaling. Endocr Rev 15:61–79

Göttlicher M, Heck S, Herrlich P 1998 Transcriptional cross-talk, the second mode of steroid hormone receptor action. J Mol Med 76:480–489

Hall BK, Miyake T 1992 The membranous skeleton: the role of cell condensations in vertebrate skeletogenesis. Anat Embryol (Berl) 186:107–124

Heller LC, Li Y, Abrams KL, Rogers MB 1999 Transcriptional regulation of the *Bmp2* gene. Retinoic acid induction in F9 embryonal carcinoma cells and *Saccharomyces cerevisiae*. J Biol Chem 274:1394–1400

Jiang H, Soprano DR, Li SW et al 1995 Modulation of limb bud chondrogenesis by retinoic acid and retinoic acid receptors. Int J Dev Biol 39:617–627

Johnson RL, Tabin CJ 1997 Molecular models for vertebrate limb development. Cell 90: 979–990

Kastner P, Messaddeq N, Mark M et al 1997 Vitamin A deficiency and mutations of RXRα, RXRβ and RARα lead to early differentiation of embryonic ventricular cardiomyocytes. Development 124:4749–4758

Kochhar DM 1973 Limb development in mouse embyros. I. Analysis of teratogenic effects of retinoic acid. Teratology 7:289–298

Koyama E, Golden EB, Kirsch T et al 1999 Retinoid signaling is required for chondrocyte maturation and endochondral bone formation during limb skeletogenesis. Dev Biol 208:375–391

Lohnes D, Mark M, Mendelsohn C et al 1994 Function of the retinoic acid receptors (RARs) during development (I). Craniofacial and skeletal abnormalities in RAR double mutants. Development 120:2723–2748

Luo G, Hofmann C, Bronckers ALJJ, Sohocki M, Bradley A, Karsenty G 1995 BMP-7 is an inducer of nephrogenesis, and is also required for eye development and skeletal patterning. Genes Dev 9:2808–2820

Lussier M, Canoun C, Ma C, Sank A, Shuler C 1993 Interdigital soft tissue separation induced by retinoic acid in mouse limbs cultured *in vitro*. Int J Dev Biol 37:555–564

Mascrez B, Mark M, Dierich A, Ghyselinck NB, Kastner P, Chambon P 1998 The RXRα ligand-dependent activation function 2 (AF-2) is important for mouse development. Development 125:4691–4707

Mendelsohn C, Lohnes D, Décimo D et al 1994 Function of the retinoic acid receptors (RARs) during development (II). Multiple abnormalities at various stages of organogenesis in RAR double mutants. Development 120:2749–2771

Niederreither K, McCaffery P, Drger UC, Chambon P, Dollé P 1997 Restricted expression and retinoic acid-induced downregulation of the retinaldehyde dehydrogenase type 2 (RALDH-2) gene during mouse development. Mech Dev 62:67–78

Rodriquez-Leon J, Merino R, Macias D, Gaffian Y, Santesteban E, Hurle JM 1999 Retinoic acid regulates programmed cell death through BMP signalling. Nat Cell Biol 1:125–126

Ruberte E, Friederich V, Morriss-Kay G, Chambon P 1992 Differential distribution patterns of CRABP I and CRABP II transcripts during mouse embryogenesis. Development 115: 973–987

Scott WJ, Walter R, Tzimas G, Sass JO, Nau H, Collins MD 1994 Endogenous status of retinoids and their cytosolic binding proteins in limb buds of chick vs mouse embryos. Dev Biol 165:397–409
Sucov HM, Izpisfla-Belmonte J-C, Gaffian Y, Evans RM 1995 Mouse embryos lacking RXRα are resistant to retinoic-acid-induced limb defects. Development 121:3997–4003
Underhill TM, Weston AD 1998 Retinoids and their receptors in skeletal development. Microsc Res Tech 43:137–155
Underhill TM, Kotch LE, Linney E 1995 Retinoids and mouse embryonic development. Vitam Horm 51:403–457
von Schroeder HP, Heersche JN 1998 Retinoic acid responsiveness of cells and tissues in developing fetal limbs evaluated in a RAREhsplacZ transgenic mouse model. J Orthop Res 16:355–364
Weston AD, Rosen V, Chandraratna RAS, Underhill TM 2000 Regulation of skeletal progenitor differentiation by the BMP and retinoid signaling pathways. J Cell Biol 148:679–690
Wolpert L 1990 Signals in limb development: STOP, GO, STAY and POSITION. J Cell Sci (suppl) 13:199–208
Zou H, Niswander L 1996 Requirement for BMP signalling in interdigital apoptosis and scale formation. Science 272:738–741
Zou H, Wieser R, Massagué J, Niswander L 1997 Distinct roles of type I bone morphogenetic protein receptors in the formation and differentiation of cartilage. Genes Dev 11:2191–2203

DISCUSSION

Kronenberg: You described how the RARs heterodimerize with the RXRs. In terms of the effects of the transgene, are the levels sufficient that you have to worry about sopping up RXRs and making thyroid hormone work less well?

Underhill: We've made dominant negatives that contain just a portion of the RAR: they are missing the N-terminal region containing the DNA binding domain. Expression of this dominant-negative in the limb under the control of the Hoxb-6 promoter does not lead to any pronounced limb defects. Under these conditions the dominant-negative is expressed to a level similar to that of the full-length RAR transgene. If the full-length RAR transgene was causing malformations due to sequestration of RXR binding partners, then one would expect that the dominant-negative which also interacts with RXRs would lead to a similar spectrum of malformations and it doesn't.

Kronenberg: Are they expressed at the same levels?

Underhill: Yes.

Hall: Could you rescue the fibula in your animals in any way? It is arrested at a mesenchyme stage: can you rescue it with BMP?

Underhill: These experiments were performed in the mouse and to better understand how the transgene alters skeletogenesis we have utilized an *in vitro* system to characterize the chondrogenic potential of the transgenic mesenchyme. We have used this as an alternative to other approaches that may involve *ex utero* surgery.

Hall: Could you rescue it conceptually?

Underhill: I don't think we could rescue fibula formation with the presence of BMP. Our experiments performed *in vitro* with transgenic limb mesenchyme demonstrate that the addition of BMPs stimulates mesenchyme condensation, but not their differentiation into chondroblasts. It appears that in this system, continued activation of the retinoid signalling pathway prevents cell differentiation and that BMPs are not able to circumvent the block in the differentiation of the transgenic mesenchyme. The primary difference here, is that in many cell systems retinoids stimulate cell differentiation, and in this system we observed that it's the actual inhibition of retinoid signalling that stimulates chondroblast differentiation. This observation does not appear to be unique to this system, as in some of the RAR knockouts, there is precocious differentiation of ventricular myocytes which contributes to a hypoplastic ventriculum. In this case, they are suggesting again that the diminution of retinoid signaling may be important in regulating cell differentiation. We are currently engaged in defining the molecular mechanisms by which transgene expression blocks chondroblast differentiation and how addition of retinoid antagonists stimulate chondroblast differentiation. So we are looking downstream of the receptors to attempt to resolve how continued activation of this pathway with the transgene suppresses chondroblast differentiation. We are initially focusing on activating protein 1 (AP1) and whether expression of the transgene interferes with the activity of this complex. If we can define downstream targets of the receptors, then manipulation of these genes may allow us to rescue transgene induced skeletal defects.

Tickle: Why is it just the fibula and that part of the limb that's affected? Also, it looked as though one of your embryos had an extra digit in the limb.

Underhill: That is a very good question, and we are not sure why the fibula is most affected, but it might be due in part to its size. The fibula is the smallest long bone in the hind limb, thus inhibition of chondrogenesis by transgene expression may affect this element to a greater extent then the other elements. Another possibility, which we have not looked into, is that transgene expression is higher or sustained for a longer period of time in the region of the hind limb that gives rise to the fibula. With respect to the periodic duplications we observe, we have found that if you look at the limb early on, we can actually determine which embryos are transgenic based on limb morphology. The limbs of transgenic embryos look very similar to what you observe if you put a bead of RA at the anterior margin in the chick limb; in other words we observe a very pronounced outgrowth along the anterior margin. I think the transgene has an affect early on, possibly by slightly modifying patterning cues, and this leads to expansion of the anterior margin and the formation of an extra digit.

Ornitz: Have you looked at sonic hedgehog expression?

Underhill: Yes, we have looked at Sonic hedgehog expression in the transgenic embryos using whole-mount *in situ* hybridization and it appears that it might be slightly elevated in the anterior margin, but it is really difficult to detect.

Hall: Is the extra digit always on the anterior margin?

Underhill: Yes, the extra digit is always observed on the anterior margin in the hind limb. On the forelimb we see polysyndactyly of the fourth digit. This phenotype is not 100% penetrant, showing up in approximately 40% of the animals.

Kronenberg: What sort of level of RAR have you increased to, to get these effects?

Underhill: That is difficult to determine, but I would say at least a fivefold increase in expression across the limb.

Kronenberg: How does that compare to the levels of RXR?

Underhill: We have not compared the level of the transgene to that of RXRs.

Newman: We have looked at cultures of chicken forelimb and hindlimb for response to retinoic acid. We see tremendous differences (Downie & Newman 1994). In the mouse the fore and hindlimb are much more similar than they are in chicken. Are the responses of the tissues consistent with this?

Underhill: We do not observe any major differences in the response of forelimb and hindlimb mesenchyme to RA or transgene expression. In both the fore and hindlimb expression of the transgene appears to inhibit chondrogenesis. The differences in the severity of the malformations observed between the fore and hindlimbs of transgenic animals can be attributed to differences in transgene expression. The transgene is expressed to a much higher level in the hindlimb than the forelimb mesenchyme.

Beresford: You have used this transgene to unmask this interesting effect with the RAR overexpression, but can you make sense of what you observe in your model if you go back and map out the expression domain of the endogenous gene?

Underhill: Yes, I think the transgene can provide us with some information on the function of endogenous retinoid signaling in skeletal development. During the course of limb development in the mouse, RARγ becomes abundantly expressed in chondroblasts, while RARα, expression is decreased in this population. Thus, in the transgenic animals the RARα transgene is expressed at much higher levels in condensing mesenchyme than normally. If down-regulation of RARα is important in chondroblast differentiation, then continued expression of the RARα transgene would interfere with this process. Another possibility is that increased expression of the RARα transgene upsets the balance between RARα and RARγ expression and this prevents chondroblast differentiation. In addressing receptor activity it is also important to examine ligand availability. There are two key enzymes that are involved in ligand bioavailability: one is retinaldehyde dehydrogenase 2, which is involved in the synthesis of retinoic acid, and the other is a molecule that is involved in the degradation of retinoic acid, P450RA or CYP26. It has been recently reported by de Roos et al (1999) that *Cyp26* is expressed in precartilaginous condensations, both in the limb cartilages as well as in other cartilages within the developing embyro. Retinaldelhyde dehydrogenase 2 is not expressed in these regions. In this respect, changes in receptor expression are coupled to a reduction in ligand bioavailability and together these contribute to chondroblast differentiation. Stimulation of chondroblast differentiation through addition of an RAR antagonist to wild-type cultures has confirmed, in part, this model.

References

de Roos K, Sonneveld E, Compaan B, ten Berge D, Durston AJ, van der Saag PT 1999 Expression of retinoic acid 4-hydroxylast (CYP26) during mouse and *Xenopus laevis* embryogenesis. Mech Dev 82:205–211

Downie SA, Newman SA 1994 Morphogenetic differences between fore and hind limb precartilage mesenchyme: relation to mechanisms of skeletal pattern formation. Dev Biol 162:195–208

General discussion II

Cartilaginous condensations and the events taking place within them

Bard: Although most of the work of this symposium rightly focuses on the signals that make mesenchyme form condensations and the subsequent regulatory pathways that lead to bone formation and growth, I don't think that we are paying enough attention to the events taking place within the condensation itself.

There are three obvious stages in the formation of endochondral bones: (1) making the initial condensation, (2) establishing the future pattern of the bone, and (3) the subsequent steps of differentiation, morphogenesis and growth. In the case of the forelimb, for example, the initial proximal–distal set of condensations that form only demarcates the basic pattern of a single humerus, a radius–ulnar doublet, a condensation that will form the carpal bones, and the five pre-digit condensations. It should also be pointed out that some of these condensations merge with proximal and distal ones. The next step is the exact delineation of the bones and this seems to take place within each condensation in an autonomous way. Only after this has been done can the subsequent events of differentiation, morphogenesis and growth take place.

The purpose of this brief comment is to take a simple look at the events taking place within the condensation, the least understood part of bone development. The formation and differentiation of mesenchymal condensations is not, of course, a process limited to the limb; it is the normal mode of development for a wide variety of mesenchyme-based tissues that include muscles, dermal derivatives (e.g. feather and hair papillae), teeth and nephrons, as well as bones. It should however be pointed out that, although condensation formation is a key stage in the differentiation of these tissues, there is not a single case where we understand how bringing cells close together (by loss of matrix, migration, the production of adhesion molecules) changes their co-operative properties and so allows them to set out on a new developmental pathway.

An interesting example is the formation of the nephron: here, signalling from the epithelial duct instructs a small group of metanephric mesenchyme cells to condense (Davies & Bard 1998). Once this has happened, a series of apparently condensation-autonomous events then take place which result in the formation of the nephron. The cells within the condensation undergo a mesenchyme-to-epithelial transition and start to form a tubule that fuses to the duct at one end, and makes a glomerulus at the other, with the tubule as a whole undergoing the extensive differentiation and elongation that marks the nephron. The response is so sophisticated that it is hard to see how the condensation-forming signal can do any more than activate a pathway which is self-controlled and self-sustained and whose molecular underpinnings are only beginning to be understood (Kispert et al 1998).

I wish to suggest that autonomous spatial patterning events of similar complexity take place within the chondrogenic condensation and it is these that specify the map that defines the phenotype of a particular bone. Controlling the downstream cassettes of gene expression that leads to osteogenesis is then as straightforward here as anywhere else; the difficult problem is to work out how patterning takes place within a chondrogenic aggregate.

Investigating the emergence of pattern in limb condensations

It is worth trying, just as an exercise, to consider just how much of the future development of a bone is established at this condensation stage. A good example is the formation in the third digit of the metacarpal and three phalanges with their intervening joints that all arise from a single condensation in the early handplate. Pattern formation here will clearly have to specify the following processes:

- Proximal–distal polarity
- Partitioning of the condensation into four cartilaginous subdivisions with intervening joint regions
- Initiating and co-ordinating the mobilization of the appropriate cassettes of genes
- Specifying the individual features of each bone. These include its pattern of growth (and hence shape) and the location of bone–ligament and bone–tendon adhesions.

Little is known about any of these processes (other than the first, which clearly derives from events taking place in the sub-apical ectodermal ridge progress zone) to the extent that we do not even know whether the condensations of each bone are the same size when they first form.

The obvious approach to elucidating what is going on here is to look for informative mutants. A particularly intriguing example is that of the mouse *brachypody* mutant where the second and third metacarpals fail to form. It is now clear that a key event underlying morphogenesis here is the behaviour of GD5, a member of the transforming growth factor (TGF)β family of signalling proteins: this early marker (and mediator) of joint formation fails to be expressed in that part of the developing condensation which would normally give rise to the missing joint (Storm et al 1994, Storm & Kingsley 1999), although the genetic pathway here is still to be elucidated.

We of course need more such mutants so that their underlying genetic lesion can be elucidated and integrated into developmental pathways. With luck, the mouse mutant screens currently under way at the Jackson Laboratory and at the MRC Unit at Harwell will produce further helpful skeletal abnormalities that illuminate the events taking place within chondrogenic and other condensations. But we also want *in vitro* systems so that we can experimentally manipulate condensation formation and development — there is a great deal to do be done in teasing out what is going on here.

I appreciate that it is always easy to say how little we understand of some developmental event, and how much more work needs to be done. Nevertheless, I believe that the pattern-forming events taking place within chondrogenic and other mesenchymal condensations are so important that they merit all the effort that we can put into working out what is going on in these small aggregates.

Hall: If we were all sitting here writing an exam paper, and the question was, 'List the genes which affect the pattern of a skeleton rather than differentiation of the skeleton', we should be able to come up with a list of genes. But it seems to be difficult to do that: to disentangle those genes from the ones that are involved in differentiation. The retinoids seem to be prime candidates that have been around for quite some time as patterning genes for the skeleton.

Karsenty: Things are very complex, but complexity is made of details. There is nothing wrong in working out complexity molecule by molecule. I think the genetic approach is still the most valid one to ask the most precise and narrow-minded questions to make progress.

Bard: Sometimes it helps to sort of say, 'I would like to be there: how do I get there?' You have to do both.

Kingsley: You mentioned at the end that there must be interesting suggestions from clinical genetics of abnormalities that affect patterning. In large part, although it didn't come from clinical genetics, the way we originally got interested in mouse mutations was by looking at the classical skeletal material that had been described, and saying, of the mutant phenotypes that have been described, which of them look like they might affect the fundamental behaviour of condensation, formation, branching and segmentation? These are the genes we've essentially chosen to study and isolate. Two of the mutants we chose for this reason turned out to be bone morphogenetic proteins. That may be disappointing to you in the sense that it gets right back to the details, but I actually think BMPs are an interesting class of molecules that sit at an interface between patterning and differentiation. They are molecules that can stimulate key events in skeletal formation, but they are deployed in limbs in a specific pattern. I think that they bridge patterning events and differentiation events.

Bard: But what sets the pattern of BMP expression?

Kingsley: That's a challenging problem that we're now working on. But our strategy is to take the molecules that have come from the mutants and attempt to push this back to the next stage, which is that these are the molecules that are setting and affecting the patterns. The key to solving the link between earlier events and patterning is to figure out what is deploying them in the specific patterns in which they turn on. We are currently wrestling with this difficult problem.

An analogy can be drawn here with the stripes in the *Drosophila* embryo. The early fly embryo goes through a beautiful stripy pattern, and there was a phase before many of the mutants had been isolated when the theoreticians had an explanation for how it was possible to generate these beautiful periodic patterns, and many models were actually based on a Turing-type mechanism. The molecular biologists then went in and cloned the genes and bashed away on the promoters that were actually setting up the stripes. When all the dust settled, the answer was that the stripes are made by very inelegant mechanisms. In fact, I remember the title of the *Nature* news and views on this paper was 'Making stripes inelegantly', because instead of generating the stripes by a nice uniform theoretical mechanism, they were built up by the summation of a terrible number of details (Akam 1989). And what controlled stripe three was different from stripe four, and different from stripe five and so on. There was a whole series of different regulatory elements that is a summation of inhibition and repression based on other molecules that were present in gradients in the fly embryo. Although the answer wasn't elegant, the answer was there and it was there in details, and it came from finding the molecules and then doing the hard work starting with the molecules to define how the patterns were being generated.

Bard: I can't get the same feeling in the limb, because there are so many segments and transcription factors before the patterns are set up. We have all the bricks there for a complicated answer.

Wilkins: We've hardly heard about the *Hox* genes in this meeting. These do affect patterning in the limb in fairly specific ways, but we don't know how they do it. There is a huge gap between the expression of the *Hox* genes, which are transcription factors, and the formation of condensations and the ultimate patterns. In the literature there is very little discussion of what fills that gap.

Hall: There is a Hox pattern along the axis that correlates with where the limbs are going to develop, but there's a very big gap between those two patterns.

Wilkins: Somehow the Hox expression patterns translate into the sorts of processes that we want to understand, that set the size of condensations and so on.

Newman: One perspective that is missing from this discussion is the question of evolution. In *Drosophila* there are these seven identical-looking even-skipped stripes that are generated in modern organisms in a genetically ornate fashion, in which each stripe seems to be specified by separate promoters and molecular cues. But it's hard to imagine that the seven-stripe pattern emerged all at once with this cumbersome mechanism. In fact, if you look at those genes with striped expression patterns, such as *even-skipped* and *fushi tarazu*, you find that they encode transcription factors which diffuse in a syncitium. In addition, there are positively autoregulatory promoter elements for each of these genes (Harding et al 1989, Ish-Horowicz et al 1989). Thus, even though there are stripe-specific promoter elements and cues for the *Drosophila* stripes, there is also the remnant of a positively autoregulatory diffusive mechanism, which could have provided the originating pattern and acted as a template for subsequent evolution. Just because you see complexity in a modern organism, it doesn't mean that the originating pattern-forming process was equally complex — it almost certainly wasn't.

Kingsley: If I understand correctly, you are saying that perhaps a billion years ago there was a simple Turing model, and since then organisms have elaborated a series of mechanisms that generate the same pattern that the Turing model generated but now in a much uglier fashion.

Newman: Exactly. For some things, this elaboration might have really taken hold, so that there is barely a remnant of the original mechanism. But in other systems, such as the limb the originating mechanism might still persist. If you take limb cells and put them in culture you get beautiful spot-like patterns: every limb is made up of rods and spots of cartilage. There are no limbs generated in the course of evolution that lack this same general pattern. It seems that natural selection is playing with a basic process that it can fine-tune, elaborate and reinforce, but underlying it all there's a generic mechanism that gives you stripes and spots.

Bard: You can get your rods and stripes in two ways. One is by a Turing-type mechanism; the other is from a unique pattern that will arise by the interplay of diffusion gradients acting on an existing substratum which is non-uniform.

Newman: You can do it both ways, but it is easier to arrive at the second way if you start with the first one.

References

Akam M 1989 *Drosophila* development: making stripes inelegantly. Nature 341:282–283

Davies JA, Bard JBL 1998 The development of the kidney. Curr Topic Dev Biol 39:245–301

Harding K, Hoey T, Warrior R, Levine M 1989 Autoregulatory and gap gene response elements of the even-skipped promoter of *Drosophila*. EMBO J 8:1205–1212

Ish-Horowicz D, Pinchin SM, Ingham PW, Gyurkovics HG 1989 Autocatalytic ftz activation and instability induced by ectopic ftz expression. Cell. 57:223–232

Kispert A, Vainio S, McMahon AP 1998 Wnt-4 is a mesenchymal signal for epithelial transformation of metanephric mesenchyme in the developing kidney. Development 125:4225–4234

Storm EE, Kingsley DM 1999 GDF5 coordinates bone and joint formation during digit development. Dev Biol 209:11–27

Storm EE, Huynh TV, Copeland NG, Jenkins NA, Kingsley DM, Lee SJ 1994 Limb alterations in *brachypodism* mice due to mutations in a new member of the TGFβ-superfamily. Nature 368:639–643

Defects in extracellular matrix structural proteins in the osteochondrodysplasias

Daniel H. Cohn

Medical Genetics, Steven Spielberg Pediatric Research Center, Ahmanson Department of Pediatrics, Cedars-Sinai Medical Center, and Departments of Human Genetics and Pediatrics, UCLA School of Medicine, Los Angeles, CA 90048, USA

Abstract. Mutations in the genes that encode structural proteins of the extracellular matrix affect one or more steps in the diverse set of coordinated events necessary for ordered skeletal development. Depending on the role of the gene product and the severity of the defect, disruption of endochondral ossification and linear growth, the structural integrity and stability of articular cartilage, and/or mineralization can occur. Several themes have emerged from the molecular dissection of these disorders; most of the osteochondrodysplasias that result from defects in structural proteins are inherited in an autosomal dominant fashion; a spectrum of related clinical phenotypes can be produced by distinct mutations in the same gene; haploinsufficiency for the gene product usually produces a milder clinical phenotype than do mutations resulting in synthesis of structurally abnormal proteins. For structural defects, a dominant-negative effect resulting from presence of the abnormal protein in the matrix appears to be the primary determinant of phenotype. Secondary effects on extracellular matrix protein structure can result from defects in post-translational maturation, including hydroxylation, sulfation and proteolytic cleavage, and produce distinct osteochondrodysplasias. Overall, the inherited disorders of skeletogenesis have revealed the exquisite sensitivity of the architecture of the extracellular matrix to the quantity and quality of matrix molecules.

2001 The molecular basis of skeletogenesis. Wiley, Chichester (Novartis Foundation Symposium 232) p 195–212

The last decade has witnessed an explosion in our understanding of the molecular basis of human osteochondrodysplasias (Mundlos & Olsen 1997a,b). The identities of the osteochondrodysplasia disease genes reflect the diverse set of coordinated events necessary for ordered skeletal development. Depending on the specific gene product involved, the molecular defects disrupt one or more steps in skeletogenesis, including (a) effects on mesenchymal condensation and establishment of the size and shape of the cartilaginous skeletal primordia; (b) disarray in the growth plate with consequent defects in endochondral ossification and linear growth; (c) degradation of the structural integrity and stability of articular cartilage leading to degenerative processes in the joints; and (d) disrupted mineralization and bone (re)modelling.

Many of the osteochondrodysplasia disease genes encode structural proteins of the extracellular matrix. Despite the expression of these proteins early in skeletal development, establishment of the cartilage anlagen in most of these disorders proceeds normally. More commonly, defects in the structural proteins of the matrix disrupt the ordered process of differentiation, proliferation, hypertrophy and mineralization at the growth plate. These alterations lead to disordered linear growth and often produce deformity in addition to short stature. For articular cartilage, a structurally altered cartilage extracellular matrix is inherently unstable and is subject to degradation, leading to early onset degenerative joint disease and osteoarthritis. This is the major cause of morbidity in these conditions and is particularly evident in the main weight bearing joints, the hips and knees. Not surprisingly, the need for joint replacement is a common consequence in many osteochondrodysplasias.

The most recent classification of this clinically and genetically heterogeneous group of disorders recognizes over 150 distinct osteochondrodysplasia phenotypes (International Working Group on the Constitutional Diseases of Bone 1998). The classification is primarily radiographic, with the different disorders divided on the basis of the specific skeletal elements that are affected and the severity of the defect. The characterization of the disease genes in many of these conditions is leading to an emerging molecular classification, separate but parallel to the clinical and radiographic nosology, that is useful in defining the conditions in biomolecular terms.

Primary defects in extracellular matrix structural proteins

Molecular analysis of the osteochondrodysplasias has revealed a number of themes for the mechanisms and consequences of defects in extracellular matrix structural proteins. Rather than providing a cursory survey of each gene and disorder, several examples will be presented that illustrate the major themes and attempt to integrate what has been learned from the analysis of the distinct disorders into a coherent whole.

The type II collagenopathies — a spectrum of osteochondrodysplasia phenotypes

Type II collagen, the homotrimeric product of the *COL2A1* gene, is distributed in cartilage, the nucleus pulposus of the vertebral bodies, and the vitreous of the eye. This distribution suggested the hypothesis that clinical phenotypes with abnormalities in these tissues could result from mutations in the *COL2A1* gene.

TABLE 1 Primary structural protein defects in the human osteochondrodyplasias

		Phenotype	
Protein	*Gene*	*Haploinsufficiency*	*Structural mutation*
Type I collagen	COL1A1	OI type I	OI (all types) EDS (arthrochalasis type)
	COL1A2		OI (all types) EDS (arthrochalasis type)
Type II collagen	COL2A1	Stickler syndrome	SED SEMD Strudwick Kniest dysplasia Hypochondrogenesis Achondrogenesis type II
Type X collagen	COL10A1	Schmid metaphyseal dysplasia	
Type XI collagen	COL11A1		Stickler syndrome
	COL11A2	*	Stickler syndrome
Type IX collagen	COL9A1		
	COL9A2		MED
	COL9A3		MED
COMP	COMP		Pseudoachondroplasia MED

*Homozysosity for COL11A2 null or structural mutations produces OSMED.
OI, osteogenesis imperfecta; EDS, SED, spondyloepiphyseal dysplasia; SEMD, spondyloepimetaphyseal dysplasia.

The hypothesis was first proved in a family with spondyloepiphyseal dysplasia congenita, with the identification of a small in-frame deletion in the *COL2A1* gene (Lee et al 1989). Subsequent studies in patients with this and related phenotypes featuring the combination of radiographic abnormalities in the spine and epiphyses along with abnormalities of the eyes ranging from myopia to vitreous liquefaction to retinal detachment, established that a spectrum of clinical conditions across a broad range of severity could all result from dominant *COL2A1* mutations (Spranger et al 1994). The clinical spectrum of the type II collagenopathies (term coined by Francesco Ramirez) now includes Stickler syndrome, spondyloepiphyseal dysplasia congenita, the Strudwick form of spondyloepimetaphyseal dysplasia, Kniest dysplasia, hypochondrogenesis, and achondrogenesis type II. The definition of the type II collagen disorders validated the concept of bone dysplasia families, groups of conditions unified by distinct mutations in the same gene, originally suggested by Spranger (Spranger 1989).

Despite the large number of *COL2A1* mutations that have been characterized, there is surprisingly little correlation between genotype and phenotype. The most consistent finding is that a phenotype at the mild end of the type II collagenopathy spectrum, Stickler syndrome, usually results from *COL2A1* null mutations, primarily frameshift or premature termination codon mutations (Snead & Yates 1999). In the few bone dysplasia families in which both null mutations and structural defects in the same gene have been described, it is consistently observed that null alleles produce phenotypes at the milder end of the spectrum (Table 1).

All other known phenotypes within the type II collagen disorders result from structural alteration of the molecule. Most of the mutations are point mutations that lead to single amino acid substitutions for glycine codons within the greater than 1000-residue triple helical domain of the molecule. This reflects both the essential requirement of glycine in every third position for proper triple helix assembly and the large mutational target presented by the first two nucleotides of each glycine codon. Exon skipping mutations or other in-frame deletions also lead to phenotypes within this spectrum. Kniest dysplasia is the only phenotype within the group for which the nature and location of the mutations cluster. In most cases, small deletions toward the N-terminal end of the triple helical domain are seen (Wilkin et al 1999). From a mechanistic viewpoint, there is evidence that the normal chains loop out in the region of the deletion in heterotrimeric molecules containing both normal and abnormal chains (Weis et al 1998). For the remaining disorders, no such correlation has been seen. For example, in a recent series of twelve consecutive cases of achondrogenesis type II/hypochondrogenesis, mutations of a variety of types distributed throughout the triple helix were characterized. The data serve to illustrate how little we really understand about the functional domains within the type II collagen molecule and suggest that approaches other than mutation analysis will be required to understand how the mutations exert their effect on phenotype.

At the cellular level, type II procollagen molecules containing a chain with a structural defect (7/8 of the molecules in an individual heterozygous for a *COL2A1* mutation), are poorly secreted and a proportion of them accumulate in the rough endoplasmic reticulum (RER) (Fig. 1). This implies a quantitative component to the effect of mutation on phenotype that results from the secretion defect. For the lethal forms within this spectrum, biochemical analysis of cartilage has shown a decrease in the amount of type II collagen in the matrix and an increased amount of type I collagen. There are few data describing the effect of retention of the type II procollagen in the RER on chondrocyte metabolism, but this remains as a possible component of phenotypic expression.

Not all of the abnormal type II procollagen molecules are retained within the chondrocyte, as type II collagen molecules containing chains derived from the

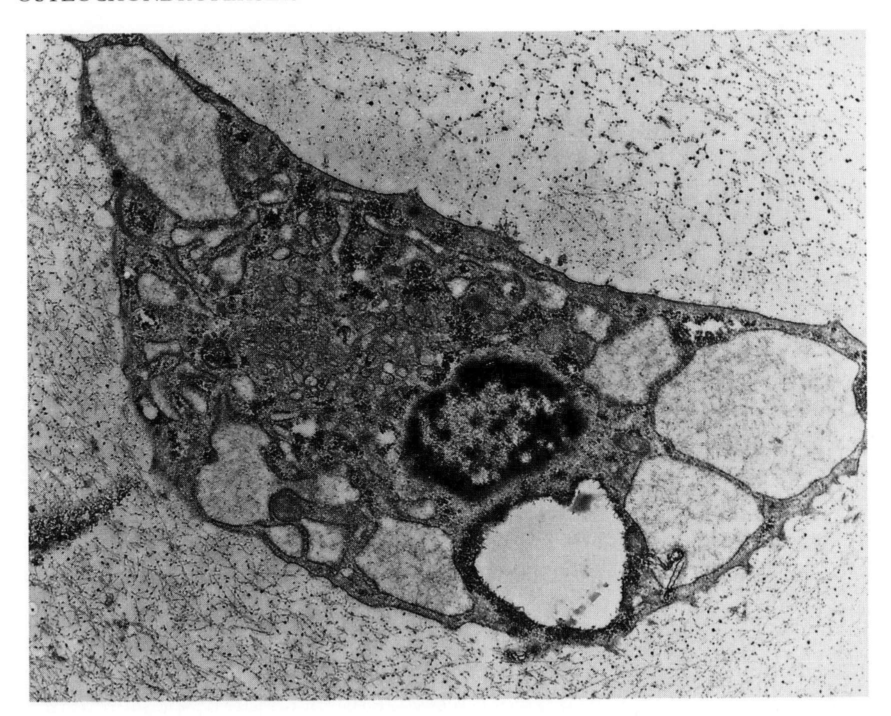

FIG. 1. Cartilage defects in the type II collagen disorders. Retention of type II collagen in the RER in a case (R97-177B) of hypochondrogenesis.

mutant allele are found in the cartilage extracellular matrix. In the matrix, the abnormal molecules are incorporated and cross-linked into the functional unit, the collagen fibril, which also contains type XI and type IX collagen molecules (Fig. 2). The multimeric nature of the cartilage collagen fibril provides the opportunity for dramatic magnification of the dominant negative effect of abnormal molecules on normal molecules in the matrix, and suggests that even a small proportion of abnormal molecules may be sufficient to disrupt the function and stability of the cartilage. In addition, as the ratios among the cartilage collagens can regulate collagen fibril diameter, altered collagen fibril structure is a common observation within this group of conditions (Fig. 3B).

Finally, as these disorders have primary effects on growth, including short stature and deformity, effects at the level of the growth plate are consistently seen. Linear growth is driven by the extent of chondrocyte proliferation (number of chondrocytes) and by their expansion in size during hypertrophy. Directional growth relies on the regular columnar arrangement of growth plate chondrocytes. In the growth plates of patients within the type II collagen disorders, both shortened and irregular columns of chondrocytes are observed. Thus the nature of the matrix elaborated by the growing chondrocyte can have an effect on both proliferation and the orientation of growth.

Defects in minor cartilage collagens produce distinct osteochondrodysplasias

Mutations in the genes that encode the minor cartilage collagens, types IX, X, and XI, also produce osteochondrodysplasia phenotypes (Table 1). Mutations in the *COL10A1* gene, which is expressed exclusively by growth plate chondrocytes, result in the relatively mild Schmid type of metaphyseal chondrodysplasia (Warman et al 1993). The mutations appear to result in haploinsufficiency for the protein (Chan et al 1999), and lead to a short growth plate. Structural mutations in type X collagen have not been described in patients. However, an engineered structural mutation in *COL10A1* in the mouse produces a moderately severe skeletal phenotype (Jacenko et al 1993), suggesting that we have yet to recognize the clinical consequences of such mutations in humans.

Type XI collagen is the heterotrimeric product of the *COL2A1*, *COL11A1* and *COL11A2* genes. Along with type II collagen, it is incorporated within the main collagenous fibril of the cartilage extracellular matrix. Heterozygosity for structural mutations in *COL11A1* or *COL11A2* produces forms of Stickler syndrome (Vikkula et al 1995, Richards et al 1996). Homozygosity for such defects provides a unique example of a recessive osteochondrodysplasia resulting from a structural mutation in an extracellular matrix structural protein, otospondylomegaepiphyseal dysplasia (OSMED). Although similar in their effects on skeletogenesis, patients with mutations in the *COL11A2* gene, which is not expressed in the vitreous, do not have eye abnormalities, while mild vitreous anomalies (less severe than the type II collagenopathies) are present in patients with *COL11A1* mutations (Snead & Yates 1999).

Type IX collagen (Olsen 1997) is cross-linked to the surface of the type II/XI collagen fibril, and may play a role in regulating fibril diameter. The N-terminal end of the molecule, which is encoded by *COL9A1*, *COL9A2* and *COL9A3*, projects out into the perifibrillar space (Fig. 2) and is postulated to mediate interactions between the fibril and other matrix molecules. Mutations in *COL9A2* and *COL9A3* have been identified in patients with multiple epiphyseal dysplasia (MED) (Muragaki et al 1996, Paassilta et al 1999). This relatively mild osteochondrodysplasia is characterized by mild short stature, a variable degree of brachydactyly, and epiphyseal abnormalities. The spine is usually unaffected. The phenotype usually comes to attention in mid-childhood, primarily because of joint pain. Joint replacement in the third or fourth decade is the most common complication.

The identified mutations all resulted in skipping of exon 3 of *COL9A2* or *COL9A3*, yielding a protein with a 12 amino acid deletion within the amino-

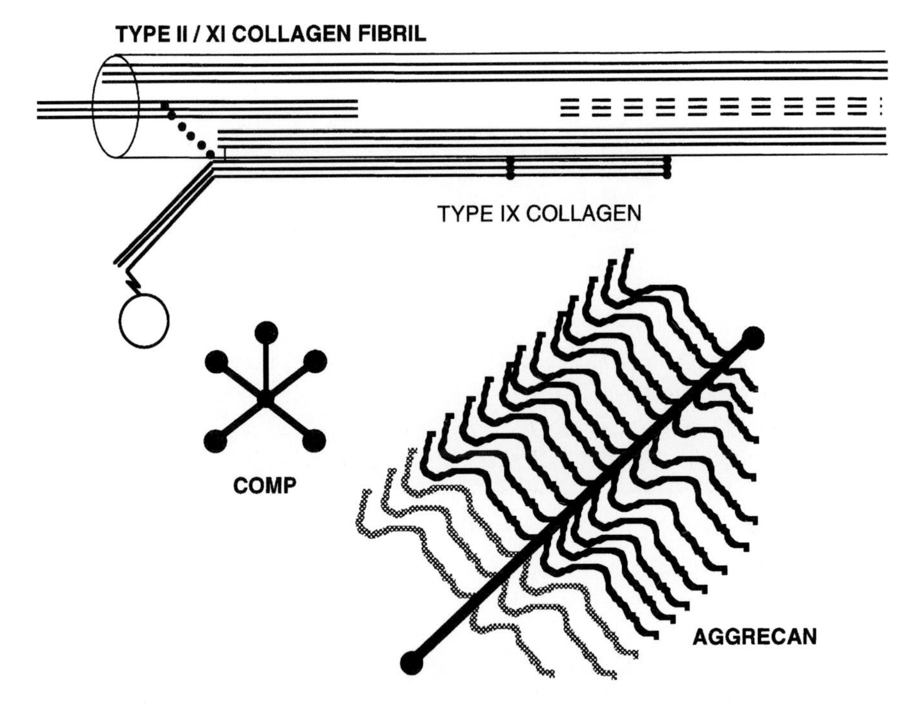

FIG. 2. The heterotypic cartilage collagen fibril. The fibril is assembled with collagen types II and XI within the fibril and type IX collagen cross-linked to the surface.

terminal collagenous domain, a portion of the molecule that extends away from the collagen fibril. The abnormal protein is not retained in the RER (van Mourik et al 1998) and the amount of type IX collagen in the matrix appears to be normal, suggesting that the abnormal molecules are incorporated into the matrix. The hypothesis has thus been raised that the MED phenotype results from disruption of an interaction between type IX collagen and one or more non-collagenous proteins of the matrix, interactions that are critical for matrix stability.

Cartilage oligomeric matrix protein — a link between the collagen fibril and non-collagenous proteins of the matrix?

A prime candidate for a protein that interacts with type IX collagen is cartilage oligomeric matrix protein (COMP) a homopentameric glycoprotein in the thrombospondin protein family (Hedbom et al 1992). Like the type IX procollagen genes, mutations in *COMP* can produce MED (Briggs et al 1995), suggesting the hypothesis that the two proteins interact and that functionally compromising either interacting molecule yields a similar phenotypic outcome.

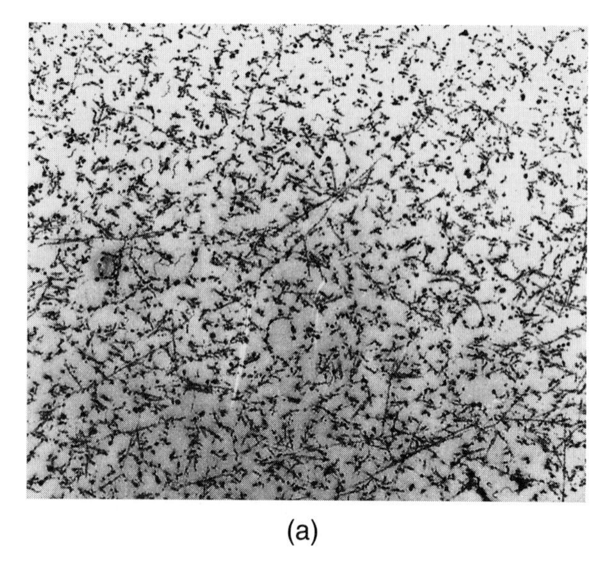

(a)

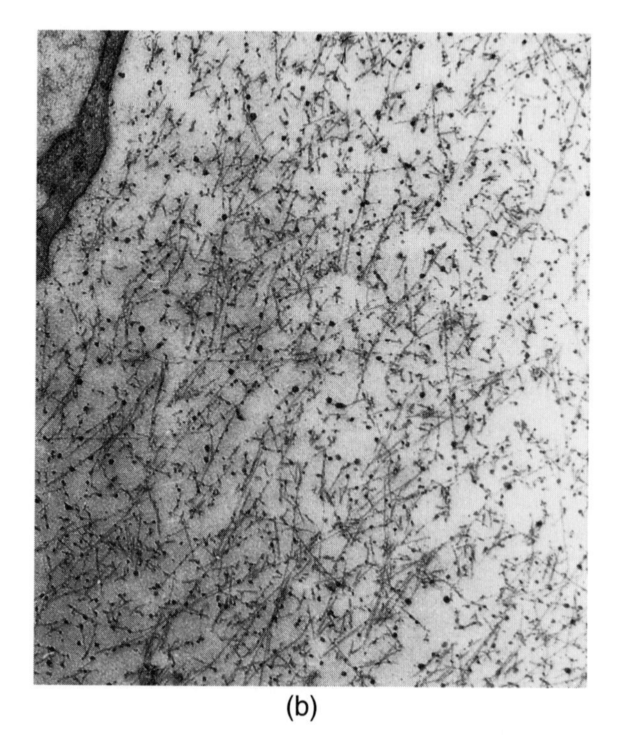

(b)

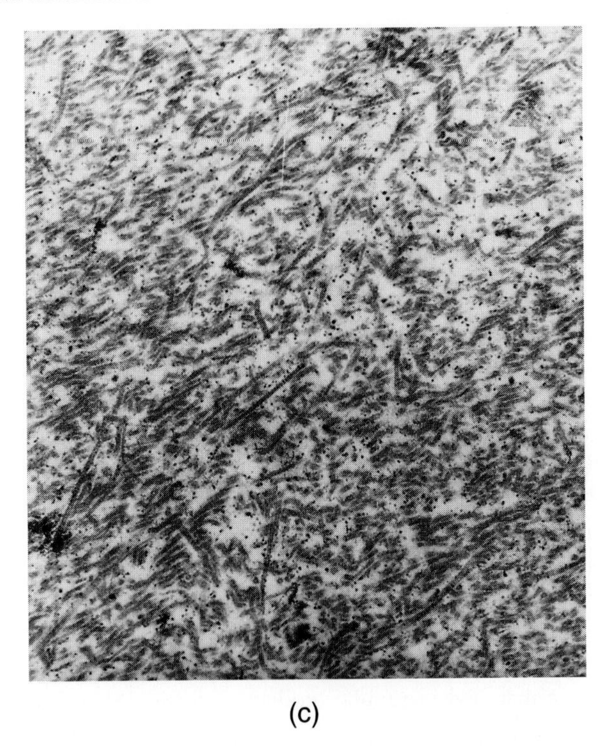

(c)

FIG. 3. Cartilage extracellular matrix in (a) a control, (b) a case of hypochondrogenesis (R97-177B) and (c) a case of pseudoachondroplasia (R68-029A). Collagen fibrils are sparser and thinner in the type II collagen disorders (b), while in the COMP disorders they are denser and thicker (c).

COMP mutations can also produce pseudoachondroplasia (PSACH) (Briggs et al 1995, Hecht et al 1995), a more severe disorder with markedly reduced stature and increased deformity. Radiographs demonstrate that pseudoachondroplasia is a severe spondyloepimetaphyseal dysplasia, with marked abnormalities at the spine, hips, knees and hands. *COMP* is also expressed in tendon and ligament, perhaps explaining the loose joints that also characterize this disorder.

Most of the mutations that produce either PSACH or MED fall within the calcium binding domain of COMP (Briggs et al 1998). As with the type II collagen disorders, a clear understanding of the relationship between genotype and phenotype has not emerged from mutation analysis. Speculation has centred on the effect of reduced calcium binding on the structure of the COMP monomer, with the degree of disruption of the structure determining the phenotype produced. Testing this hypothesis will require structural studies of the various mutant proteins.

COMP pentamers are cotranslationally assembled in the RER. Chondrocytes from PSACH and MED patients with *COMP* mutations show retention of COMP in this compartment (Maynard et al 1972), attesting to a form of quality control likely related to monitoring the folded structure of COMP. Interestingly, the RER inclusions stain with antibodies to COMP as well as antibodies to type IX collagen (Maddox et al 1997, Délot et al 1998) and the cartilage-specific proteoglycan aggrecan (Stanescu et al 1984). These data support the hypothesis that COMP can bind these two molecules, and that their retention in the RER is secondary to the retention of COMP.

Cultured tendon and ligament cells from patients with *COMP* mutations show that synthesis and secretion of COMP is difficult to distinguish from control cells, implying that abnormal COMP molecules contribute to the pool of extracellular matrix molecules and exert their phenotypic effect through a dominant-negative mechanism (Délot et al 1998). Examination of the cells by electron microscopy has shown that the cultured cells have RER inclusions similar to those in cartilage, but only after a considerable time in culture. This observation suggests that tissue-specific mechanisms of disease are at least superficially similar among tissues that express *COMP*.

COMP is expressed throughout the cartilage, codistributed with type II collagen. At the tissue level, cartilage from patients with *COMP* mutations (Fig. 3C) show thickened, irregular collagen fibrils (Stöss & Pesch 1985), pointing to a role in regulating the structure of the fibril, perhaps compatible with an interaction mediated by type IX collagen. The major morbidity associated with these conditions is degenerative joint disease, particularly at the hip, reflecting the consequences of an altered cartilage matrix structure. The alteration in structure is also evident at the growth plate, with shortened and irregular growth plate chondrocyte columnization (Sillence et al 1979).

Finally, bovine chondrocytes have been shown to bind COMP (DiCesare et al 1994). By analogy with interactions between thrombospondin 1 and cells that express this molecule, the interaction may be mediated through a sequence in the carboxyl-terminal domain of COMP similar to a thrombospondin 1 sequence that binds an integrin-associated protein. The intriguing hypothesis suggested by these data, that COMP may mediate signalling between the matrix and the chondrocyte, could then constitute one of the regulatory signals used by chondrocytes to monitor and react to their environment.

Summarizing to this point, a molecular understanding of the consequences of abnormalities in structural proteins of the cartilage extracellular matrix has yielded the following insights and hypotheses. First, a spectrum of conditions, from mild to lethal, can result from different types of mutations in a single gene product. Second, structural abnormalities generally yield a more severe phenotypic effect than haploinsufficiency. Third, the structurally abnormal proteins are poorly secreted by chondrocytes and accumulate in the RER, perhaps contributing to phenotypic expression. Fourth, abnormal molecules do contribute to the pool of matrix proteins and likely exert a dominant negative effect on matrix function, particularly at the level of the collagen fibril. Fifth, specific intermolecular interactions among matrix molecules may be compromised by structural abnormalities in the interacting partners. Sixth, the abnormal structure of the matrix can affect both the ordered developmental sequence of chondrocytes at the growth plate and the stability of articular cartilage.

Secondary defects in extracellular matrix protein structure—defects in post-translational maturation

The structural proteins of the extracellular matrix undergo a variety of post-translational modifications, including hydroxylation, glycosylation, sulfation and proteolysis. These steps serve to mature the proteins and confer on them their functional properties. Skeletal abnormalities resulting from defects in the hydroxylation of the collagens are seen in the ocular form of Ehlers-Danlos syndrome (EDS type VI) and defects in proteolytic processing of the collagens have been described in the arthrochalasis and dermatosparaxis forms of EDS (EDS types VIIA, VIIB and VIIC) (Byers 1996). The proteoglycans that make up the bulk of the non-collagenous proteins of cartilage are heavily modified with sulfate-containing sugars, so of particular interest in the biology of cartilage are the steps that lead to post-translational sulfation of these molecules.

Sulfation pathway defects in the osteochondrodysplasias

In cartilage, the first step in the sulfation pathway (Fig. 4) involves transport of sulfate into the chondrocyte, primarily mediated by the diastrophic dysplasia sulfate transporter (DTDST). The essential role of sulfation in skeletogenesis was established by the identification of DTDST as the diastrophic dysplasia (DTD) disease gene (Hästbacka et al 1994). Diastrophic dysplasia is an autosomal recessive, severe osteochondrodysplasia characterized by markedly short, deformed limbs, kyphoscoliosis, significant brachydactyly with a characteristic hitchhiker thumb, and club feet. Based on similarities in radiographic presentation and cartilage histomorphology, identification of the DTDST disease gene allowed confirmation of the hypothesis that the lethal disorders atelosteogenesis type II (Hästbacka et al 1996) and achondrogenesis type IB (Superti-Furga et al 1996) also resulted from mutations in the diastrophic dysplasia disease gene. More recently, it has been shown that a recessive form of MED can result from homozygosity for a DTDST mutation (Superti-Furga et al 1999). Thus, as with some of the dominant osteochondrodysplasias, a range of

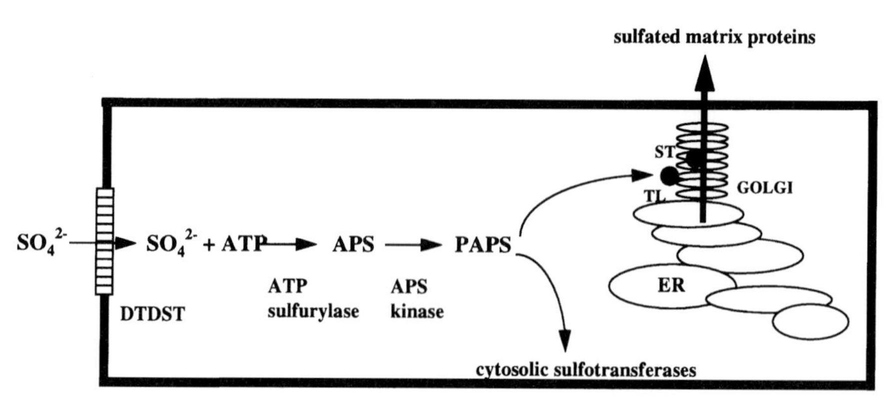

FIG. 4. The sulfation pathway.

phenotypes, from mild to lethal, can result from different mutations, or in this case combinations of mutations, in the same gene. For the DTDST disorders, there is a direct relationship between the severity of the transport deficit and phenotype, with the most severe phenotype produced by homozygosity for DTDST null alleles. Carriers of DTDST mutations do not show skeletal abnormalities, implying that 50% of the normal level of sulfate transport is compatible with normal skeletal development.

The second step in the sulfation pathway, activation of inorganic sulfate to the useable form, PAPS (peroxidase–antiperoxidase), is carried out in two steps: sulfation of ATP to produce APS (adenosine-5′-phosphosulfate) followed by phosphorylation of APS to produce PAPS (Geller et al 1987). In higher animals, there are currently two known PAPS synthetase genes, PAPSS1 and PAPSS2, each a bifunctional gene that encodes both enzyme activities (Li et al 1995, Kurima et al 1998, ul Haque et al 1998). PAPS can be directly used by cytosolic sulfotransferases. For secreted proteins, sulfation occurs in the Golgi. Translocation of PAPS to the Golgi is mediated by one or more PAPS translocases, the identities of which have yet to be firmly established. Within the Golgi, a variety of substrate-specific sulfotransferases utilize PAPS to transfer sulfate to specific positions on specific sugar molecules.

Reduced sulfate activation, due to homozygosity for a null mutation in *PAPSS2*, produces a recessive form of spondyloepimetaphyseal dysplasia, SEMD Pakistani type (ul Haque et al 1998). The phenotype consists of short stature with epiphyseal and mild metaphyseal irregularity, platyspondyly, and mild brachydactyly (Ahmad et al 1998) and is intermediate in severity between two of the DTDST disorders, MED and DTD. The relatively mild nature of the phenotype, at least as compared with DTD, likely results from PAPSS1-derived PAPS synthesis in cartilage. It can be predicted that milder PAPSS2 mutant alleles could produce a distinct osteochondrodysplasia, perhaps similar to MED, as might defects in PAPSS1. This possibility, together with the observation that less than half of MED cases are accounted for by the known loci, suggest that MED may be the most genetically heterogeneous of the osteochondrodysplasias.

At the tissue level, as observed by alcian blue staining, the DTDST disorders exhibit a global decrease in the level of sulfate in cartilage. A pericellular ring of condensed collagen fibrils surrounds the chondrocytes, suggesting that sulfate groups on the sugar side-chains of the proteoglycans serve in part to define the architecture of the matrix by preventing collagen aggregation. Cartilage has not been available from patients with the milder sulfation disorders, but analysis of cartilage from the brachymorphic mouse, which has deficient PAPSS2-derived sulfate activation (Kurima et al 1998, ul Haque et al 1998), shows reduced proteoglycan granule size and an increased density of collagen fibrils (Orkin et al 1977), also compatible with a role for sulfated proteoglycans in separating collagen fibrils.

It has been suggested that the phenotypes of the sulfation disorders may be in part determined by decreased response to growth factors resulting from undersulfated heparan sulfate proteoglycans, which are required for growth factor presentation to their receptors. This hypothesis is supported by the skeletal phenotype of a mouse in which the *heparan sulfate 2-sulfotransferase* gene has been knocked out (Bullock et al 1998). In this mouse, the long bones exhibit proximal shortening and widening at the growth plate, and the ribs are short, widened and prematurely fused, possibly reflecting increased skeletal maturation. Also compatible with this idea is the observation that growth plate chondrocytes from the brachymorphic mouse are less numerous and smaller than normal chondrocytes, indicating that decreased sulfation affects both proliferation and differentiation of these cells.

Conclusions

The architecture of the cartilage extracellular matrix is exquisitely sensitive to the relative abundances of matrix molecules, the primary structures of the proteins, and their state of post-translational maturation. Abnormalities in the structural proteins have distinct clinical outcomes, depending on the specific role of each protein in skeletogenesis. However, common to most of these conditions is alteration of chondrocyte alignment and proliferation at the growth plate, leading to short stature and deformity. For articular cartilage, disrupting the structure and intermolecular interactions of cartilage extracellular matrix structural proteins compromises the ability of the joint surfaces to withstand compressive loads, leading to degradation of the matrix and osteoarthrosis.

The molecular basis for many osteochondrodysplasias remains undiscovered. In addition, defects in some known cartilage extracellular matrix structural proteins have yet to be associated with an inherited disease phenotype. For some of these, skeletal phenotypes in the mouse have been described (Li & Olsen 1997). Thus one current challenge is identifying the appropriate human cognate, a process that may be obscured by interspecific biological differences. It is also likely that novel structural proteins important for skeletogenesis will be identified as a more complete set of genes expressed in skeletal tissues is found by genome-based approaches. As revealed by analysis of the sulfation pathway disorders, description of the molecular basis of some osteochondrodysplasias may identify an unrecognized role for known biochemical pathways that are of particular importance in skeletal development and homeostasis.

From a practical viewpoint, recent identification of the molecular basis of many osteochondrodysplasias has greatly enhanced our ability to provide diagnostic studies to families. The remaining challenge is to translate these findings to therapy. For the defects in the genes that encode the structural proteins, enhanced gene expression of the normal allele is an obvious strategy for those phenotypes that result from haploinsufficiency for the gene product. For the structural defects, repopulation of the articular cartilage with chondrocytes that synthesize a normal matrix may be the best approach. For the disorders of post-translational modification, a deeper understanding of each biochemical pathway may suggest ways to detour the cellular machinery around the defect. For some pathways, including sulfation, even small increases in the level of modification may greatly improve cartilage function.

Acknowledgements

I thank my many colleagues and collaborators, particularly those within my laboratory, upon whose work this article is based. I appreciate the critical review of this manuscript by D. Krakow and W.R. Wilcox. I also gratefully acknowledge the enthusiastic participation of osteochondrodysplasia patients and their families in our research projects. Work in the laboratory is supported by grants from the National Institutes of Health.

References

Ahmad M, Haque MF, Ahmad W et al 1998 Distinct, autosomal recessive form of spondyloepimetaphyseal dysplasia segregating in an inbred Pakistani kindred. Am J Med Genet 78:468–473

Briggs MD, Hoffman SM, King LM et al 1995 Pseudoachondroplasia and multiple epiphyseal dysplasia due to mutations in the cartilage oligomeric matrix protein gene. Nat Genet 10: 330–336

Briggs MD, Mortier GR, Cole WG et al 1998 Diverse mutations in the gene for cartilage oligomeric matrix protein in the pseudoachondroplasia-multiple epiphyseal dysplasia disease spectrum. Am J Hum Genet 1998 62:311–319

Bullock SL, Fletcher JM, Beddington RS, Wilson VA 1998 Renal agenesis in mice homozygous for a gene trap mutation in the gene encoding heparan sulfate 2-sulfotransferase. Genes Dev 12:1894–1906

Byers PH 1996 Ehlers–Danlos syndrome. In: Rimoin DL, Connor JM, Pyeritz RE (eds) Principles and practice of medical genetics. Churchill Livingstone, New York, p 1067–1082

Chan D, Freddi S, Weng YM, Bateman JF 1999 Interaction of collagen α1(X) containing engineered NC1 mutations with normal α1(X) *in vitro*. Implications for the molecular basis of schmid metaphyseal chondrodysplasia. J Biol Chem 274:13091–13097

Délot E, Brodie SG, King LM, Wilcox WR, Cohn DH 1998 Physiological and pathological secretion of cartilage oligomeric matrix protein by cells in culture. J Biol Chem 273:26692–26697

DiCesare PE, Morgelin M, Mann K, Paulsson M 1994 Cartilage oligomeric matrix protein and thrombospondin 1. Purification from articular cartilage, electron microscopic structure and chondrocyte binding. Eur J Biochem 223:927–937

Geller DH, Henry JG, Belch J, Schwartz NB 1987 Co-purification and characterization of ATP-sulfurylase and adenosine-5′-phosphosulfate kinase from rat chondrosarcoma. J Biol Chem 262:7374–7382.

Hästbacka J, de la Chapelle A, Mahtani MM et al 1994 The diastrophic dyslplasia gene encodes a novel sulfate transporter: positional cloning by fine-structure linkage disequilibrium mapping. Cell 78:1073–1087

Hästbacka J, Superti-Furga A, Wilcox WR, Rimoin DL, Cohn DH, Lander ES 1996 Atelosteogenesis type II is caused by mutations in the diastrophic dysplasia sulfate-transporter gene (DTDST): evidence for a phenotypic series involving three chondrodysplasias. Am J Hum Genet 58:255–262

Hecht JT, Nelson LD, Crowder E et al 1995 Mutations in exon 17B of cartilage oligomeric matrix protein (COMP) cause pseudoachondroplasia. Nat Genet 10:325–329

Hedbom E, Antonsson P, Hjerpe A et al 1992 Cartilage matrix proteins. An acidic oligomeric protein (COMP) detected only in cartilage. J Biol Chem 267:6132–6136

International Working Group on Constitutional Diseases of Bone 1998 International nomenclature and classification of the osteochondrodysplasias (1997). Am J Med Genet 79:376–382

Jacenko O, LuValle PA, Olsen BR 1993 Spondylometaphyseal dysplasia in mice carrying a dominant negative mutation in a matrix protein specific for cartilage-to-bone transition. Nature 365:56–61

Kurima K, Warman ML, Krishnan S et al 1998 A member of a family of sulfate-activating enzymes causes murine brachymorphism. Proc Natl Acad Sci USA 95:8681–8685

Lee B, Vissing H, Ramirez F, Rogers D, Rimoin D 1989 Identification of the molecular defect in a family with spondyloepiphyseal dysplasia. Science 244:978–980

Li H, Deyrup A, Mensch JR Jr, Domowicz M, Konstantinidis AK, Schwartz NB 1995 The isolation and characterization of cDNA encoding the mouse bifunctional ATP sulfurylase-adenosine 5′-phosphosulfate kinase. J Biol Chem 270:29453–29459

Li Y, Olsen BR 1997 Murine models of human genetic skeletal disorders. Matrix Biol 16:49–52

Maddox BK, Keene DR, Sakai LY et al 1997 The fate of cartilage oligomeric matrix protein is determined by the cell type in the case of a novel mutation in pseudoachondroplasia. J Biol Chem 272:30993–30997

Maynard JA, Cooper RR, Ponseti IV 1972 A unique rough surfaced endoplasmic reticulum inclusion in pseudoachondroplasia. Lab Invest 26:40–44

Mundlos S, Olsen BR 1997 Heritable diseases of the skeleton. Part I: molecular insights into skeletal development-transcription factors and signaling pathways. FASEB J 11:125–132

Mundlos S, Olsen BR 1997 Heritable diseases of the skeleton. Part II: molecular insights into skeletal development-matrix components and their homeostasis. FASEB J 11:227–233

Muragaki Y, Mariman EC, van Beersum SE et al 1996 A mutation in the gene encoding the α2 chain of the fibril-associated collagen IX, COL9A2, causes multiple epiphyseal dysplasia (EDM2). Nat Genet 12:103–105
Olsen BR 1997 Collagen IX. Int J Biochem Cell Biol 29:555–558
Orkin RW, Williams BR, Cranley RE, Poppke DC, Brown KS 1977 Defects in the cartilaginous growth plates of brachymorphic mice. J Cell Biol 73:287–299
Paassilta P, Lohiniva J, Annunen S et al 1999 *COL9A3*: a third locus for multiple epiphyseal dysplasia. Am J Hum Genet 64:1036–1044
Richards AJ, Yates JR, Williams R et al 1996 A family with Stickler syndrome type 2 has a mutation in the COL11A1 gene resulting in the substitution of glycine 97 by valine in α1 (XI) collagen. Hum Mol Genet 5:1339–1343
Sillence DO, Horton WA, Rimoin DL 1979 Morphologic studies in the skeletal dysplasias. Am J Pathol 96:813–870
Snead MP, Yates JR 1999 Clinical and molecular genetics of Stickler syndrome. J Med Genet 36:353–359
Spranger J 1989 Radiologic nosology of bone dysplasias. Am J Med Genet 34:96–104
Spranger J, Winterpacht A, Zabel B 1994 The type II collagenopathies: a spectrum of chondrodysplasias. Eur J Pediatr 153:56–65
Stanescu V, Stanescu R, Maroteaux P 1984 Pathogenic mechanisms in osteochondrodysplasias. J Bone Joint Surg (Am) 66:817–836
Stöss H, Pesch HJ 1985 Structural changes of collagen fibrils in skeletal dysplasias. Ultrastructural findings in the iliac crest. Virchows Arch A Pathol Anat Histopathol 405:341–364
Superti-Furga A, Hästbacka J, Wilcox WR et al 1996 Achondrogenesis type IB is caused by mutations in the diastrophic dysplasia sulphate transporter gene. Nat Genet 12:100–102
Superti-Furga A, Neumann L, Riebel T et al 1999 Recessively inherited multiple epiphyseal dysplasia with normal stature, club foot, and double layered patella caused by a DTDST mutation. J Med Genet 36:621–624
ul Haque MF, King LM, Krakow D et al 1998 Mutations in orthologous genes in human spondyloepimetaphyseal dysplasia and the brachymorphic mouse. Nat Genet 20:157–162
Vikkula M, Mariman EC, Lui VC et al 1995 Autosomal dominant and recessive osteochondrodysplasias associated with the COL11A2 locus. Cell 80:431–437
van Mourik JB, Buma P, Wilcox WR 1998 Electron microscopical study in multiple epiphyseal dysplasia type II. Ultrastruct Pathol 22:249–251
Warman ML, Abbott M, Apte SS et al 1993 A type X collagen mutation causes Schmid metaphyseal chondrodysplasia. Nat Genet 5:79–82
Weis MA, Wilkin DJ, Kim HJ et al 1998 Structurally abnormal type II collagen in a severe form of Kniest dysplasia caused by an exon 24 skipping mutation. J Biol Chem 273:4761–4768
Wilkin DJ, Artz AS, South S et al 1999 Small deletions in the type II collagen triple helix produce Kniest dysplasia. Am J Med Genet 85:105–112

DISCUSSION

Kingsley: You mentioned that achondrogenesis 1B is the clinical name for a condition that results from homozygosity for null mutations in the sulfate transport. What does that look like? Is there any cartilage formed?

D. Cohn: These mice are extremely small, and they die because their chest doesn't grow properly. The cartilage looks very similar to that in diastrophic dysplasia: histologically we see rings of collapsed collagen around the chondrocytes. The growth plates are extremely disorganized, there are zones of degenerating cartilage, and the chondrocytes have very little ability to align at the growth plate. Having said that, if you measure the level of sulfation of aggrecan, for example, in tissue, there is still a level of sulfate on aggrecan that remains. So there are other ways—perhaps only salvage pathways—of getting sulfate in. Transport of sulfated amino acids may be up-regulated, or there could be secondary accessory pathways for the transport of sulfate into those cells. We don't know the precise level of the defect in post-translational sulfation that produces achondrogenesis 1B: is it that the level of sulfation of the extracellular matrix molecules is 5% of normal or 10% of normal? It is not known what level of decreased sulfation results in each of these phenotypes.

Russell: With regard to the COMP story, Dick Heinegard makes a big point about the distribution through cartilage being non-homogeneous. I was wondering whether in the syndromes you were describing anyone has looked at that.

D. Cohn: His data suggest that there is a higher concentration of COMP in the immediate pericellular region around the chondrocyte, and that there is then a diffuse distribution of COMP throughout the rest of the matrix. We have repeated some of these experiments with lower concentrations of antibody, and as we reduce the concentration of antibody we get a much more homogeneous picture. I wonder to what extent there really is more COMP immediately adjacent to the chondrocytes. In pseudoachondroplasia chondrocytes, whether there is a clearing of COMP around the chondrocyte is not known. We haven't had much cartilage from patients with which to test that idea. But we are doing direct experiments looking at which specific amino acid residues of COMP might be involved in binding chondrocytes.

Bard: One of your electron micrographs caught my eye. It was in COMP. It showed type II collagen fibrils showing normal 640 Å periodicity. Is this correct?

D. Cohn: It seems that type II collagen is able to assemble in a proper way, but the fibril diameter is just a little bit bigger.

Bard: Normally it's very thin.

D. Cohn: Yes, so the abnormality is in the lateral assembly of the type II collagen.

Bard: Do you have any idea why this was happening?

D. Cohn: Not really. This is speculation, but perhaps it is because of a disrupted interaction with type IX collagen.

Chen: I'm struck by the concentration of the mutations in COMP within the calmodulin domains. It is known that the extracellular growth factor (EGF) domain also binds calcium, and thus may be important for the function of COMP. Why are there no mutations identified in EGF domains?

D. Cohn: It surprised us as well. We certainly expect that there are phenotypes that will result from mutations in the EGF domain of COMP. We have looked at a number of other spondyloepiphyseal dysplasias and spondyloepimetaphyseal dysplasias to try to identify them. As yet, no mutations in that domain have been identified in any phenotype. Having said that, another way you can look for them is to ask whether in osteochondrodysplasias in which there are inclusions in the RER, you can identify any others in which the inclusions stain with COMP antibody. We have now looked at a couple of dozen different phenotypes with inclusions in the rough endoplasmic reticulum and we don't see any others that stain with antibodies to COMP. Perhaps the best way to address that problem is to construct a transgenic mouse with a mutation in that domain to explore the possibility that a skeletal phenotype would result. This may then bring into play the ability to detect things that we wouldn't detect by just clinically ascertaining patients. This would include everything from lethal phenotypes to those that are really very mild.

Newman: Some of these chondrocytes making the aberrant proteins look pretty sick. Can you attribute some of the symptomology to altered chondrocyte function apart from matrix occlusion? In other words, perhaps it is not so much that the matrix is compromised as that the growth of the chondrocytes is.

D. Cohn: One of the questions would be, is the accumulation of so much of this abnormal protein within the rough endoplasmic reticulum affecting cellular metabolism in some more general way? A number of people have looked to see whether there is increased apoptosis in chondrocytes from these patients. There are some limited data on pseudoachondroplasia, which suggests that there may be, but in general that's not what we see. The distinct clinical outcomes of mutations in each of the different proteins, despite the fact that they all accumulate a tremendous amount of abnormal protein within the rough endoplasmic reticulum, I think argues a little bit against that.

Hall: Presumably, you're also getting disrupted feedback from the extracellular matrix to the cells. Thus you are not maintaining differentiation in the normal way.

D. Cohn: That makes a lot of sense and is likely to be true, although there are very few data which say that it is. Now knowing the underlying biology of some of these conditions, we have the ability to look at this, with our eyes more open.

Genetic control of bone and joint formation

David M. Kingsley

Howard Hughes Medical Institute and Department of Developmental Biology, Beckman Center B300, Stanford University School of Medicine, Stanford CA 94305-5329, USA

Abstract. The form and pattern of the vertebrate skeleton is thought to be strongly influenced by several fundamental morphogenetic behaviours of mesenchymal cells during embryonic development. Recent genetic and developmental studies have identified some of the genes that play an important role in controlling both the aggregation of mesenchymal cells into rough outlines of future skeletal elements (condensations), and in controlling where skeletal precursors cleave or segment to produce separate skeletal elements connected by joints. Members of the bone morphogenetic protein (BMP) family appear to play an important role in both processes. Mouse and human mutations in these genes lead to defects in formation of specific bones and joints, with striking specificity for particular anatomical locations. Results from a range of experiments suggest that these molecules may have multiple functions during normal skeletal development and patterning. A major challenge for the future is to identify genes and pathways that can maintain, repair, or stimulate the regeneration of bone and joint structures at later developmental stages.

2001 The molecular basis of skeletogenesis. Wiley, Chichester (Novartis Foundation Symposium 232) p 213–234

My laboratory has been interested in the genetic mechanisms that control skeletal patterning during vertebrate development. More than a decade ago, Shubin, Alberch, Oster and Murray proposed that skeletal patterns emerge from a few fundamental behaviours of mesenchyme cells (Shubin & Alberch 1986, Oster et al 1988). One of these fundamental behaviours is the aggregation of mesenchymal cells into regions of higher cell density called condensations. The condensations represent rough outlines of future skeletal elements and the cells within them will later differentiate and elaborate extracellular matrix to differentiate into cartilage, which is usually replaced by bone to give rise to adult skeletal elements. Some of these condensations undergo a second fundamental behaviour, elongating and branching at their terminal ends to generate a Y-shaped bifurcation. Finally, some skeletal elements undergo transverse cleavage events that give rise to multiple daughter elements from an initial precursor. The cleavage or segmentation sites represent the sites of future joints between skeletal elements. Combinations of these three proposed fundamental behaviours, formation of *de novo* condensations, branching of condensations, and segmentation of condensations, can then be combined in various ways to produce the sort of patterns seen in vertebrate skeletons (Shubin & Alberch 1986, Oster et al 1988).

What are the mechanisms and molecules that may control these basic behaviours of mesenchyme cells during normal development? My laboratory has been taking a genetic approach to that problem, taking advantage of the large number of classical mouse mutations that have been described that disrupt various aspects of skeletal development. Over 150 such mutations have been described, many of them mapped to particular chromosome regions (Grüneberg 1963, Green 1989). Here I will review our study of two of these classical mouse mutations, *short ear* and *brachypodism*, both of which are known to disrupt the formation and behaviour of early skeletal condensations.

Short ear, *Bmp5*, and genetic control of *de novo* condensations

Mutations at the mouse *short ear* locus were first recognized because of their obvious effects on the size and shape of the external ear (Lynch 1921). Although named for the pronounced shortening they cause of the ear pinna, mutations at this locus also disrupt a number of internal skeletal structures (Green & Green 1946, Green 1951, 1968). For example, *short ear* mutants are typically missing one of the 13 pairs of ribs, show loss or alterations of bones at both ends of the sternum, and are missing the characteristic ventral processes normally found only on the sixth cervical vertebrae. Developmental studies done many years ago by Margaret Green show that these changes arise at the early condensation stage of normal development (Green & Green 1942). At particular times and places, like the ear, or the sternum, or regions adjacent to the sixth cervical vertebrae, condensations either fail to form in *short ear* mice, or form with altered shapes and sizes. These data suggested that the gene acts at the earliest stages of skeletal development, and might provide new insights into the mechanisms that control the key process of condensation formation in particular body regions.

Using a chromosome walking and positional cloning approach, we showed that the product of the *short ear* gene is one of the large number of secreted proteins in the bone morphogenetic protein (BMP) or transforming growth factor β (TGFβ) family (Kingsley et al 1992, King et al 1994). Many of the BMP members of this family were originally isolated on the basis of their remarkable ability to induce the entire process of endochondral bone formation when implanted at ectopic sites in animals, beginning with local aggregation of cells, differentiation into cartilage and later replacement by bone (Reddi & Huggins 1972, Wozney et al 1988, Luyten et al 1989, Sampath et al 1990). The *Bmp5* mutations in *short ear* mice provided the first genetic evidence that members of this gene family are also required for normal condensation and skeletal formation during embryonic development.

How is the normal expression of the *Bmp5* gene related to the skeletal defects in *short ear* mice? We have carried out a series of *in situ* hybridization studies that show that at specific sites in embryos, *Bmp5* turns on at the earliest stages of skeletal formation, in shapes that prefigure specific skeletal condensations (King et al 1994, 1996). In *short ear* mice that cannot express *Bmp5*, these condensations fail to form (for example, the condensation for the ventral process on the sixth cervical vertebrate). These results suggest that embryos may be using BMPs at least in part much as biochemists have been, turning on their expression at particular times and places to induce the formation of condensations, cartilage and bone during normal skeletal development (Kingsley 1994). The combined expression patterns of BMPs, their known ability to induce the entire process of endochondral bone formation, and the early condensation defects seen in mice missing particular family members, all suggest that BMPs may be the endogenous molecules that normally stimulate one of the three fundamental behaviours of mesenchymal cells, the formation of early skeletal condensations.

Brachypodism, GDFs and genetic control of condensation and segmentation

What mechanisms may control the other postulated fundamental behaviours of condensations during skeletal formation? We have become quite interested in the segmentation process through our studies of another classical mouse mutation called *brachypodism*. This spontaneous mouse mutation has very little effect on the axial skeleton, but decreases the length of the long bones in the limb, and has a marked effect on the length of the feet (Landauer 1952, Grüneberg & Lee 1973). The length effect comes both from decreasing the size of the long arch of the foot, and by reducing the number of phalanges in the digits. The effect on the phalanges attracted a great deal of interest from developmental biologists because the number of bones in the digits is frequently taken as measure of digit identity in embryological experiments, and is the sort of trait that varies between organisms with limbs adapted to different life styles (Hinchliffe & Johnson 1980). In the mouse, the trait is controlled by a single Mendelian gene, again providing an opportunity to study the mechanisms that control early skeletal development using a phenotype-driven, positional cloning approach.

Several years ago, we showed that the normal product of the *brachypodism* gene was a new member of the BMP family, called growth/differentiation factor 5 (GDF5) (Storm et al 1994). This molecule had originally been isolated by Se-Jin Lee's laboratory at Johns Hopkins University in a screen for new TGFβ superfamily members. *Gdf5* is closely related to two other genes called *Gdf6* and *Gdf7*, and forms a new subgroup within the TGFβ superfamily (Storm et al 1994).

By sequence comparison, *Gdf5,6,7* are much more closely related to BMPs than to other family members. In fact, *Gdf5,6,7* and are as closely related to *Bmp2,4* and *Bmp5,6,7* as those molecules are to each other. Mutations at the *brachypodism* locus completely inactivate the function of GDF5, producing a null phenotype of shortened bones and elements missing from the fingers (Storm et al 1994). Mutations in the human version of the *Gdf5* gene (also called *CDMP1*) have subsequently been found by Frank Luyten's group and Matt Warman's group, in patients with Hunter–Thompson-type (Thomas et al 1996) and Grebe-type acromesomelic chondrodysplasia (Thomas et al 1997), or in patients with autosomal dominant brachydactyly type C (Polinkovsky et al 1997).

Elaine Storm in my lab has carried out detailed expression studies of the mouse *Gdf5* gene during normal limb development (Storm & Kingsley 1996). The gene is expressed in early skeletal condensations, along the edges of skeletal elements, and in interdigit mesenchyme at early limb stages. At later stages the most striking aspect of its expression is a dramatic series of stripes seen across developing skeletal condensations. These stripes correspond to the sites where joints will form in the developing limb, with one stripe for the shoulder, one for the elbow, a series of stripes in the wrist region and a series of stripes across the digits. We were immediately struck by the correspondence between this pattern and the normal process of joint formation (Storm & Kingsley 1996). The first histological sign of joint formation is normally the appearance of a stripe of cells called the interzone. Cells within the interzone later undergo programmed cell death, producing a region of decreased cell density called the three-layered interzone. A variety of matrix remodelling and other events later lead to accumulation of fluid in this region, eventually leading to two separate skeletal elements connected by a synovial joint (Mitrovic 1977, 1978, Craig et al 1987, 1990, Pitsillides et al 1995). The *Gdf5* gene is one of the first markers known to mark the sites of joint formation, and is expressed even before the interzone can be recognized at the histological level. Because of this striking pattern, it is now becoming widely used as a joint marker in a other studies of limb development (Brunet et al 1998).

We think that the members of the *Gdf5,6,7* family are not just useful markers for joint formation, but also play a key role in the process. The first suggestion that this might be true came from the genetic defects in joint development seen in *brachypodism* mice with null mutations in the *Gdf5* gene. Classical studies on these mice by Grüneberg and Lee suggested that the animals showed an altered pattern of joint development in the digit region, and milder joint abnormalities in the hip, knee, wrist, and ankle regions. The digit rays in mutant mice are somewhat thinner than normal, suggesting a possible early function in condensation formation (Grüneberg & Lee 1973). Subsequently, the middle portion of the digit ray fails to develop normally, a fault proposed to be a defect in segmentation or joint formation (Hinchliffe & Johnson 1980).

Is expression of GDF5 protein sufficient to trigger the process of joint formation at new sites? To test this we have used recombinant GDF5 protein provided by Genetics Institute to express GDF5 at new sites in developing limbs (Storm & Kingsley 1999). Surprisingly, beads soaked in recombinant GDF5 protein actually inhibit the process of joint formation when implanted alongside developing chick digits. Extra growth of cartilage is seen in existing skeletal elements towards the beads, suggesting that GDF5 protein can stimulate cartilage growth in existing skeletal elements.

We have recently done similar experiments in mouse limbs grown in organ culture (Storm & Kingsley 1999). Unsegmented mouse digit rays present on the day of explant will subdivide during several days of development *in vitro* using this system. Joint formation is easily recognized by the formation of dark stripes of dying cells across the digits. Implants of GDF5 protein beads induce a halo of cartilage surrounding the implant. The cells surrounding the bead strongly express markers of cartilage development (*collagen II*, Ihh). Implants of GDF5 protein also suppress expression of endogenous *Gdf5* message in developing interzones. These results suggest that one function of GDF5 protein may normally be to stimulate cartilage development and restrict expression of *Gdf5* itself to the joint region.

Ectopic expression experiments are subject to many possible artefacts, including the provision of unnaturally large amounts of protein that stimulate a broader range of receptors than those activated by the endogenous ligand. To test whether the paradoxical results obtained with recombinant GDF5 protein reflect endogenous roles of GDF5, we have re-examined the development of skeletal defects in *brachypodism* mice using both histology and molecular markers. As expected if GDF5 protein normally stimulates cartilage development and inhibits its own expression in the developing digit rays, limbs of *brachypodism* mutant mice show a failure of cartilage development in the central portion of the digit ray (recognized by a failure of expression of *collagen II*). Interestingly, a variety of markers of joint development expand in the central region, including *Gdf5* itself, *Gli3* and another marker called *Egr*. The results of overexpressing GDF or removing its function are thus reciprocal and complementary, providing a strong argument that GDF5 is required both for cartilage development and for restricting the location of joint development in the developing digit (Storm & Kingsley 1999).

In addition to these early functions, we believe that GDF5 plays an essential role in later segmentation events as well. The developing digit may not be the best location to see both functions, because the earlier defects in condensation and cartilage development make it difficult to determine whether the subsequent failure in segmentation that occurs in the digits is merely secondary to the earlier skeletal defects. However, it is worth noting that *brachypodism* mice show defects in joint formation at other locations as well, including hips, knees, wrist, and ankles (Grüneberg & Lee 1973, Storm & Kingsley 1996). At these sites we do not see the unusual failure of *collagen II* expression and expansion of joint markers that are seen in the digits.

In addition, we have recently inactivated two genes closely related to *Gdf5*, *Gdf6* and *Gdf7*, using mouse embryonic stem cell technology (Settle & D. M. Kingsley, unpublished results 2000). The *Gdf6* knockout mice show defects in joint formation at new locations in the limbs. When we look at early expression of molecular markers at these locations, we see a situation that is much simpler than the complex changes observed in the digits of *brachypodism* mutant animals. Early skeletal condensations in wild-type mice express collagen II diffusely, then segment into separate regions divided by stripes of cells expressing joint markers like *Gdf5* and *PTHrP*. The early skeletal condensations in *Gdf6* mutant mice fail to segment, and show expanded expression of *collagen II*, and reduced expression of joint markers. No contraction of *collagen II* expression or expansion of joint markers is seen in the *Gdf6* mutant mice, in striking contrast to the unusual situation in the digits of *brachypodism* mice. These results are consistent with a key role for GDF6 in controlling the early formation or maintenance of joints at particular anatomical locations.

The combined studies of the *short ear*, *brachypodism* and *Gdf6* mutant phenotypes suggest that members of the BMP family may play an important role in at least two of the fundamental behaviours of mesenchyme cells earlier postulated by Shubin and Alberch. The studies of *short ear* and *brachypodism* mice show that *Bmp5* and *Gdf5* are both required for normal formation of skeletal condensations in particular body regions. Some of the joint defects in *Gdf5* and *Gdf6* mutant mice suggest that members of the BMP family are also required for the segmentation process that generates joints between skeletal structures. The role in joint development could involve multiple stages, including regulation of other genes expressed in joints, stimulation of programmed cell death in the interzone region, and stimulation of cell proliferation or differentiation at the ends of developing skeletal elements. Evidence for many of these functions has also been reported in parallel studies from other labs, using either recombinant GDF5 protein or ectopic expression of *Gdf5* sequences in developing chicks and mice (Francis-West et al 1999, Merino et al 1999, Tsumaki et al 1999).

Skeletal patterning and BMP gene regulation

Studies of different BMPs during normal development have revealed striking differences in the expression patterns of different family members in different skeletal structures. The *Bmp5* gene, for example, is expressed in condensations inside and at the base of the ear, a site where we do not see expression of

Bmp2,4,6,7,8 (King et al 1996). The *Gdf5,6,7* genes also show striking differences in their expression in different subsets of joints of the developing limb (Settle & Kingsley 2000). Mutations in different BMP genes also disrupt skeletal development in surprisingly specific skeletal regions. These expression and mutant data both suggest that individual anatomical features in the vertebrate skeleton may be encoded by the expression patterns of specific members of the BMP family. Gene duplication within the BMP family, followed by accumulation of specific regulatory elements within each BMP family member, may provide a simple mechanism for achieving independent genetic control over the size and shape of individual skeletal elements in the vertebrate skeleton (Kingsley 1994).

To test this model, we recently began detailed studies of the regulatory sequences that determine where and when BMPs are expressed during embryogenesis. The most detailed work has been so far on the *Bmp5* gene, where a large number of classical mutant alleles were available to help identify the location of key regulatory regions. A combination of gene expression studies in these unique regulatory mutations, and analysis of a series of transgenic mouse constructs containing different *Bmp5* gene regions fused to a *lacZ* reporter construct, have helped identify a large series of different control elements that drive expression in different *Bmp5* expression domains (DiLeone et al 1998, 2000). Interestingly, we do not find general control elements for expression in developing skeletal structures. Instead, we have strong evidence for a whole series of separate elements that each drive expression at very specific anatomical locations within the skeleton. For example, one element drives expression only at the top of the sternum, but not in the rest of the sternal bands or the developing ribs. These studies suggest a surprisingly direct relationship between *Bmp* control elements and the formation of specific skeletal structures. We are currently expanding these studies to the *Gdf* genes in order to further study the mechanisms that control the process of joint formation at different anatomical locations. Using a similar approach we have been able to develop constructs that drive consistent expression of reporter genes in the developing joints of embryos (Schoor & Kingsley, unpublished results 2000). Although we have not yet narrowed any of the elements sufficiently to be able to use them for identification of upstream regulatory factors, we think these elements can already provide a useful tool for driving the expression of other genes at particular locations in the skeleton, or for conditional gene inactivation in developing joints.

Genetic studies of skeletal maintenance and disease

Many of the same genetic techniques we have been using to study early skeletal patterning mutations can also be used to isolate genes that control important skeletal diseases. For example, we have recently been using positional cloning techniques to identify a novel gene disrupted by a classical mouse mutation called progressive ankylosis (*Ank*) (Ho et al 2000). *Ank/Ank* mutant mice are born with normal appearing skeletons, but develop a severe, progressive form of arthritis and mineral deposition that begins several weeks after birth (Sweet & Green 1981). Studies of genes like *Ank* may help identify novel pathways that control maintenance of joints long after they have formed in embryos (Ho et al 2000). We believe that genetic approaches in mice have great potential for future studies of many different areas of bone and joint maintenance, in addition to their important role in studying the fundamental mechanisms that govern the condensation, and segmentation of mesenchyme cells during embryonic development.

Acknowledgement

This work was supported in part by grants from the National Institutes of Health. D. M. K. is an assistant investigator of the Howard Hughes Medical Institute.

References

Brunet LJ, McMahon JA, McMahon AP, Harland RM 1998 Noggin, cartilage morphogenesis, and joint formation in the mammalian skeleton. Science 280:1455–1457

Craig FM, Bentley G, Archer CW 1987 The spatial and temporal pattern of collagens I and II and keratan sulphate in the developing chick metatarsophalangeal joint. Development 99:383–391

Craig FM, Bayliss MT, Bentley G, Archer CW 1990 A role for hyaluronan in joint development. J Anat 171:17–23

DiLeone RJ, Russell LB, Kingsley DM 1998 An extensive 3′ regulatory region controls expression of *Bmp5* in specific anatomical locations of the mouse embryo. Genetics 148: 401–408

DiLeone RJ, Marcus GA, Johnson MD, Kingsley DM 2000 Efficient studies of long-distance *Bmp5* gene regulation using bacterial artificial chromosomes. Proc Natl Acad Sci USA 97:1612–1617

Francis-West PH, Abdelfattah A, Chen et al 1999 Mechanisms of GDF-5 action during skeletal development. Development 126:1305–1315

Green EL, Green MC 1942 The development of three manifestations of the *short ear* gene in the mouse. J Morphol 70:1–19

Green EL, Green MC 1946 Effect of the *short ear* gene on number of ribs and presacral vertebrae in the house mouse. Am Nat 80:619–625

Green MC 1951 Further morphological effects of the *short ear* gene in the house mouse. J Morphol 88:1–22

Green MC 1968 Mechanism of the pleiotropic effects of the *short ear* mutant gene in the mouse. J Exp Zool 167:129–150

Green MC 1989 Catalog of mutant genes and polymorphic loci. In: Lyon MF, Searle AG (eds) Genetic variants and strains of the laboratory mouse. Oxford University Press, Oxford, p 12–403

Grüneberg H 1963 The pathology of development. Blackwell Scientific Publications, Oxford

Grüneberg H, Lee AJ 1973 The anatomy and development of *brachypodism* in the mouse. J Embryol Exp Morphol 30:119–141

Hinchliffe JR, Johnson DR 1980 The development of the vertebrate limb. Clarendon Press, Oxford

Ho AM, Johnson MD, Kingsley DM 2000 Role of the mouse *ank* gene in control of tissue calcification and arthritis. Science 289:265–270

King JA, Marker PC, Seung KJ, Kingsley DM 1994 BMP5 and the molecular, skeletal, and soft-tissue alterations in *short ear* mice. Dev Biol 166:112–122

King JA, Storm EE, Marker PC, DiLeone R, Kingsley DM 1996 The role of BMPs and GDFs in development of region-specific skeletal structures. Ann NY Acad Sci 785:70–79

Kingsley DM 1994 What do BMPs do in mammals? Clues from the mouse *short ear* mutation. Trends Genet 10:16–21

Kingsley DM, Bland AE, Grubber JM et al 1992 The mouse short ear skeletal morphogenesis locus is associated with defects in a bone morphogenetic member of the TGFβ superfamily. Cell 71:399–410

Landauer W 1952 Brachypodism, a recessive mutation of house mice. J Hered 43:293–298

Luyten FP, Cunningham NS, Ma S et al 1989 Purification and partial amino acid sequence of osteogenin, a protein initiating bone differentiation. J Biol Chem 264:13377–13380

Lynch CJ 1921 Short ears, an autosomal mutation in the house mouse. Am Nat 55:421–426

Merino R, Macias D, Gaffian Y et al 1999 Expression and function of *Gdf-5* during digit skeletogenesis in the embryonic chick leg bud. Dev Biol 206:33–45

Mitrovic DR 1977 Development of the metatarsophalangeal joint of the chick embryo: morphological, ultrastructural and histochemical studies. Am J Anat 150:333–347

Mitrovic D 1978 Development of the diarthrodial joints in the rat embryo. Am J Anat 151:475–85

Oster GF, Shubin N, Murray JD, Alberch P 1988 Evolution and morphogenetic rules: the shape of the vertebrate limb in ontogeny and phylogeny. Evolution 42:862–884

Pitsillides AA, Archer CW, Prehm P, Bayliss MT, Edwards JC 1995 Alterations in hyaluronan synthesis during developing joint cavitation. J Histochem Cytochem 43:263–273

Polinkovsky A, Robin NH, Thomas JT et al 1997 Mutations in CDMP1 cause autosomal dominant brachydactyly type C. Nat Genet 17:18–19

Reddi AH, Huggins C 1972 Biochemical sequences in the transformation of normal fibroblasts in adolescent rats. Proc Natl Acad Sci USA 69:1601–1605

Sampath TK, Coughlin JE, Whetstone RM et al 1990 Bovine osteogenic protein is composed of dimers of OP-1 and BMP-2A, two members of the transforming growth factor-β superfamily. J Biol Chem 265:13198–13205

Shubin NH, Alberch P 1986 A morphogenetic approach to the origin and basic organization of the tetrapod limb. Evol Biol 20:319–387

Storm EE, Kingsley DM 1996 Joint patterning defects caused by single and double mutations in members of the bone morphogenetic protein (BMP) family. Development 122:3969–3979

Storm EE, Kingsley DM 1999 GDF5 coordinates bone and joint formation during digit development. Dev Biol 209:11–27

Storm EE, Huynh TV, Copeland NG, Jenkins NA, Kingsley DM, Lee S-L 1994 Limb alterations in *brachypodism* mice due to mutations in a new member of the TGF-β superfamily. Nature 368:639–643

Sweet HO, Green MC 1981 Progressive ankylosis, a new skeletal mutation in the mouse. J Hered 72:87–93

Thomas JT, Lin K, Nandedkar M, Camargo M, Cervenka J, Luyten FP 1996 A human chondrodysplasia due to a mutation in a TGF-β superfamily member. Nat Genet 12:315–317

Thomas JT, Kilpatrick MW, Lin K et al 1997 Disruption of human limb morphogenesis by a dominant negative mutation in CDMP1. Nat Genet 17:58–64

Tsumaki N, Tanaka K, Arikawa-Hirasawa E et al 1999 Role of CDMP-1 in skeletal morphogenesis: promotion of mesenchymal cell recruitment and chondrocyte differentiation. J Cell Biol 144:161–73
Wozney JM, Rosen V, Celeste AJ et al 1988 Novel regulators of bone formation: molecular clones and activities. Science 242:1528–1534

DISCUSSION

Burger: Why do all these different joints need different genes?

Kingsley: When you compare skeletons from different animals, one of the most obvious features is that the morphology and structure of both joints and bones have to be controlled independently. The size and shape of the different joints and bone is highly specific to different organisms. One of the mechanisms that may be involved in doing that, is to take families of molecules, each of whose members have similar activity, but which can be controlled independently from each other. This is essentially what we see when we do the regulatory or promoter-bashing work on these genes: each of them appear to have accumulated very specific sets of regulatory elements that lay out the expression of them but not others. This may well be the tip of a very complicated iceberg. This is probably one method that evolution uses to independently control bone and joint formation in different body regions, to help generate the highly specific patterns and shapes and sizes that are seen throughout the skeleton.

Burger: Then it would be interesting to compare mouse with chicken, where the function of the arm is so different.

Kingsley: One of the predictions of this sort of model is that if you compare the regulatory apparatus surrounding the *BMP* genes in different organisms, you may find differences in the arrays of the anatomy elements that are related to the differences in the morphologies of the different organisms. We've actually seriously considered trying to do that. In one study, we decided to take an even more general approach, which was simply to do honest genetic mapping of the morphological differences in the skeletons of different organisms, so that we could compare the location of genes that control skeletal morphology to the location of the *BMPs* and also everyone else's favourite candidate genes. The basic idea is to take all of the genes identified by developmental biologists and compare the mapped location of those genes to the mapped locations of the traits that are varying between organisms in order to produce morphological changes. I think this is an approach that will answer the question in the long term. We were using 3-spine sticklebacks as a model system to make the genetic work possible.

Hall: We happily recognize kidney and heart as separate organs, because they are made of different sorts of cells and tissues, but we don't recognize vertebrae and limbs as different organs, because they happen to be made of the same tissue. Perhaps we should start thinking of them as different organs that happen to share a common cell type, and not necessarily look for the same controls in the vertebrae as one would find in the limbs.

Kingsley: I agree. One of the almost frightening results from this sort of study, is that the mechanisms that are used to generate morphology or skeletal structures in one location may be quite different from those used in another location. Conceptually, they may be similar in that they may involve different members of the same classes of signalling molecules, but we have seen cases where, if you look at the genes that control cartilage over there, you get a different answer than if you look over here. This is frustrating in the sense that we can't always generalize the conclusions from one location to the other. But it's also exciting in the sense that it can be related to the whole issue of how skeletal patterns are created in the first place. This may be to use members of families deployed independently under separate genetic control to lay out events in different body features.

Karsenty: This relates to patterning. There are different mechanisms for different elements, but what has been illustrated during this meeting with PTHrP and FGFR mutations, is that when you come to cell differentiation, very often the mechanism is more general. In this sense, this is part of the same organ that is made up of several hundred elements.

Hall: So as a tissue undergoing differentiation we are looking at common control, whereas in an organ system we are looking at different controls.

Karsenty: One emphasis that I really appreciated in your paper, and I think was slow to appear in this meeting, is that the future of genetics and molecular biology is in the physiology of the skeleton, because we don't know how it works. Not only do we not know how it works, we don't know how to mineralize bone diseases such as arthritis and oesteoporosis, which are diseases that affect physiology. I think it is very important to use mouse or chicken or human to look at these issues.

Kronenberg: I want to probe David a bit more about this puzzle: you take away GDF5 or GDF6 and you don't get that cleavage step, but you can make it happen by giving it back. When you gave back GDF5 on a bead, you said that the effect on condensation dominates to the extent that you would have missed the joint formations. Now that you have showed us the GDF6 story, it seems to be simpler. A naïve prediction is that a GDF6 bead in the right place in the wrist might bring back the cleavage. Have you done that experiment?

Kingsley: No, we have not. Technically this is harder because the elements are really small. What we know of these molecules in joint formation is that they're expressed in patterns clearly related to joint formation and are required for the process, but no one has been able to trigger the whole process of joint formation using either recombinant protein or viral overexpression of GDFs. They are molecules that are necessary, but they're not sufficient.

Kronenberg: The expectation is that if you put GDF6 any old place, you will not get anything interesting — you will just get more chondrocytes. But if you put it in exactly the right place, one would hope to be able to trigger a special cascade that then could be further studied. There must be something special about the competence of those cells.

Kingsley: There are multiple possibilities. One is the competence of the cells. Another is other factors that may be in the environment. The recombinant protein experiments that we've done have all been done with GDF5 homodimers, for example, whereas these are molecules that may normally function also as heterodimers. There are other BMPs present in the limb some of which have activities that can be more closely related to segmentation. For example, Juan Hurle's lab has done experiments putting BMP2 and 4 beads into developing chick limbs. This results in pretty halos of cell death around the beads, and if you put them at the ends of skeletal elements, the skeletal will bifurcate and branch around the implanted bead. This is much more the sort of activity that we were expecting on the basis of the expression patterns and mutant phenotypes of *Gdf5*. One possibility is that the GDFs normally act as heterodimers with molecules of that class. This is a very hard thing to try to recapitulate using a homodimer of protein. Yet another possibility is exactly what you said: there may be a whole array of a kind of response pre-programming, such that if the cells are not in the right state then GDF5 will not initiate the joint cascade.

Blair: Can you tell us a little bit more about these anatomically specific promoter sites that appear to be binding sites for known promoters?

Kingsley: Unfortunately, we can't tell you much more about them yet. This is partly related to the scale of the problem. For the gene where we have studied the most, the *Bmp5* gene, some of the control elements are located more than 270 kb away from the promoter. Therefore the procedures that we had to use to find the genes essentially involved DNA constructs that were two orders of magnitude bigger than most people normally do when they are trying to find this regulatory element. Still, after several years of working on the problem, we're down to constructs of somewhere between 2 and 10 kb for the regulatory elements, which is still further away than we need to be to answer the question you've asked.

Newman: I am not sure if you are suggesting that the various anatomically specific promoter elements and the various BMPs that play different anatomical roles are all derived from some primordial single factor that organized the entire skeleton originally. Or rather, do you believe that these different skeletal elements came into being when the different genes and promoters for them emerged during the course of evolution?

Kingsley: Since the different members of the BMP family are expressed in different patterns, what I have always thought happened is that there were ancestral versions of the genes that were duplicated, and then the various copies in the vertebrate genome have subsequently gained and lost all sorts of specific control elements that have recruited the expression of that particular family member into a little shape or location, that, if it turned out to be useful in creating an aspect of skeletal morphology, was saved, and if not it was lost. After a few hundred million years of doing this, there is now this very complicated situation with dozens of genes controlled by complicated arrays of *cis*-acting regulatory elements, different from each other for the most part, and responsible for creating different aspects of skeletal morphology.

Newman: That would also be my interpretation. Can you then look at the different sequences of the different factors and trace them back? Do they converge at some ancestral vertebrate?

Kingsley: These sorts of data will be one of the nice things that will come out of the various genome projects. Most of the sequence comparisons that we have now are limited to the coding regions of the genes. The work of getting the *cis*-acting regulatory sequences is going to be an enormous project, because in this gene the regulatory sequences are literally spread out over hundreds of kilobases of DNA of exactly the type people normally don't bother sequencing. The genome projects should prove to be a real treasure trove in this respect.

Reddi: Gdf5 has some human mutations which result in chondrodysplasia.

There is one question that has puzzled me. Limb development on the left and right hand sides of the body is controlled by exactly the same genes, yet the limbs on one side are a mirror image of those on the other. There are genes which specify asymmetry in organs such as the heart, for example. Is there any theoretical explanation for the left–right differences in joint formation?

Kingsley: We have not found any expression patterns in these genes that are related left–right differences. I suspect that there are mechanisms that deploy these genes in specific locations within limbs, and the left–right aspect of it is probably governed by the mechanisms that take whole cassettes and do it on the left-hand side or the right-hand side.

Reddi: The topic of this meeting is the molecular control of skeletogenesis. We have mostly concerned ourselves with cartilage bone, but at least in the joints we cannot ignore the ligaments and tendons, and the bone–ligament continuum. In this regard, people have implanted GDF7 ectopically and claimed that it might induce ligament-like structures. What do you make of this in relation to some of the bead experiments you have done? Have you had occasion to implant beads of GDF6 and GDF7 in the same experiment in which you implanted GDF5, where you saw a halo effect of cell death? If you had put in GDF6 or GDF7, would you have got a different effect? Is this a reasonable experiment?

Kingsley: It's a reasonable experiment that we haven't done. The tendon induction results are controversial: different groups trying what seem to be similar experiments have reported different results. On the genetic side, in addition to the skeletal phenotypes that are reported in brachypodism animals, they do have defects in tendon pattern. This was work originally done in the classical phenotypic description of the animals by Grüneberg and Lee, which showed that the number of tendons and their insertions are disrupted. Many people hear that and think that it's likely to be secondary to the skeletal defects: that if the bones are wrong, then the tendons that normally insert in them will also be wrong. Grüneberg had a fairly good argument that that probably wasn't true—some of the tendon defects are seen even in heterozygous mice that don't show the skeletal abnormality. These are still the only *genetic* data in the literature to my mind that suggest that the members of this family may be doing something related to tendon patterning.

Wilkie: Something that confused me is that you were talking about *brachypodism* in mice and doing *in situ* hybridizations with *Gdf5*. Is *brachypodism* a true null allele?

Kingsley: When we did that we were using an *in situ* probe that recognizes a message encoding a non-functional gene product. Thus you can still produce the transcript, even though that particular allele encodes a null mutation that would be unable to produce protein. Thus you can still ask what's happening to the gene expression.

Wilkie: As you know, the equivalent mutation to *Gdf5* in the human affects the *CDMP1* gene, which causes a heterozygous loss of function phenotype termed type C brachydactyly (Polinkovsky et al 1997). One of the features that you can get is hypersegmentation of some of the phalanges, which seems to fit rather better with the predictions of the consequences of a modest reduction in *Gdf5*.

Kingsley: In the particular clinical cases where it has been shown, is there hypersegmentation? I don't remember hypersegmentation as a prominent phenotype, at least in those families where brachydactyly type C had actually been tied to mutations in CBMP1. I thought it was usually shortening and occasionally fusion of phalanges and digits two and five without the hypersegment. Just to explain for those who are unfamiliar with this work, there have been several clinical diseases in humans that have been tied to mutations in the human version of *Gdf5*, which is also called *CDMP1*. This is work that Frank Luyten's lab and Matt Warman's lab have done (Thomas et al 1996, 1997, Polinkovsky et al 1997). The disease comes in several flavours. There is a recessive form that has a loss-of-function severe skeletal phenotype, and there is a heterozygous form which is generally associated with mild shortening of the digits and a phenotype called brachydactyly C.

Mundlos: We recently had a family in which the parents were affected by brachydactyly type C and a son with a greater dysplasia and was homozygous.

Kingsley: There is one family described who have severe phenotypes that are heterozygous, and in that case the explanation is that they are dominant negative.

Karsenty: It is not the only protein in the TGFβ family that has no phenotype in heterozygous mice yet in human the phenotype is present in heterozygote individuals. I don't know what this means about the biology of the coupling, but it means there is something specifically different between mouse and humans when it comes to BMPs.

Mundlos: I am still confused about how GDF5 is having its effect. If I have got it right, you first have an effect when it is expressed at an early stage in the developing anlage, and it is making cartilage. But later on it is obviously expressed in a place where there is no cartilage, and it is obviously not making cartilage. Could this be a dosage effect? How could you by just one day apart have something completely different?

Kingsley: This is the fundamental puzzle. Everyone is going to have to get used to having their molecules do more than one thing in different places in the embryo. I think people are still surprised when they hear this, because the two things both seem to be in skeletal tissue and only a day apart. Despite that, I think the logical problem is not that different than having the result that many people found, that the molecules do more than one thing in different places.

Mundlos: I was also struck by that picture where you showed that in P2, there are basically more chondrocytes: the condensation is very small. This is exactly what it looks like in our mice. There is nothing there. The two phalanges are reduced to one phalanx. This is not really making joints, but rather it is taking away one piece, which reduces three to two. Is GDF5 really involved in making joints? The effect is not making a joint; the effect is taking away P2.

Kingsley: I don't know what the histology says in the short digit mice. In the brachypodism mice, people have compared the length of the combined P1 and P2 with what is seen is wild-type, and it doesn't look like there is just a deletion of an element.

Mundlos: Instead of two it is one and a half.

Kingsley: It starts out as the length of two, but the two don't develop and segment the way that they are supposed to.

Mundlos: That is not the time point when chondrocytes condense.

Kingsley: It occurs later. That is why I said that I think the digit region of the developing brachypodism animal is a very complicated place to try and unravel the combined effects on condensation and digit segmentation. The simpler place to look is the ankle region of the *Gdf6* mutant animal, where there isn't an early effect on condensation. Here you are looking at collagen II and GDF5, and it appears that a precursor fails to segment in the absence of the GDF6 product, without earlier abnormality in cartilage differentiation. If we only had the brachypodial phenotype, I'd still be wondering exactly what was going on. I should say that the brachypodism animals, in addition to having the phenotype in the fingers, have joint defects at other locations. They have fusions in the wrists and ankles at different locations.

Mundlos: So the effect you see is not just a missing joint.

Kingsley: No, in the digits of brachypodism it is not just a missing joint, it is abnormal cartilage differentiation that we think reflects an earlier role for the gene in controlling condensation and cartilage formation.

Bard: Is the condensation shorter in this digit?

Kingsley: Not when it starts.

Bard: At the time you would expect the new joint to form in the condensation in the wild-type, has the condensation not grown adequately in the mutant?

Kingsley: At the time when the wild-type joint forms, the mutant digit ray is just as long but is showing abnormal differentiation.

Hall: It looks almost like a switch; if these cells are not being programmed to become chondrogenic, they are becoming a joint.

Pizette: Doesn't collagen type II come back on again after a while?

Kingsley: There is a failure to thrive in the middle of the digit array. When we look late enough, we do see a cartilage element. It appears that is formed by appositional cartilage growth: we see cartilage being laid down from the edges of the element. It starts out hollow looking and with time the cartilage grows in from the side.

Chen: There is another possibility from the data you have shown. You have shown *in situ* hybridization which detects the message. This means that this is where *Gdf5* is made, but it doesn't necessarily mean that it is where it acts. For example, *Indian hedgehog* is expressed in the pre-hypertrophic middle zone, but it affects the adjacent zones of proliferation and hypertrophy. Even though *Gdf5* mRNA is made between cartilage elements, the protein could still affect cartilage elements.

Kingsley: I agree; that is a good point. We have looked at mRNA patterns and not protein patterns. This is not for lack of trying, it is just our antibodies have not worked. Certainly in the case of digit phenotype in brachypodism, which is wider than just the stripe of expression, diffusion affecting the surrounding skeletal elements is a good possibility. In the more localized defects that are seen in the wrists and ankles of the *Gdf6* mutant, we're not really seeing phenotypes that go much beyond the stripe of expression that we see by RNA *in situ* hybridization. This could still be an effect on the cartilage immediately surrounding the stripe, but it doesn't appear to spread very far.

Hall: It sounds as if what you're doing is suppressing chondrogenesis within the condensation, but you are not suppressing the development of the perichondrium around the boundary of that condensation, which later on makes cartilage which comes in from the side. You seem to have a boundary layer there which is escaping the suppression.

Kingsley: That is possible.

Ornitz: So *Gdf5* is expressed in every joint, but the major phenotype you see affects a single joint. Do you think that there are other members of the family that are overlapping in expression in all of the joints?

Kingsley: This idea was the motivation for us doing the *Gdf6* and *Gdf7* knockouts. Their expression patterns are broadly consistent with what you have just said: they are molecules of the same families that are expressed in some joints but not others. The part of that story that remains to be unravelled is to build in all the doubles and the triples to see what new functions are uncovered by stripping out the various members.

Ornitz: But even *Gdf6* and *Gdf7* don't account for the proximal joints.

Kingsley: Gdf6 is made in elbows, knees, wrists and ankles; *Gdf7* is made in one stripe pattern in the digit and in a patch near the shoulder. We are not missing that many joints.

Ornitz: So it's possible that double knockout of *Gdf5* and *Gdf6* would affect many more joints.

Kingsley: We'll see.

Burger: I'm trying to grasp this idea about the morphology of the joint. Even in the phalanges, which have simple joints, one is a ball and the other is a cup, and that is how they fit in each other. The fact that the same molecule occurs just in one stripe makes it difficult for me to envisage how it could produce these morphologies.

Kingsley: There are many aspects of joint morphology that are not going to be determined just by these sorts of gene products. There's been a lot of modelling done to suggest that the kind of cup and ball arrangement seen in joints may actually be a mechanical response of the joint.

Burger: It is present very early on, and will even develop in culture.

Kingsley: I think there are studies showing that if developing embryos are paralysed, they will not develop that joint structure, which suggests that it is dependent upon mechanical influences (Drachman & Sokoloff 1966).

Reddi: A problem concerns which cells give rise to both the growth plate cartilage and the articular cartilage. Do you think there is a common precursor for both articular cartilage and growth plate cartilage? If there is, we are confronted with a paradox. The life of the growth plate chondrocytes seems to be ephemeral, compared with articular cartilage, which seems to have a permanent phenotype unless it is ravaged by rheumatoid arthritis or osteoarthritis with ageing. Are both of these from the same cell type?

Kingsley: It's an interesting question. This is why I asked Henry Kronenberg earlier about the PTHrP expression around the articular cartilage. There is articular cartilage which you want to hang on to for as long as possible, versus growth plate which is programmed to go. These same blue cell chimera experiments that Henry Kronenberg was using to look at PTHrP phenotypes also provide a way of asking lineage questions in developing skeletal elements. Have you have seen any interesting columns of blue to suggest that articular and growth rate cartilage may be related?

Kronenberg: I think there are some really interesting lineage experiments that need doing, but we haven't attempted them yet. When Schipani and Jüppner rescue the PTHrP knockout mice by using the *collagen II*-driven overactive PTH/PTHrP receptor transgene, what happens is that the animals live for a few months because their bones are extended, but the transgene seems to become less active over time, as *collagen II* transgenes often seem to. The growth plates fuse and disappear, but the articular cartilage doesn't disappear. There may be some abnormalities; these mice are still being evaluated. But broadly speaking, you run out of the stem cells that are necessary for growth plate chondrocytes, but you haven't run out of the cells that are at the articular surfaces. So there must be some point, if they start out perhaps from the same general pool, at which they start following different rules. There are a number of groups of chondrocytes that behave differently in the *PTHrP* knockout mice. The contrast between the tracheal cartilage, on the one hand, and Meckle's cartilage and the cartilage in the ribs near the sternum on the other is instructive. If you take away PTHrP, these two latter cartilages become bone, but the tracheal cartilage isn't touched. The articular chondrocytes at the top of the growth plate and in the trachea don't have PTH/PTHrP receptors—this may explain their insensitivity to knockout of PTHrP.

Poole: It's worth bearing in mind that although a cell may form articular cartilage and thus not go through a hypertrophic pathway, in osteoarthritis cells can often express the hypertrophic phenotype as part of the pathology. This is associated with the mineralization of extracellular matrix. Type X collagen is expressed in these cells, there is up-regulation of annexin V, up-regulation of PTHrP, and increased apoptosis. These changes are all associated with hypertrophy in endochondral ossification and with articular chondrocytes in osteoarthritis that tend to be near the articular surface. We should remember that the potential to express the hypertrophic phenotype is there even in a cell in articular cartilage.

Kingsley: What do you think normally holds the articular cartilage?

Poole: I think there is a regulatory suppressive mechanism. It is either going one way or the other: either there is a facilitating process or a suppressing process or both. I think it is also environmental, and it comes back to the unique position that the growth plate chondrocytes find themselves in with respect to the surrounding tissues. There are some very important issues with vascular invasion, for example—preliminary work is starting to indicate that endothelial blood vessel cells can perhaps also regulate differentiation of chondrocytes suggesting further cross-talk.

Burger: There is a theory developed by Dennis Carter's group that explains the maintenance of articular cartilage, that is based on the mechanical stimuli that they get. It is difficult to prove that of course.

Karsenty: There is at least one gene, expressed in resting chondrocytes and in articular chondrocytes only. This suggests that there is a commonality, which makes sense.

Mundlos: There is another striking pattern, that of cartilage matrix protein. This gene is expressed everywhere in the cartilage except in the presumptive joint area.

Poole: Well, the interesting thing about CMP is that it reappears in arthritis. This again is an association with the expression of the hypertrophic phenotype.

Kronenberg: Is it certain that those hypertrophic chondrocytes came from articular chondrocytes, as opposed to mesenchymal cells nearby?

Poole: If we look at osteoarthritic cartilage, there is no evidence that a mesenchymal cell could penetrate cartilage and gain entry into this exclusive club.

Kronenberg: But inflammatory cells got there.

Poole: No, not into cartilage, only on it. They are usually very remote in osteoarthritis. This is different to rheumatoid arthritis.

Meikle: Articular cartilage is not totally inert. It can be replaced. Mankin (1962) carried out come classical experiments in the rabbit with tritiated thymidine labelling, and demonstrated a proliferative zone within the articular cartilage.

Poole: Yes, the chondrocytes formed the so-called clusters or clones, and this is very much the feature of the pathology. When you probe those cells, they are often the cells that tend to express type X collagen and so on.

Chen: Even though chondrocytes in normal adult articular cartilage do not make type X collagen or CMP, if they are isolated and cultured they will make these markers. One interpretation is that there is a suppressive environment in the matrix that prevents them from becoming hypertrophic.

Poole: I would like to ask about the *Ank/Ank* mouse. The calcification of the articular cartilage is abnormal and unlike anything quite like that in arthritis, and yet you're saying that you see it in development. Do you see any change in expression in the articular cartilage in these mice compared with what you might see in a healthy mouse?

Kingsley: That is something we are currently interested in. Unfortunately, the *in situ* protocols that we used in the lab work well at prenatal stages but don't work well with postnatal stages. This is particularly frustrating in this case because the phenotype is not seen in the joints until about 4–5 weeks of age. We don't know yet whether the development of that phenotype arises because of interesting changes in the expression of the genes.

Poole: In view of this linkage that you're seeing in the *Ank/Ank* mouse, can anybody comment on what happens to those cells when they undergo this process of programmed cell death to form the joint cavity? Is there any evidence of any hypertrophy related to apoptosis?

Kingsley: The stage at which the cell death occurs going from inter zone to a three-layer interzone is way before chondrocytes show any maturity of hypertrophy. There is no evidence for the role of hypertrophy in that cell death.

Reddi: I suspect this discussion will have provided a lot of food for thought for Marty Cohn, because in his work on pythons he may have to look at why the joints disappear. Perhaps GDFs play a role there too!

M. Cohn: Some of those mutant skeletal patterns look like reptilian ankles. The increased complexity of compartmentalization during evolution of the carpus and tarsus is interesting, and if skeletal development is as modular as it appears to be in these mice, it points to some interesting candidate mechanisms.

Kingsley: We're quite interested in the molecular basis of morphological changes during vertebrate in evolution, because it's so appealing to extend the sorts of models that have come out of the work on mouse mutants. I'm also fully aware of how complicated evolutionary changes are likely to be. This is why in thinking about how to approach the problem experimentally, we've decided not to bet just on BMP genes, but to set up a general genetic approach that will map the loci responsible for morphological changes, and allow us to compare these loci with the location of BMP genes and everyone else's favourite candidate genes too.

M. Cohn: In your talk, you mentioned Neil Shubin's and Pere Alberch's paper on branching and segmentation during limb development (Shubin & Alberch 1986). From your data, it is clear that segmentation is an important part of limb morphogenesis. I am less sure about branching, The Shubin and Alberch paper was based on a stained series of developing limbs, and an inference of a developmental process was made from the Y-shaped pattern of the early humerus, ulna and radius. There is good experimental evidence against the idea that distal structures branch from proximal structures. For example, when the distal tip of a limb bud is transplanted, prior to skeletal condensation, to another site in the embryo, the distal limb bud cells give rise to digits in the absence of any proximal structures. Other evidence which argues against the idea that distal structures branch from proximal structures comes from barrier experiments, in which an impermeable barrier is implanted between the presumptive humerus and the ulna and radius. Although the barrier prevents mechanical branching, distal structures develop normally. My inclination is not to accept branching as a part of limb morphogenesis. Do you know of evidence supporting branching as a processes in development of the limb skeleton?

Kingsley: You're quite right. The original hypothesis was based on simply looking at histology sections of developing embryos without an experimental component at the same time.

Kronenberg: You could argue that the pre-pattern might branch. It is conceivable that there could be something that could branch before the condensations are visibly distinct.

Bard: The experimental data suggest that branching is not needed.

Kronenberg: A counter-argument would be that you are putting the barrier in too late, or that you are taking away this region after that branch formed. But I guess that kind of argument is already taking away a lot of what the branching is presumably all about, which would be cell proliferation.

M. Cohn: Lewis Wolpert has shown that the skeletal pattern is specified long before the appearance of skeletal condensations. I think that the branching and segmentation model of limb development may conflate an anatomical pattern with a developmental process.

Kronenberg: It could be signalling molecules that get streamed, then.

Newman: There is an interesting theoretical point here. The branching picture was not only based on histological observations, but also on a mechanochemical model (Oster et al 1983) that interestingly had the same formal properties as the Turing reaction–diffusion model (Newman & Frisch 1979). However, the embodiment of that model in the mechanochemical and tension-producing fields led to a kind of natural inference that there would be mechanical continuity between the sequential elements, even if they were more numerous. In contrast, the original reaction–diffusion model was able to generate elements that would be binary or multiple without physical continuity at the branch point. Since then, evidence has mounted against the mechanochemical idea (Miura & Shiota 2000a), whereas the reaction–diffusion idea is still viable (Miura & Shiota 2000a,b).

Hall: I think Martin Cohn is right: I've always been concerned about the branching model because I don't see any evidence for branching as a mechanism, despite the evidence for branching as a pattern.

References

Drachman DB, Sokoloff L 1966 The role of movement in embryonic joint development. Dev Biol 14:401–420

Mankin HJJ 1962 Localization of tritiated thymidine in articular cartilage of rabbits. I. Growth in immature cartilage. J Bone Joint Surg 44A:682–698

Miura T, Shiota K 2000a Extracellular matrix environment influences chondrogenic pattern formation in limb bud micromass culture: experimental verification of theoretical models. Anat Rec 258:100–107

Miura T, Shiota K 2000b TGFβ2 acts as an 'activator' molecule in reaction–diffusion model and is involved in cell sorting phenomenon in mouse limb micromass culture. Dev Dyn, in press

Newman SA, Frisch HL 1979 Dynamics of skeletal pattern formation in developing chick limb. Science. 205:662–668

Oster GF, Murray JD, Harris AK 1983 Mechanical aspects of mesenchymal morphogenesis. J Embryol Exp Morphol 78:83–125

Polinkovsky A, Robin NH, Thomas JT et al 1997 Mutations in CDMP1 cause autosomal dominant brachydactyly type C. Nat Genet 17:18–19

Shubin N, Alberch P 1986 A morphogenetic approach to the origin and basic organization of the tetrapod limb. Evol Biol 18:319–387

Thomas JT, Lin K, Nandedkar M, Camargo M, Cervenka J, Luyten FP 1996 A human chondrodysplasia due to a mutation in a TGFβ superfamily member. Nat Genet 12:315–317

Thomas JT, Kilpatrick MW, Lin K et al 1997 Disruption of human limb morphogenesis by a dominant negative mutation in CDMP1. Nat Genet 17:58–64

The molecular basis of osteoclast differentiation and activation

Tatsuo Suda, Kanichiro Kobayashi, Eijiro Jimi, Nobuyuki Udagawa and Naoyuki Takahashi

Department of Biochemistry, School of Dentistry, Showa University, Tokyo 142-8555, Japan

Abstract. Osteoclasts develop from haemopoietic cells of the monocyte–macrophage lineage. Osteoblasts or stromal cells are essentially involved in osteoclastogenesis through cell–cell interaction with osteoclast progenitor cells. Recent findings indicate that osteoblasts/stromal cells express osteoclast differentiation factor (ODF, also called RANKL, TRANCE and OPGL) as a membrane-associated factor in response to several osteotropic factors to support osteoclast differentiation. ODF is a new member of the tumour necrosis factor (TNF) ligand family. Osteoclast precursors, which express RANK, a TNF receptor family member, recognize ODF through cell–cell interactions with osteoblasts/stromal cells, and differentiate into osteoclasts in the presence of macrophage colony-stimulating factor (M-CSF). Osteoclastogenesis inhibitory factor (OCIF, also called OPG) is a secreted TNF receptor, which acts as a decoy receptor for ODF. ODF is responsible for inducing not only differentiation, but also activation of osteoclasts. Interleukin 1α (IL-1α) can be substituted for ODF in inducing the activation of osteoclasts. Recently, it was shown that mouse TNFα stimulated the differentiation of M-CSF-dependent mouse bone marrow macrophages into osteoclasts in the presence of M-CSF without any help of osteoblasts/stromal cells. Osteoclast formation induced by TNFα was inhibited by antibodies against TNF type 1 receptor (TNFR1) or TNFR2, but not by OCIF. Osteoclasts induced by TNFα formed resorption pits on dentine slices only in the presence of IL-1α. These results demonstrate that TNFα stimulates osteoclast differentiation through a mechanism independent of the ODF–RANK interaction. TNFα and IL-1α may play an important role in pathological bone resorption due to inflammation.

2001 The molecular basis of skeletogenesis. Wiley, Chichester (Novartis Foundation Symposium 232) p 235–250

Osteoclasts, the multinucleated giant cells that are present only in bone, develop from haemopoietic cells of the monocyte–macrophage lineage. Osteoblasts or bone marrow stromal cells are involved in osteoclastic bone resorption, which consists of two processes; one is the recruitment of new osteoclasts and the other is the activation of mature osteoclasts. We developed a co-culture system of mouse osteoblasts/stromal cells and haemopoietic cells (spleen cells or bone marrow cells) (Takahashi et al 1988), in which osteoclasts were formed in response to several osteotropic factors including 1α,25-dihydroxyvitamin D3 [1α,25$(OH)_2$D3], parathyroid hormone (PTH), interleukin 1 (IL-1) and IL-11 (Suda et al 1992). Our results suggest that osteoblasts/stromal cells are crucial for osteoclast formation. Macrophage colony-stimulating factor (M-CSF, also called CSF-1), a soluble factor produced by osteoblasts/stromal cells, is essential for osteoclast development (Yoshida et al 1990, Felix et al 1990). Cell–cell contact between osteoblasts/stromal cells and haematopoietic osteoclast progenitors is also critical for osteoclast development (Takahashi et al 1988, Suda et al 1992). Subsequent experiments have established that the target cells of osteotropic factors for inducing osteoclast formation in the co-culture are osteoblasts/stromal cells (Suda et al 1995). From these experimental results, we proposed a hypothesis that osteoblasts/stromal cells induce osteoclast differentiation factor (ODF) as a membrane-associated cytokine in response to various osteotropic factors (Fig. 1; Suda et al 1995). Osteoclast progenitors recognize ODF through cell–cell interaction with osteoblasts/stromal cells and differentiate into osteoclasts. Chambers et al (1993) also proposed that SOFA (stromal osteoclast forming activity) expressed by osteoblasts/stromal cells is essential for osteoclast differentiation (Fig. 1). SOFA appeared to be identical to ODF.

Discovery and identification of OPG/OCIF

In 1997, Simonet et al cloned a new member of the tumour necrosis factor (TNF) receptor family, termed osteoprotegerin (OPG). Interestingly, OPG lacked a transmembrane domain and represented a secreted TNF receptor (Fig. 2). Hepatic expression of *Opg* in transgenic mice resulted in osteopetrosis (Simonet et al 1997). Tsuda et al (1997) independently isolated a novel protein termed osteoclastogenesis inhibitory factor (OCIF) from conditioned media of human fibroblast culture. The cDNA sequence of OCIF is identical to that of OPG.

OPG/OCIF is a 401 amino acid protein with four cysteine-rich domains and two death domain homologous regions (Fig. 2). The death domain homologous regions share structural features with 'death domains' of TNF receptor p55, Fas and TRAIL receptors, which mediate apoptotic signals. OPG/OCIF strongly inhibited osteoclast formation induced by either 1α,25$(OH)_2$D3, PTH, prostaglandin E2 (PGE2) or IL-11 in mouse co-culture (Simonet et al 1997, Tsuda et al 1997, Yasuda et al 1998a). Analyses of transgenic mice expressing *Opg/Ocif* and animals injected with OPG/OCIF have demonstrated that this factor suppresses osteoclastic bone resorption, resulting in increased bone mass (Simonet et al 1997, Yasuda et al 1998a). In contrast, *Opg/Ocif* knockout mice exhibited severe osteoporosis due to enhanced osteoclastogenesis (Bucay et al 1998, Mizuno et al 1998). These results suggest that OPG/OCIF is a physiologically important inhibitor of osteoclastic bone resorption.

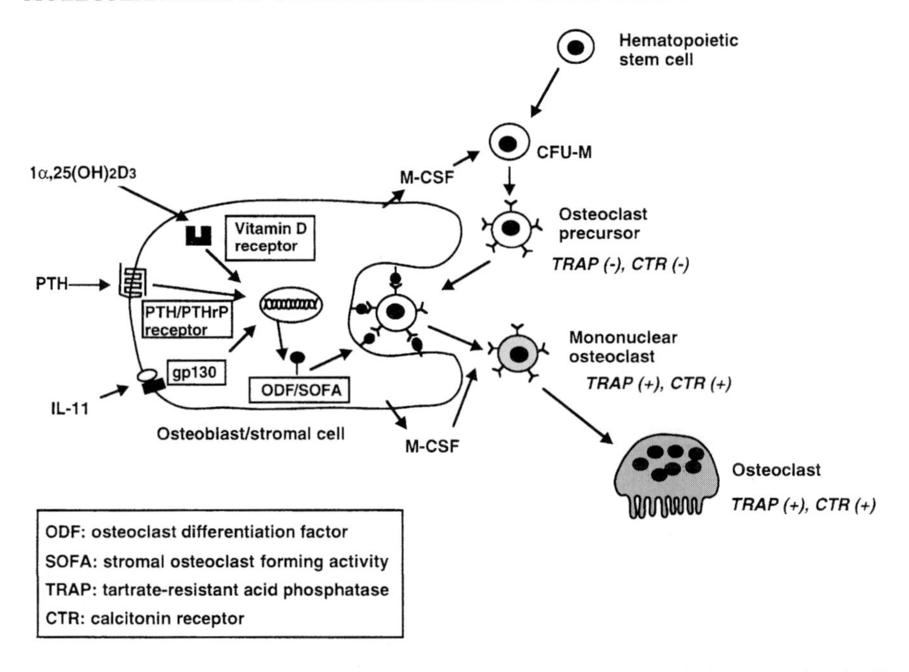

FIG. 1. A hypothetical concept of osteoclast differentiation. Osteotropic factors such as 1α,25 $(OH)_2D3$, PTH and IL-11 stimulate osteoclast formation in co-cultures of osteoblasts/stromal cells and haematopoietic cells. Target cells for these factors are osteoblasts/stromal cells. Three different signalling pathways mediated by 1α,25$(OH)_2D3$ receptor (vitamin D receptor), PTH/PTHrP receptor, and gp130 similarly induce ODF (also called stromal osteoclast forming activity, SOFA) as a membrane associated factor in osteoblasts/stromal cells. Osteoclast progenitors of the monocyte–macrophage lineage recognize ODF bound to the cell membrane of osteoblasts/stromal cells through cell–cell interaction, then differentiate into osteoclasts which express TRAP and calcitonin receptors (CTR). M-CSF produced by osteoblasts/stromal cells is a prerequisite for both proliferation and differentiation of osteoclast progenitors.

Molecular cloning of *Odf*

The mouse bone marrow-derived stromal cell line ST2 supports osteoclast formation from mouse spleen cells in the presence of 1α,25$(OH)_2D3$ and dexamethasone (Udagawa et al 1989). Radioactive OPG/OCIF bound to a single class of high affinity binding sites on ST2 cells induced by 1α,25$(OH)_2D3$ and dexamethasone. Using OPG/OCIF as a probe, Yasuda et al (1998b) cloned a cDNA with an open reading frame encoding 316 amino acid residues from an expression library of ST2 cells. The OPG/OCIF-binding molecule was a type H transmembrane protein of the TNF ligand family (Fig. 2). Since the OPG/OCIF-binding molecule satisfied major criteria of ODF, this molecule was renamed ODF (Yasuda et al 1998b). Lacey et al (1998) also cloned a ligand for OPG (OPGL) from

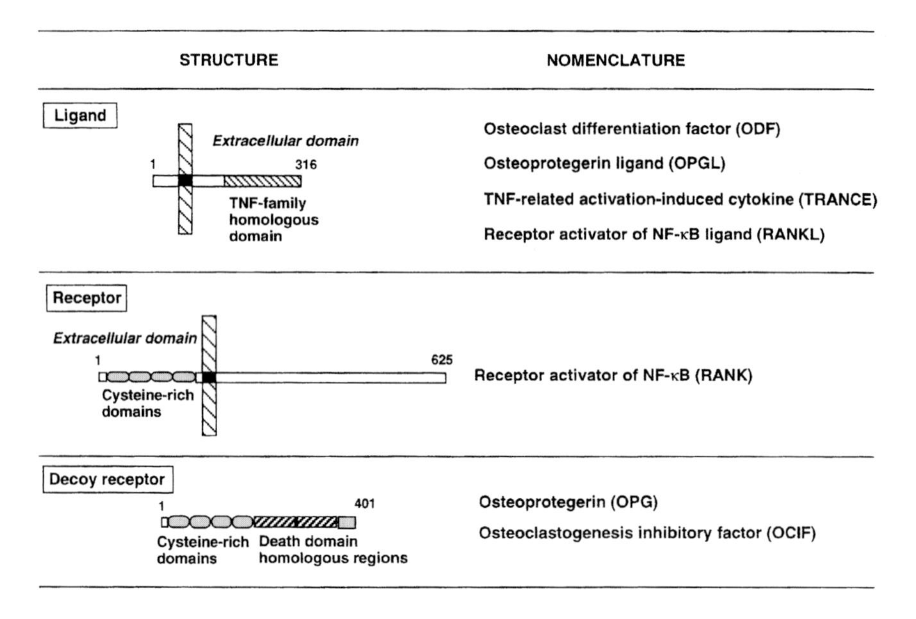

FIG. 2. A schematic representation of the ligand, receptor and decoy receptor of the new TNF receptor-ligand family involved in osteoclast formation. Different nomenclatures for the same ligand, receptor, and decoy receptor of the new TNF receptor-ligand family involved in osteoclast formation are listed.

an expression library of the murine myelomonocytic cell line 32D. OPGL was identical to ODF. Molecular cloning of ODF/OPGL demonstrated that it is identical to TRANCE (TNF-related activation-induced cytokine) cloned from murine T cell hybridomas (Wong et al 1997) and RANKL (receptor activator of NF-κB ligand) cloned from a cDNA library of murine thymoma EL40.5 cells (Anderson et al 1997). Thus, ODF, OPGL, TRANCE and RANKL are different names for the same molecule (Fig. 2) (Suda et al 1999).

RANK is a signalling receptor for ODF

The research group who cloned RANKL cloned a new member of the TNF receptor family termed 'RANK' from a cDNA library of human dendritic cells (Anderson et al 1997). The mouse RANK cDNA encoded a type 1 transmembrane protein of 625 amino acid residues (Fig. 2). A genetically engineered soluble form of RANK, an extracellular domain of RANK, inhibited ODF-mediated osteoclast formation (Nakagawa et al 1998). Polyclonal antibodies against the extracellular domain of RANK induced osteoclast formation in pure spleen cell culture in the presence of M-CSF, probably through clustering of

RANK (Nakagawa et al 1998). In contrast, the Fab fragment of the antibody blocked the binding of ODF to RANK, causing the inhibition of the ODF-mediated osteoclastogenesis (Nakagawa et al 1998). These results suggest that RANK is the sole signalling receptor for ODF in inducing osteoclast differentiation (Suda et al 1999).

Role for ODF in osteoclast differentiation and activation

A genetically engineered soluble form of ODF (sODF) together with M-CSF induced osteoclast formation from spleen cells in the absence of osteoblasts/stromal cells, which was abolished completely by simultaneous addition of OPG/OCIF (Yasuda et al 1998b). Human osteoclasts were also formed in cultures of human peripheral blood mononuclear cells in the presence of sODF and human M-CSF (Matsuzaki et al 1998). This suggests that the regulatory mechanisms of human osteoclast formation are essentially the same as those of mouse osteoclast formation. Recently, Kong et al (1999) reported that *Opgl* (*Odf*) knockout mice exhibited typical osteopetrosis with no osteoclasts and total occlusion of bone marrow space within endosteal bone. This suggests that ODF is an absolute requirement for osteoclast development.

Survival, fusion and pit-forming activity of osteoclasts were also induced by adding sODF (Jimi et al 1999a). Treatment of osteoclasts with OPG/OCIF suppressed their survival, fusion and pit-forming activity supported by sODF. These results indicate that ODF induces not only osteoclast differentiation but also osteoclast activation (Fig. 3).

IL-1 directly induces osteoclast activation

We have reported that IL-1α stimulates osteoclast formation in co-culture of osteoblasts and spleen cells (Akatsu et al 1991). Osteoclast formation induced by IL-1α in the co-culture was completely inhibited by adding OPG/OCIF. Thus, the effect of IL-1α on osteoclast differentiation is an indirect effect through ODF production by osteoblastic cells. Besides its indirect action in osteoclast differentiation, we recently found that IL-1α directly acts on mature osteoclasts to stimulate their function (Jimi et al 1999b). When IL-1 was added to purified osteoclasts, this inflammatory cytokine prolonged the survival and stimulated pit-forming activity of purified osteoclasts in the absence of osteoblastic cells. In fact, osteoclasts expressed IL-1 type 1 receptor (IL-1R) (Jimi et al 1999b) as well as RANK (Jimi et al 1999a). Treatment of purified osteoclasts with either IL-1 or sODF similarly activated NF-κB within 30 min. IL-1 receptor antagonist (IL-1ra) inhibited IL-1-induced NF-κB activation, but it did not inhibit ODF-induced activation of purified osteoclasts (Suda et al 1999). In contrast, OPG/OCIF

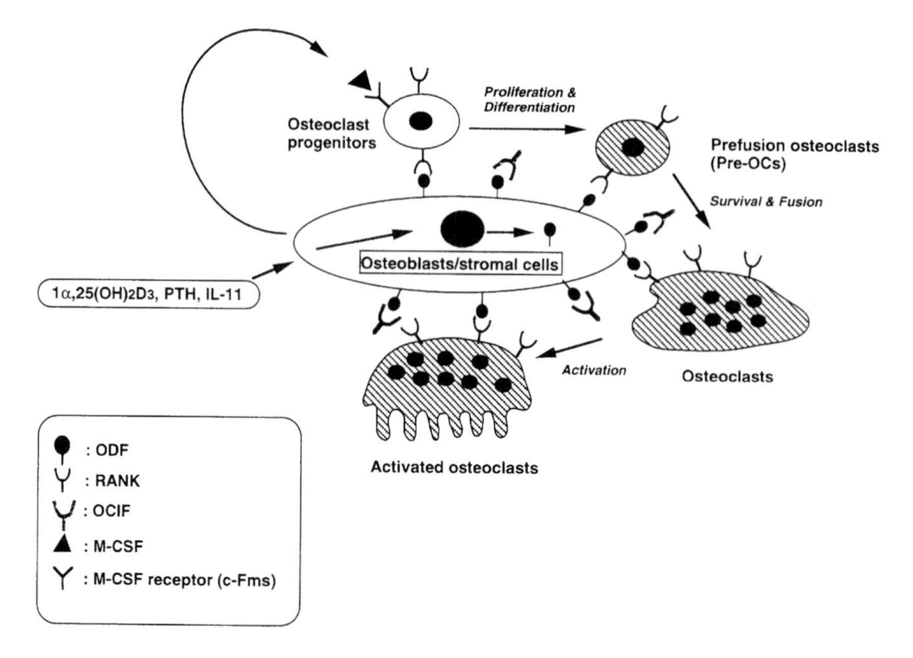

FIG. 3. A schematic representation of osteoclast differentiation and function supported by osteoblasts/stromal cells.

inhibited ODF-induced NF-κB activation, but it did not inhibit IL-1-induced NF-κB activation (Jimi et al 1999a). Thus, it is likely that ODF and IL-1 share a common intracellular signalling pathway of NF-κB in purified osteoclasts (Fig. 4).

Signal transduction of TNF family members

Recently, much attention has been focused on the role of TNFα in osteoclastic bone resorption, in particular in inflammatory bone diseases (Abu-Amer et al 1997, Assuma et al 1998, Merkel et al 1999). TNFα induces its biological response through two cell surface receptors termed TNF receptor type 1 (TNFR1, p55) and TNF receptor type 2 (TNFR2, p75) (Tartaglia & Goeddel 1992). TNF-receptor associated factor 2 (TRAF2) has been shown to be an adaptor protein for mediating the signals induced by TNFR1 and TNFR2 (Fig. 5) (Yeh et al 1997, Arch et al 1998). In addition, the binding of TNFα to TNFR1 triggers apoptosis in many types of cells.

ODF induces its signals through RANK. Recent studies have demonstrated that the cytoplasmic tail of RANK interacts with TRAFs 1, 2, 3, 5 and 6 (Fig. 5) (Wong et al 1999). The interaction of these TRAFs with RANK appears to be involved in osteoclast differentiation and activation. IL-1R is shown to use TRAF6 as a signal

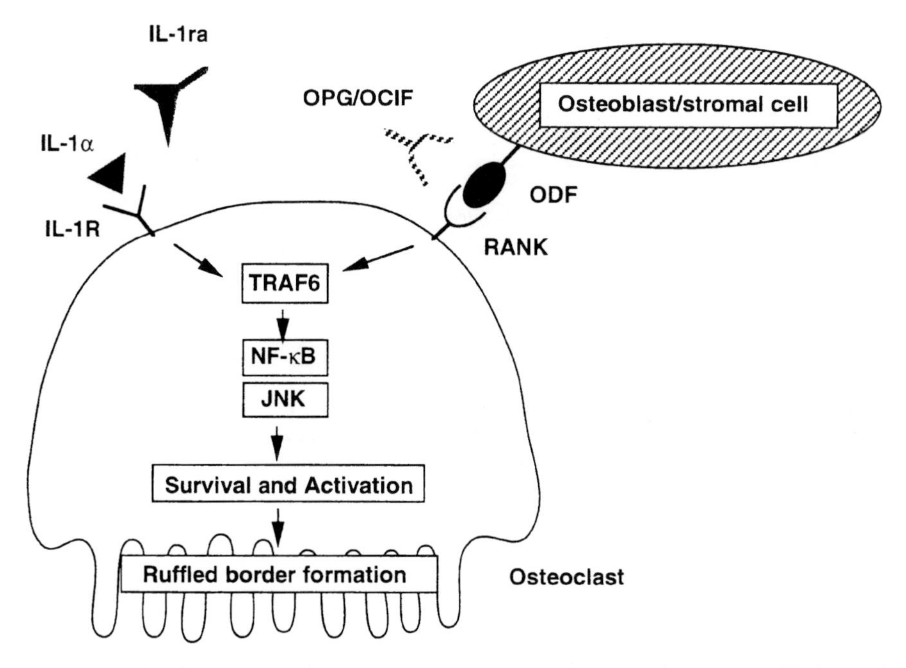

FIG. 4. Signals induced by IL-1α and ODF in osteoclasts. Osteoclasts express IL-1 type 1 receptor (IL-1R) and RANK. IL-1α and ODF similarly induce the survival, fusion and activation of osteoclasts in the absence of osteoblasts/stromal cells. Both IL-1α and ODF activate NF-κB and c-Jun N-terminal kinase (JNK) through their respective receptors. TRAF6 appears to be a common signalling molecule involved in osteoclast function induced by ODF and IL-1α. IL-1 receptor antagonist and OPG/OCIF inhibit IL-1α- and ODF-induced signals, respectively.

transducing molecule (Fig. 5) (Cao et al 1996). Recently, Lomaga et al (1999) and Naito et al (1999) independently reported that *Traf6* knockout mice developed severe osteopetrosis. It was shown that in spite of the presence of similar numbers of osteoclasts between TRAF6 knockout mice and wild-type mice, osteoclasts in *Traf6* knockout mice failed to form ruffled borders (Lomaga et al 1999). This suggests that TRAF6-induced signals are involved in osteoclast activation, which confirms our previous findings that IL-1 directly induces activation of mature osteoclasts (Fig. 4).

TNFα induces osteoclast differentiation by a mechanism independent of the ODF–RANK signalling system

Very recently, we found that TNFα induced osteoclast formation by a mechanism independent of the ODF–RANK interaction (Kobayashi et al 2000). When mouse bone marrow cells were cultured with M-CSF for 3 days and non-adherent cells

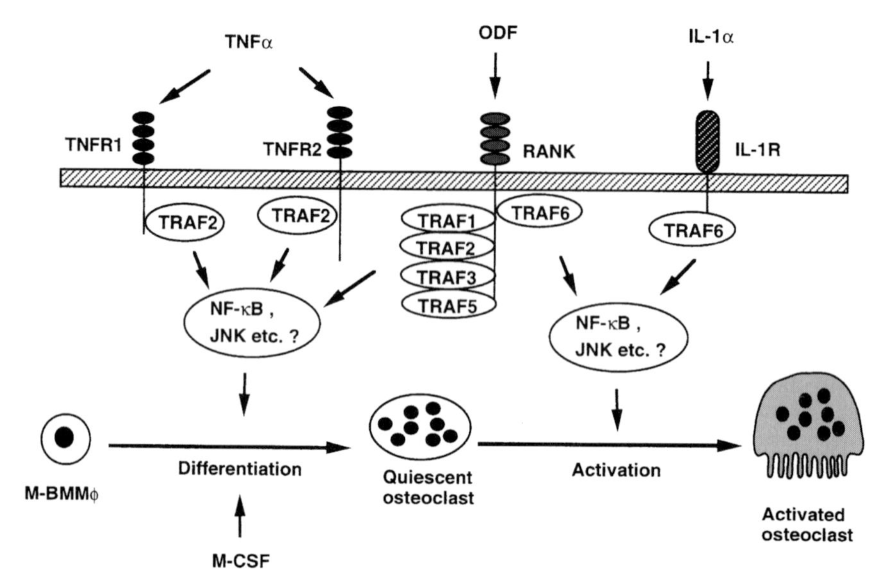

FIG. 5. Schematic representation of ligand-receptor systems in osteoclast differentiation and function regulated by TNFα, ODF and IL-1α. TNFα and ODF independently stimulate osteoclast differentiation. Osteoclast differentiation induced by TNFα occurs via TNFR1 and TNFR2 expressed by osteoclast precursors. ODF induces osteoclast differentiation through RANK-mediated signals. M-CSF is a common factor required for both TNFα- and ODF-induced osteoclast differentiation. Activation of osteoclasts is induced by ODF and IL-1α through RANK and type 1 IL-1α receptor (IL-1R), respectively. TRAF6 appears to be involved commonly in osteoclast activation induced by ODF and IL-1α. Common signalling cascades such as NF-κB and JNK activation may be involved in the differentiation and activation of osteoclasts induced by TNFα, ODF/RANKL, and IL-1α.

were removed, adherent cells of uniform size and shape remained on the culture dish. Almost all of the adherent cells strongly expressed macrophage-specific antigens such as Mac-1, Moma-2 and F4/80. We call these adherent cells M-CSF-dependent bone marrow macrophages (M-BMMϕ), which contained seldom alkaline phosphatase-positive osteoblasts/stromal cells.

ODF induced differentiation of M-BMMϕ into tartrate-resistant acid phosphatase (TRAP)-positive mononuclear and multinucleated osteoclasts within 3 days, in the presence of M-CSF (Kobayashi et al 2000). Mouse TNFα similarly induced osteoclast formation in the presence of M-CSF. Human TNFα induced only a few osteoclasts at a high concentration in the presence of M-CSF. IL-1α failed to induce osteoclast formation in M-BMMϕ cultures even in the presence of M-CSF. Mouse TNFα binds to both mouse TNFR1 and TNFR2, whereas human TNFα binds only to mouse TNFR1 (Tartaglia & Goeddel 1992). This suggests that both TNFR1- and TNFR2-mediated signals are required for the differentiation of osteoclasts induced by TNFα (Fig. 5).

Reverse transcriptase (RT)-PCR indicated that M-BMMϕ expressed not only RANK and c-Fms (M-CSF receptor) but also TNFR1 and TNFR2 (Kobayashi et al 2000). Osteoclast formation induced by sODF was completely inhibited by adding OPG/OCIF, but that induced by mouse TNFα was not. Adding antibodies against TNFR1 and TNFR2 blocked osteoclast formation induced by mouse TNFα but not by sODF. These results indicate that TNFα induces osteoclast differentiation through TNF receptors but not RANK, suggesting that TNFα stimulates osteoclast formation by a mechanism independent of the ODF–RANK interaction (Kobayashi et al 2000).

To further confirm this possibility, we prepared M-BMMϕ from both *Tnfr1* and *Tnfr2* knockout mice. M-BMMϕ obtained from the wild-type mice differentiated into osteoclasts in response to sODF or TNFα in the presence of M-CSF. However, M-BMMϕ prepared from *Tnfr1* knockout mice differentiated into osteoclasts in response to sODF, but they did not differentiate into osteoclasts in response to mouse TNFα. Similarly, *Tnfr2* knockout mouse-derived M-BMMϕ differentiated into osteoclasts in response to sODF, but osteoclast differentiation induced by mouse TNFα was markedly decreased in *Tnfr2* knockout mouse-derived M-BMMϕ cultures (Kobayashi et al 2000). These results confirm that both TNFR1- and TNFR2-mediated signals are important for TNFα-induced osteoclast differentiation.

Autoradiography using labelled calcitonin showed that numerous grains due to the binding of labelled calcitonin were observed on TRAP-positive cells induced by mouse TNFα together with M-CSF even in the presence of OPG/OCIF. Numerous grains were also observed on TRAP-positive cells induced by sODF plus M-CSF. No calcitonin receptor-positive cells appeared in the cultures treated with ODF in the presence of OPG/OCIF. These results indicate that TRAP-positive cells induced by mouse TNFα as well as sODF expressed calcitonin receptors, but that TNFα-induced osteoclast formation occurs by a mechanism independent of the ODF-RANK interaction (Kobayashi et al 2000).

TNFα induces osteoclast differentiation but not osteoclast activation

To examine whether TNFα induces not only osteoclast differentiation but also osteoclast activation, M-BMMϕ were cultured on dentine slices in the presence of mouse TNFα, M-CSF and OPG/OCIF. Some cultures were treated with IL-1α as well. After culture for 6 days, similar numbers of TRAP-positive osteoclasts were formed on dentine slices irrespective of the presence and absence of IL-1α. However, no resorption pits were detected in M-BMMϕ cultures treated with mouse TNFα and M-CSF. Resorption pits on dentine slices were observed only in the presence of TNFα and M-CSF together with IL-1α. These results suggest that TNFα stimulates differentiation of osteoclasts, but not activation of osteoclasts. In contrast, IL-1α does not induce differentiation of osteoclasts in M-BMMϕ cultures, but stimulates pit-forming activity of osteoclasts (Kobayashi et al 2000).

Regulation of osteoclast differentiation and activation by TNF ligand family members

Figure 5 summarizes the regulation of osteoclast differentiation and activation by TNF ligand family members. TNFα induces osteoclast differentiation through TNFR1 and TNFR2 in the presence of M-CSF. TRAF2-mediated signals may be involved in osteoclast differentiation induced by TNFα. RANK-induced signals also stimulate osteoclast differentiation in the presence of M-CSF. TRAF2 or its related TRAFs appear to be involved in RANK-induced osteoclast differentiation. On the other hand, RANK- and IL-1R-mediated signals induce osteoclast activation. TRAF6 appears to be involved in osteoclast activation induced by ODF and IL-1 as a common signalling molecule (Lomaga et al 1999). NF-κB and c-Jun N-terminal kinase (JNK) activation may be involved in differentiation and activation of osteoclasts induced by TNFα, ODF and IL-1 (Fig. 5).

Conclusion and future directions

In conclusion, TNFα stimulates osteoclast differentiation by a mechanism independent of the ODF–RANK interaction. Firstly, mouse TNFα stimulates TRAP-positive osteoclast formation in M-BMMϕ cultures. The potency of human TNFα was much weaker than that of mouse TNFα. Secondly, osteoclast formation induced by mouse TNFα was inhibited by antibody against TNFR1 and TNFR2, but not inhibited by OPG/OCIF nor by the Fab fragment of anti-RANK antibody. Thirdly M-BMMϕ prepared from *Tnfr1* knockout mice and *Tnfr2* knockout mice showed low capacity to differentiate into osteoclasts in response to TNFα. TNFα-induced osteoclasts formed resorption pits only in the presence of IL-1α. From these results, we conclude that TNFα induces osteoclast differentiation but not osteoclast activation. Osteoclast differentiation induced by TNFα occurs by a mechanism independent of the ODF-RANK interaction.

Under physiological conditions, combination of ODF and M-CSF appears to play a major role in regulating osteoclast differentiation and activation. In contrast, under pathological circumstances, such as rheumatoid arthritis, periodontal diseases and postmenopausal osteoporosis, TNFα may play a major role in inducing osteoclast differentiation in the presence of M-CSF, and IL-1 induces osteoclast activation (Fig. 6). Thus, the two inflammatory cytokines,

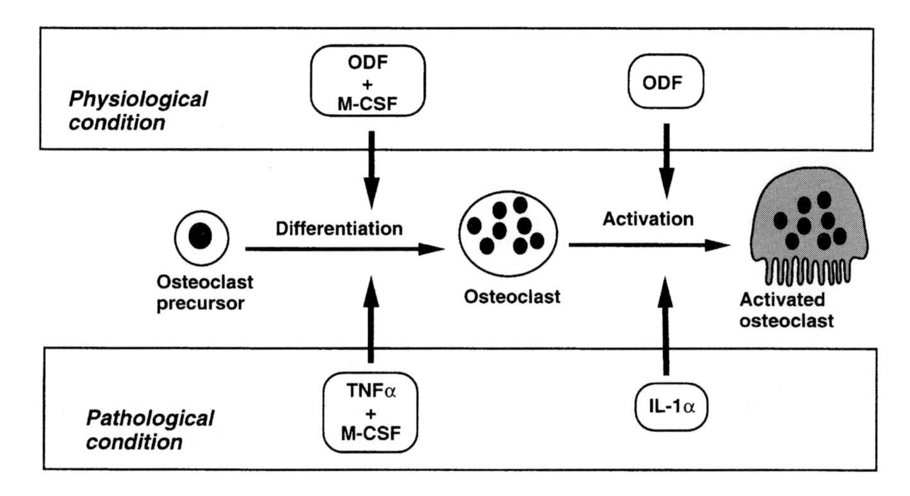

FIG. 6. Regulation of osteoclast differentiation and function under physiological and pathological conditions.

TNFα and IL-1, can be substituted for ODF in the whole process of osteoclastic bone resorption *in vitro* (Fig. 6). Further studies will elucidate the biological importance of TNFα and IL-1-induced bone resorption *in vivo*.

References

Abu-Amer Y, Ross FP, Edwards J, Teitelbaum SL 1997 Lipopolysaccharide-stimulated osteoclastogenesis is mediated by tumor necrosis factor via its p55 receptor. J Clin Invest 100:1557–1565

Akatsu T, Takahashi N, Udagawa N et al 1991 Role of prostaglandins in interleukin-1-induced bone resorption in mice *in vitro*. J Bone Miner Res 6:183–189

Anderson DM, Maraskovsky E, Billingsley WL et al 1997 A homologue of the TNF receptor and its ligand enhance T-cell growth and dendritic-cell function. Nature 390:175–179

Arch RH, Gedrich RW, Thompson CB 1998 Tumor necrosis factor receptor-associated factors (TRAFs)—a family of adapter proteins that regulates life and death. Genes Dev 12: 2821–2830

Assuma R, Oates T, Cochran D, Amar S, Graves DT 1998 IL-1 and TNF antagonists inhibit the inflammatory response and bone loss in experimental periodontitis. J Immunol 160:403–409

Bucay N, Sarosi I, Dunstan CR et al 1998 Osteoprotegerin-deficient mice develop early onset of osteoporosis and arterial calcification. Genes Dev 12:1260–1268

Cao Z, Xiong J, Takeuchi M, Kurama T, Goeddel DV 1996 TRAF6 is a signal transducer for interleukin-1. Nature 383:443–446

Chambers TJ, Owens JM., Hattersley G, Jat PS, Noble MD 1993 Generation of osteoclast-inductive and osteoclastogenic cell lines from the H-2K^{b}tsA58 transgenic mouse. Proc Natl Acad Sci USA 90:5578–5582

Felix R, Cecchini MG, Fleisch H 1990 Macrophage colony stimulating factor restores *in vivo* bone resorption in the op/op osteopetrotic mouse. Endocrinology 127:2592–2594

Jimi E, Akiyama S, Tsurukai T et al 1999a Osteoclast differentiation factor acts as a multifunctional regulator in murine osteoclast differentiation and function. J Immunol 163:434–442

Jimi E, Nakamura I, Duong LT et al 1999b Interleukin 1 induces multinucleation and bone-resorbing activity of osteoclasts in the absence of osteoblasts/stromal cells. Exp Cell Res 247:84–93

Kobayasi K, Takahashi N, Jimi E et al 2000 Tumor necrosis factor α stimulates osteoclast differentiation independent of the ODF/RANK signaling system. J Exp Med 191:275–285

Kong YY, Yoshida H, Sarosi I et al 1999 OPGL is a key regulator of osteoclastogenesis, lymphocyte development and lymph-node organogenesis. Nature 397:315–323

Lacey DL, Timms E, Tan HL et al 1998 Osteoprotegerin ligand is a cytokine that regulates osteoclast differentiation and activation. Cell 93:165–176

Lomaga NIA, Yeh WC, Sarosi I et al 1999 TRAF6 deficiency results in osteopetrosis and defective interleukin-1, CD40, and LPS signaling. Genes Dev 13:1015–1024

Matsuzaki K, Udagawa N, Takahashi N et al 1998 Osteoclast differentiation factor (ODF) induces osteoclast-like cell formation in human peripheral blood mononuclear cell cultures. Biochem Biophys Res Commun 246:199–204

Merkel KD, Erdmann JM, McHugh KP, Abu-Amer Y, Ross FP, Teitelbaum SL 1999 Tumor necrosis factor-α mediates orthopedic implant osteolysis. Am J Pathol 154:203–210

Mizuno A, Amizuka N, Irie K et al 1998 Severe osteoporosis in mice lacking osteoclastogenesis inhibitory factor/osteoprotegerin. Biochem Biophys Res Commun 247:610–615

Nakagawa N, Kinosaki M, Yamaguchi K et al 1998 RANK is the essential signaling receptor for osteoclast differentiation factor in osteoclastogenesis. Biochem Biophys Res Commun 253:395–400

Naito A, Azuma S, Tanaka S et al 1999 Severe osteopetrosis, defective interleukin-1 signalling and lymph node organogenesis in TRAF6-deficient mice. Genes Cells 4:353–362

Simonet WS, Lacey DL, Dunstan CR et al 1997 Osteoprotegerin: a novel secreted protein involved in the regulation of bone density. Cell 89:309–319

Suda T, Takahashi N, Martin TJ 1992 Modulation of osteoclast differentiation. Endocr Rev 13:66–80 (erratum: 1992 Endocr Rev 13:191)

Suda T, Takahashi N, Martin TJ 1995 Modulation of osteoclast differentiation: update 1995 Endocr Rev Monographs 4:266–270

Suda T, Takahashi N, Udagawa N, Jimi E, Gillespie MT, Martin TJ 1999 Modulation of osteoclast differentiation and function by the new members of the tumor necrosis factor receptor and ligand families. Endocr Rev 20:345–357

Takahashi N, Akatsu T, Udagawa N et al 1988 Osteoblastic cells are involved in osteoclast formation. Endocrinology 123:2600–2602

Tartaglia LA, Goeddel DV 1992 Two TNF receptors. Immunol Today 13:151–153

Tsuda E, Goto M, Mochizuki S et al 1997 Isolation of a novel cytokine from human fibroblasts that specifically inhibits osteoclastogenesis. Biochem Biophys Res Commun 234:137–142

Udagawa N, Takahashi N, Akatsu T et al 1989 The bone marrow-derived stromal cell lines MC3T3-G2/PA6 and ST2 support osteoclast-like cell differentiation in co-cultures with mouse spleen cells. Endocrinology 125:1805–1813

Yasuda H, Shima N, Nakagawa N et al 1998a Identity of osteoclastogenesis inhibitory factor (OCIF) and osteoprotegerin (OPG): a mechanism by which OPG/OCIF inhibits osteoclastogenesis *in vitro*. Endocrinology 139:1329–1337

Yasuda H, Shima N, Nakagawa N et al 1998b Osteoclast differentiation factor is a ligand for osteoprotegerin/osteoclastogenesis-inhibitory factor and is identical to TRANCE/RANKL. Proc Natl Acad Sci USA 95:3597–3602

Yeh WC, Shahinian A, Speiser D et al 1997 Early lethality, functional NF-κB activation, and increased sensitivity to TNF-induced cell death in TRAF2-deficient mice. Immunity 7:715–725

Yoshida H, Hayashi S, Kunisada T et al 1990 The murine mutation osteopetrosis is in the coding region of the macrophage colony stimulating factor gene. Nature 345:442–444

Wong BR, Rho J, Arron J et al 1997 TRANCE is a novel ligand of the tumor necrosis factor receptor family that activates c-Jun N-terminal kinase in T cells. J Biol Chem 272: 25190–25194

Wong BR, Josien R, Choi Y 1999 TRANCE is a TNF family member that regulates dendritic cell and osteoclast function. J Leukoc Biol 65:715–724

DISCUSSION

Burger: You have demonstrated nicely how the osteoblast-mediated osteoclast generation is part of the haematopoietic system — it uses the same mechanisms as developing leukocytes. You saw homologous molecules to ODF in other systems, including in dendritic cells. Is it also functioning as a decoy molecule there?

Suda: At first we thought that ODF was present only in bone, but this was not the case, because ODF was found in dendritic cells and T cells as well. ODF therefore may be important not only for osteoclast differentiation, but also for immune responses.

Beresford: In the Amgen experiment that you described, are you talking about a complete rescue of the osteopetrotic phenotype or just the appearance of a few osteoclasts?

Suda: When TNFα is administered into RANK knockout mice, they start to generate osteoclasts. These osteoclasts, however, cannot resorb bone. IL-1 is necessary for TNFa-induced osteoclasts to resorb bone.

Beresford: Following on from that, I was interested in your suggestion that there is a TNF/IL-1 dependent pathway which becomes very important in osteoclastogenesis in rheumatoid arthritis and related conditions. Have you tested that formally in an animal model? If you take an animal model of chronic rheumatoid arthritis with excessive joint destruction and you infuse OPG into those animals, you do not see any improvement. If I have understood correctly, this argues against ODF having an important role in the maintenance of chronic joint destruction; it's actually a TNF/IL-1-mediated pathway.

Suda: We postulate that the ODF-dependent pathway is important preferentially in physiologic conditions, whereas the ODF-independent pathway is important in pathological conditions. However, I cannot exclude the possibility that the ODF-dependent pathway is important in pathological conditions as well.

Russell: OPG is being developed for clinical use. It seems to work in just about everything that has been tried, including arthritis models and tumour-driven bone destruction.

Beresford: So this TNF/IL-1 'auxiliary pathway' may not be that critical in the real world setting.

Suda: We need further *in vivo* experiments to prove our hypothesis.

Russell: One of the interesting things about osteoclasts is the duration of their action and their lifespan. It seems increasingly that they have a finite lifespan which can be altered in disease states — they undergo apoptosis. What is known in relation to the RANK system about whether those factors influence lifespan? M-CSF is meant to.

Suda: The lifespan of a mature osteoclast is less than 24 h in the absence of ODF and M-CSF. When ODF or M-CSF is present, osteoclasts can survive for several days. In *in vivo* conditions many osteoblastic stromal cells are present around osteoclasts. Osteoblastic stromal cells are capable of producing ODF and M-CSF.

Burger: If this is true, how do you explain the tunnelling that osteoclasts can make in secondary osteomes. They can dig a tunnel in compact bone, which is not lined with osteoblasts or lining cells. Where does the ODF come from there?

Suda: In our hands, ODF acts as a membrane-bound factor; we do not have any evidence that ODF acts as a soluble factor. When osteoblastic cells and spleen cells are co-cultured, but separated by a membrane filter, no osteoclasts are formed.

Burger: But how do you explain this *in vivo* observation that osteoclasts are able to tunnel deep into the compact bone, and once inside they cannot encounter osteoblasts, stromal cells or lining cells?

Blair: There have been studies looking at that. The active site is full of proteinases and an acid pH. When they encounter osteoblasts, the cells are immediately apoptotic (Elmardi et al 1990, Taniwaki & Katchburian 1998). So they are unlikely to have this sort of cell surface interaction. On the other hand, at their basolateral surfaces, the osteoclasts do maintain interaction with stromal cells. The question would be whether there would be enough time in this tunnelling activity for the stromal cells to follow on. Osteoblasts are quite actively proliferative and expand to cover the available surface. I would suspect that osteoblast–osteoclast interactions at their lateral surfaces would be the major mechanism.

Suda: Osteocytes may play a similar role in stimulating osteoclast function.

Beresford: To my knowledge mice don't exhibit cortical tunnelling. Anyway in those species in which the process does occur, there is no evidence of any physical separation between the cutting cone and the cells that follow on in its wake. The idea that you have osteoblasts tunnelling through bone in isolation is actually not true: they are always in close proximity to other cells and/or their processes. I think the potential is there.

Blair: The tunnelling is mainly an artefact of hyperparathyroidism in humans with renal failure.

Burger: It is also seen in animals. The mouse is just too small and too short-lived for remodelling to take place, but it can occur if it is needed.

Blair: In humans or animals that have mature skeleton, there is a complex microarchitecture, particularly at the ends of the long bones, which is very important. If you look at where the bone is resorbed by osteoclasts, and bone is laid down by osteoblasts, this architecture is maintained. It is difficult to see how this kind of general expression of ODF/RANK ligand is managing this higher architectural behaviour.

Burger: The higher-order control can only be explained if mechanical stimuli are considered.

Suda: I would like to propose a new concept. Bone loss by osteoblast-induced osteoclasts can be followed by osteoblastic bone formation. However, TNF-α/IL-1-induced bone loss cannot be followed by osteoblastic bone formation, since ODF is induced on the plasma membrane of osteoblastic stromal cells in response to several bone resorbing factors. TNFa acts directly on osteoclast progenitors. I think that some interaction between Cbfa1 and RANK ligand/ODF may be involved in this connection.

Kronenberg: That is an interesting model. What it potentially leaves out is the other model whereby osteoclastic release of growth factors in bone matrix by itself might be able to stimulate bone formation. If that is an important stimulus in bone formation, then you would predict that TNF would release those factors during osteoclastic resorption. So if you don't see formation in inflammatory disease, either that release of growth factors is not important after all, or else inflammatory mediators can block the action. So, the relationship between bone formation and resorption may be very complicated.

Karsenty: If this mediator released by the osteoclasts that we are talking about in the *in vitro* assay is important in controlling bone formation, wouldn't you expect that in animal models where there are no osteoclasts that there will be no bone formation? We should not lose sight that in the absence of osteoclasts bone formation is not affected.

Beresford: The work that you did with the inducible osteoblast ablation model is very nice (Corral et al 1998), but it does not preclude the possibility that cells earlier in the lineage are influencing osteoclast differentiation.

Karsenty: I was thinking about the osteoclasts in terms of function, not differentiation.

Beresford: I think you have to distinguish between a skeleton that is actually growing and modelling, and one that's reached a steady-state. The interactions between the two cell populations may be fundamentally different in these two contexts.

Karsenty: Our experiment was not designed to study differentiation, but function. In the absence of bone formation, the function of the osteoclasts was not affected. This is different from differentiation. This work led us to imagine a novel regulation of bone remodelling, which *in vivo* seems to be more powerful than the alternative mechanism.

Hall: You mentioned mononuclear osteoclasts just once in your talk. Are they functional? And is this a way of cutting short the time dependence that you need for an interaction of osteoclasts with osteoblasts?

Suda: Tim Chambers has shown that mononuclear osteoclasts can also resorb bone.

Hall: Certainly, if one goes away from mammals to organisms such as fish, mononuclear resorption is the dominant situation. I had the impression that it's very rare in mammals.

Poole: It is very much a feature of rheumatoid arthritis, particularly with respect to regions at the cartilage–bone interface.

Blair: There is a question that has been puzzling me. In the normal lymph node there are T cells that bear lots of RANK ligand/ODF, there is plenty of CSF-1 and there are macrophage precursors. Why is it that there are no osteoclasts?

Suda: That's an important question. At present, we do not have any good answers to explain the exclusive localization of osteoclasts in bone.

Russell: Presumably the stem cells aren't there.

Blair: But you can get dandy osteoclasts from blood or spleen macrophages.

Burger: My question at the beginning of the discussion was related to that: is the decoy system also working to prevent this going on? Otherwise, you would have many osteoclasts.

Suda: Yes, many drug companies are interested in developing OPG/OCIF as a new drug. They also would like to get a small molecule that has the same inhibitory activity. I understand that Roussel-Uclaf is now trying to develop such a molecule.

References

Corral DA, Amling M, Priemel M et al 1998 Dissociation between bone resorption and bone formation in osteopenic transgenic mice. Proc Natl Acad Sci Usa 95:13835–13840

Elmardi AS, Katchburian MV, Katchburian E 1990 Electron microscopy of developing calvaria reveals images that suggest that osteoclasts engulf and destroy osteocytes during bone resorption. Calcif Tissue Int 46:239–245

Taniwaki NN, Katchburian E 1998 Ultrastructural and lanthanum tracer examination of rapidly resorbing rat alveolar bone suggests that osteoclasts internalize dying bone cells. Cell Tissue Res 293:173–176

Clinical disorders of bone resorption

Graham Russell, Gabrielle Mueller, Claire Shipman and Peter Croucher

Division of Biochemical and Musculoskeletal Medicine, Human Metabolism & Clinical Biochemistry, University of Sheffield Medical School, Beech Hill Road, Sheffield S10 2RX and Institute of Musculoskeletal Sciences, University of Oxford, Nuffield Orthopaedic Centre, Headington, Oxford OX3 7LD, UK

Abstract. Clinical disorders in which bone resorption is increased are very common and include Paget's disease of bone, osteoporosis, and the bone changes secondary to cancer, such as occur in myeloma and metastases from breast cancer. Clinical disorders of reduced bone resorption are less common and often have a genetic basis, e.g. in osteopetrosis, and in pycnodysostosis due to cathepsin K deficiency. Bone is metabolically active throughout life. After skeletal growth is complete, remodelling of both cortical and trabecular bone continues and results in an annual turnover of about 10% of the adult skeleton. The commonest disorder of bone resorption is osteoporosis, which affects one in three women over 50 years. Its pathophysiological basis includes genetic predisposition and subtle alterations in systemic and local hormones, coupled with environmental influences. Treatment depends mainly on drugs that inhibit bone resorption, either directly or indirectly. This includes bisphosphonates, oestrogens, synthetic oestrogen-related compounds (SERMs — selective oestrogen receptor modulators) and calcitonin. The most widely used drugs for all disorders of increased bone resorption, including osteoporosis, are the bisphosphonates. Recent elucidation of their mode of action, together with the rapidly increasing knowledge of regulatory mechanisms in bone biology, offers many opportunities for the development of new therapeutic agents.

2001 The molecular basis of skeletogenesis. Wiley, Chichester (Novartis Foundation Symposium 232) p 251–271

The physiological regulation of Ca^{2+} metabolism

The physiological regulation of Ca^{2+} homeostasis involves three main organs, the gut, the kidney, and the skeleton. The fluxes of Ca^{2+} and phosphate through these organs contribute to the integration of Ca^{2+} metabolism throughout growth and adult life.

The hormonal control of Ca^{2+} metabolism can be attributed mainly to the effects of systemic hormones, especially the Ca^{2+}-regulating hormones, parathyroid hormone (PTH), 1α,25-dihydroxy vitamin D (calcitriol) and calcitonin (CT). For example, the setting of plasma Ca^{2+} concentrations is determined mainly by the renal tubular reabsorption of Ca^{2+} and the effects of parathyroid hormone (PTH) on this process. (Fig. 1).

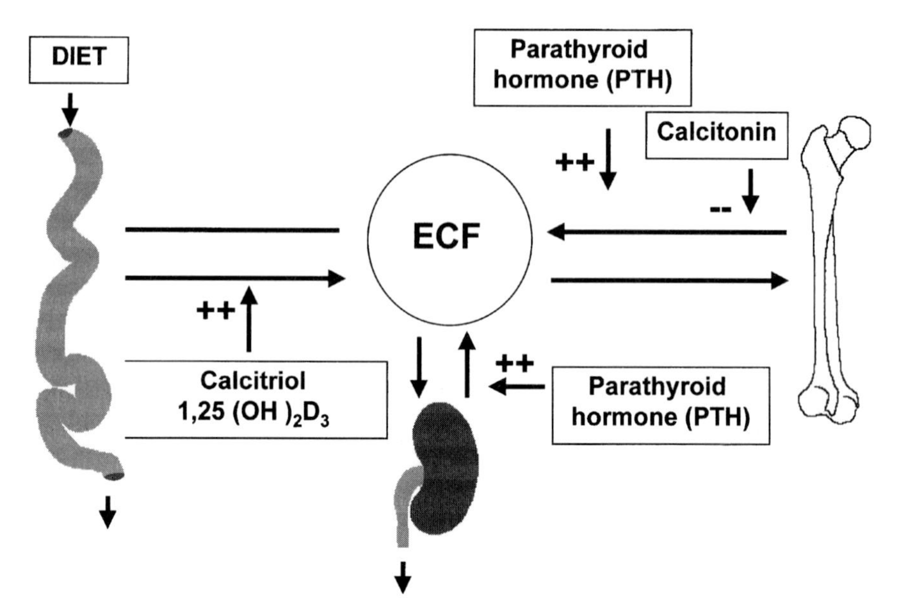

FIG. 1. The major physiological regulatory mechanisms in Ca^{2+} metabolism.

Other hormones including thyroid and pituitary hormones, and adrenal and gonadal steroids, also have major effects on the skeleton, as seen in clinical disorders in which their secretion is abnormally high or low. Many additional factors, notably cytokines and growth factors, also play a role in skeletal metabolism, in many cases by interacting locally with systemic hormones. Mechanical loading of the skeleton is also a major influence over bone remodelling.

The genetic factors that regulate skeletal development and function are gradually being identified, and recent examples include the *Cbfa1* gene for osteoblast differentiation and the RANK system for osteoclasts. Many cytokines and growth factors are involved in the induction of new bone formation, and the activation and modulation of remodelling. These and other mediators contribute not only to the physiological regulation of bone metabolism but also to the pathogenesis of skeletal diseases.

After skeletal growth is complete, remodelling of both cortical and trabecular bone continues and results in an annual turnover of about 10% of the adult skeleton. This requires the coordinated actions of osteoclasts to remove bone, and osteoblasts to replace it, and these processes may be monitored by histological means. Changes in the quality or amount of bone arise from disorders of bone modelling during growth or remodelling during adult life.

An excellent and comprehensive review of basic and clinical aspects of metabolic bone disease can be found in the Primer edited by Favus et al (1999).

The cells of bone

Osteoblasts and bone formation

Osteoblasts within trabecular bone differentiate from stromal cell precursors in bone marrow, and manufacture a complex extracellular matrix, which subsequently mineralizes. Important events for induction of new bone formation include the activation of specific 'master' genes for skeletal development, notably *Cbfa1* (Rodan & Harada 1997). Many growth factors affect bone formation (Croucher & Russell 1999). These include insulin-like growth factors (IGFs), fibroblast growth factors (FGFs), and especially members of the transforming growth factor (TGF)β family, such as the bone morphogenetic proteins (BMPs) (Wozney & Rosen 1998, Yamashita et al 1996).

There are many ways in which these and other mediators contribute to the physiological regulation of bone metabolism and to the pathogenesis of skeletal diseases.

Osteoclasts and bone resorption

Osteoclasts are multinucleated cells with a number of distinctive features (Fig. 2), and are the major cells involved in bone resorption. Osteoclasts differentiate from haematopoietic stem cell precursors under the direction of factors that include cytokines such as colony-stimulating factors (CSFs, especially M-CSF), interleukins (ILs, e.g. IL-1, IL-11), and other factors.

The recent discovery of the RANK/RANK-ligand (RANKL) system has revealed that this is one of the major regulatory systems for osteoclast recruitment and action, as discussed by Suda et al (2001, this volume) at this meeting. RANK and RANKL are members of the tumour necrosis factor (TNF) and TNF receptor families, respectively (Kong et al 1999).

Many cytokines can affect the differentiation and activity of osteoclasts and activate bone resorption. In pathological states, such as the bone destruction that occurs in rheumatoid arthritis or periodontal disease, the proinflammatory cytokines such as IL-1, IL-6 and TNF play a prominent role.

Nitric oxide (NO) is another endogenous mediator that appears to have complex effects on osteoclast function. It is likely that some of the effects of cytokines such as IL-1 and γ-interferon on bone resorption may be mediated by changes in production of NO. There are several isoforms of nitric oxide synthase (NOS). One of these (iNOS) is inducible by cytokines and can be inhibited by glucocorticosteroids.

Osteocytes and the response to mechanical effects on bone

Osteocytes lie embedded within individual lacunae in mineralized bone, and connect with each other via the canicular system. Osteocytes thus form a cellular

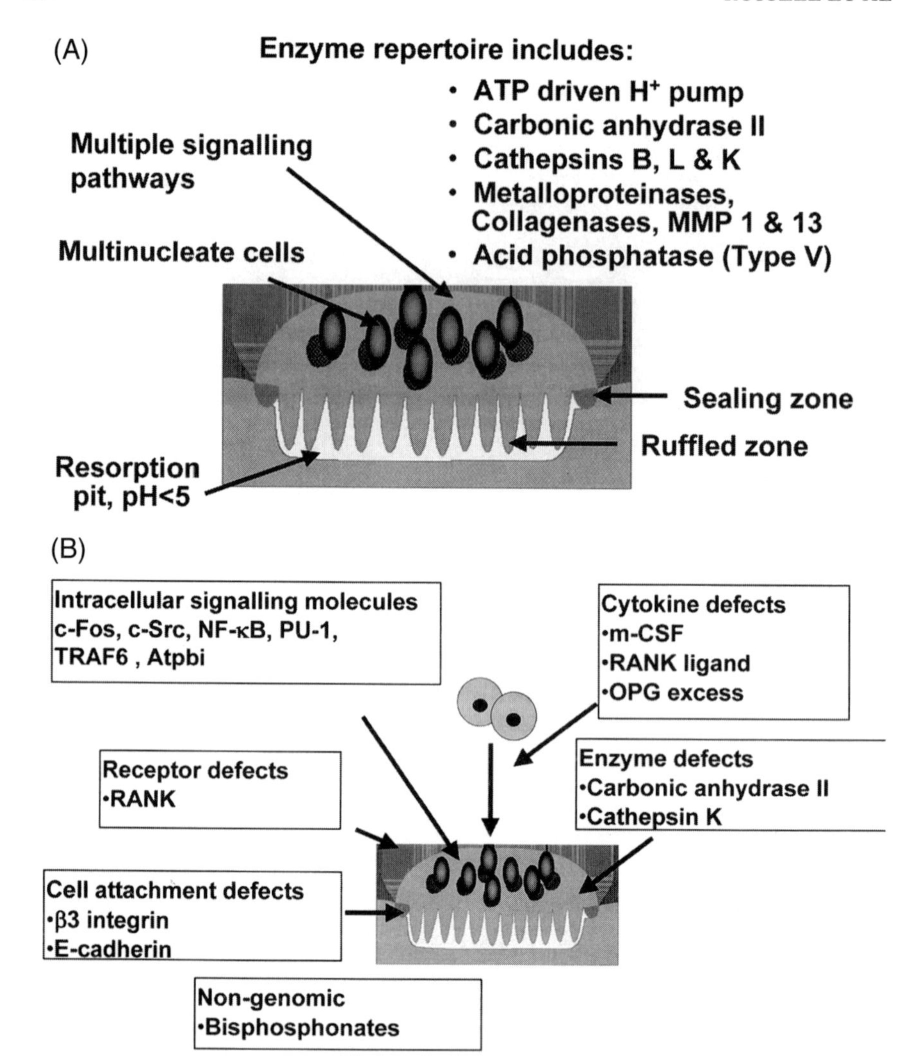

FIG. 2. (A) The main features of osteoclasts involved in bone resorption. (B) Disturbances of osteoclasts leading to impaired bone resorption and osteopetrosis.

network, much like a neural network, and are believed to be the cellular system that responds to mechanical deformation and loading in bone. Mechanical forces exert strong influences on bone shape and modelling (Lanyon 1998). Early biochemical responses to mechanical loading may include induction of prostacyclin synthesis, NO production and, later, increases in IGFs, changes in amino acid transporters and eventually increases in new bone formation (Damien et al 1998, Mason et al 1997).

The role of apoptosis in the regulation of bone turnover

Apoptosis is emerging as a major means of regulating the life span of bone cells of all lineages: osteoclasts, osteoblasts and osteocytes (Plotkin et al 1999). This may contribute to changes in bone turnover under physiological and pathological conditions. Drugs with adverse effects on bone such as glucocorticoids may induce osteocyte apoptosis, while therapeutic agents that inhibit bone resorption, including oestrogens and bisphosphonates, may shorten the lifespan of osteoclasts in this way.

Role of cytokine gene polymorphisms

There is increasing evidence for cytokine gene polymorphisms that may be linked to occurrence or severity of inflammatory on infectious disease. There are examples of such polymorphisms in the genes encoding IL-1α and β, IL-1RA, and TNFα. These genetic variants may result in differences in cytokine production.

Bone remodelling

The remodelling cycle within bone involves a similar sequence of cellular activity at both cortical and trabecular sites. An initial phase of osteoclastic resorption is followed by a more prolonged phase of bone formation mediated by osteoblasts. The amount of bone made under normal conditions corresponds very closely to the amount removed, so that in any remodelling cycle within bone, the total amount of bone tends to remain constant. Even in conditions such as Paget's disease, where there seems to be a primary acquired abnormality of osteoclasts, the subsequent formative and reparative phase of bone deposition is still closely matched to the preceding resorption. The nature of these coupling mechanisms is still poorly understood but they are very important since minor disturbances in them are likely to contribute to osteopenic or osteosclerotic states. Furthermore, any therapeutic attempts to increase bone mass, in osteoporosis for example, may be difficult to achieve unless these regulatory mechanisms can be circumvented.

The processes of bone formation and resorption may be influenced by the exposure to matrix-derived growth factors. Regulation that is achieved by factors attached to matrix, e.g. FGF attached to heparin-like glycosaminoglycans, may be a mechanism for limiting cellular responses to specific sites within bone.

A further way in which the activity of osteoblasts and osteoclasts are coordinated may be through the generation of enzymes such as plasminogen activator (PA) from osteoblasts in response to bone resorbing agents. Proteolysis of the surface matrix of bone may be an essential step in preparing it for subsequent resorption by incoming osteoclasts. Many of the agents that stimulate bone resorption (e.g. retinoids, PTH, 1,25D, IL-1) can stimulate the production of PA by osteoblast-like cells. Another mechanism may involve contraction of cells of the lining osteoblast layer in response to resorbing agents, such as PTH, thereby allowing access by osteoclasts.

Clinical assessment of bone status

Bone mass can now be measured very precisely by physical methods such as DEXA (dual energy X-ray absorptiometry), and turnover itself can be assessed using biochemical markers of bone resorption and formation.

Monitoring of bone metabolism by biochemical means depends upon measurement of enzymes and proteins released during bone formation (such as alkaline phosphatase, osteocalcin and collagen propeptides), and of degradation products produced during bone resorption. The most useful markers of bone resorption are degradation products derived from the enzymatic hydrolysis of type I collagen, particularly peptides related to regions of cross-linking with the pyridinolines.

The major diseases of bone resorption

Diseases of diminished bone resorption

The most florid examples of these are the various osteopetroses. There are now many examples of these in rodents, where the gene defects are known and in most cases have been generated in transgenic animals. Examples include mice lacking RANK or RANKL, or with defective intracellular signalling pathways such as occur after gene knockouts for *Src*, *Pu1* and other factors (Felix et al 1990, Hofbauer et al 2000, Istova 1997, Schwartzberg et al 1997). Osteopetrosis can be due to lack of differentiation of osteoclasts, or to their failure to function normally. A human disorder in which the defect is characterized is pycnodyostosis, a form of osteopetrosis that the French painter Toulouse-Lautrec is thought to have suffered from, which is associated with defective cathepsin K.

Diseases of increased bone resorption

Paget's disease. This is characterized by enlargement and deformity of bones as a result of markedly increased rates of remodelling. It is common and affects men and women with approximately equal frequency. There are marked geographic differences in prevalence, and it can affect up to 5% of more of people over 50 years of age in certain parts of the UK, such as Lancashire.

The primary defect in Paget's disease appears to reside in the osteoclast. These cells are bigger than normal, more plentiful and with more nuclei. The reason for these changes is incompletely understood. Inclusion bodies occur in the osteoclasts

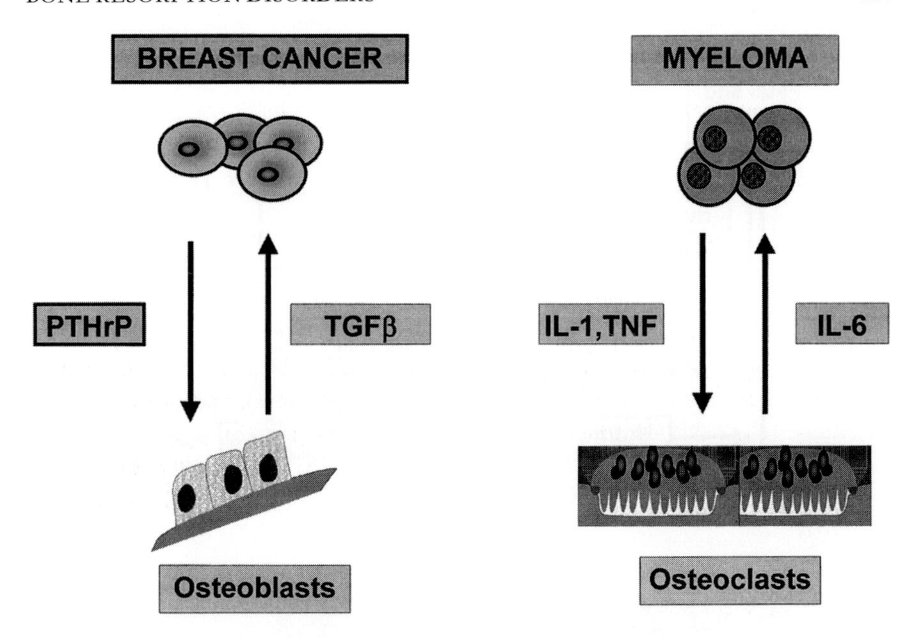

FIG. 3. Some of the regulatory mechanisms involved in bone resorption in cancer.

from Paget's disease, and also in some other disorders of bone resorption. This has led to the notion that Paget's disease has a viral origin, with members of the paramyxovirus family as putative causative agents. Measles and respiratory syncytial viruses have been implicated, and there has been much speculation about the potential involvement of dog distemper virus. The case is unproven, and current work is focusing on the role of genetic factors. In many patients there is a positive family history, and several gene loci have been associated with susceptibility. Recently a mutation in the receptor, RANK, has been found in one Paget's family, and also in a large kindred with the rare resorption disorder called familial expansile osteolysis (FEO). These mutations may cause constitutive activation of RANK, which could contribute to the increased differentiation and activity of osteoclasts.

Myeloma. This is a B cell neoplasm that produces characteristic and substantial destruction of bone with complications that include hypercalcaemia, pain and fractures. Myeloma represents a paradigm of the 'seed-soil' concept of the bi-directional interaction between tumour cells and the bone cells of the host. Thus myeloma cells may produce osteoclast-activating factors, including IL-1 and TNF, which stimulate resorption, while host bone cells produce factors such as IL-6 which stimulate the growth of myeloma cells (Fig. 3).

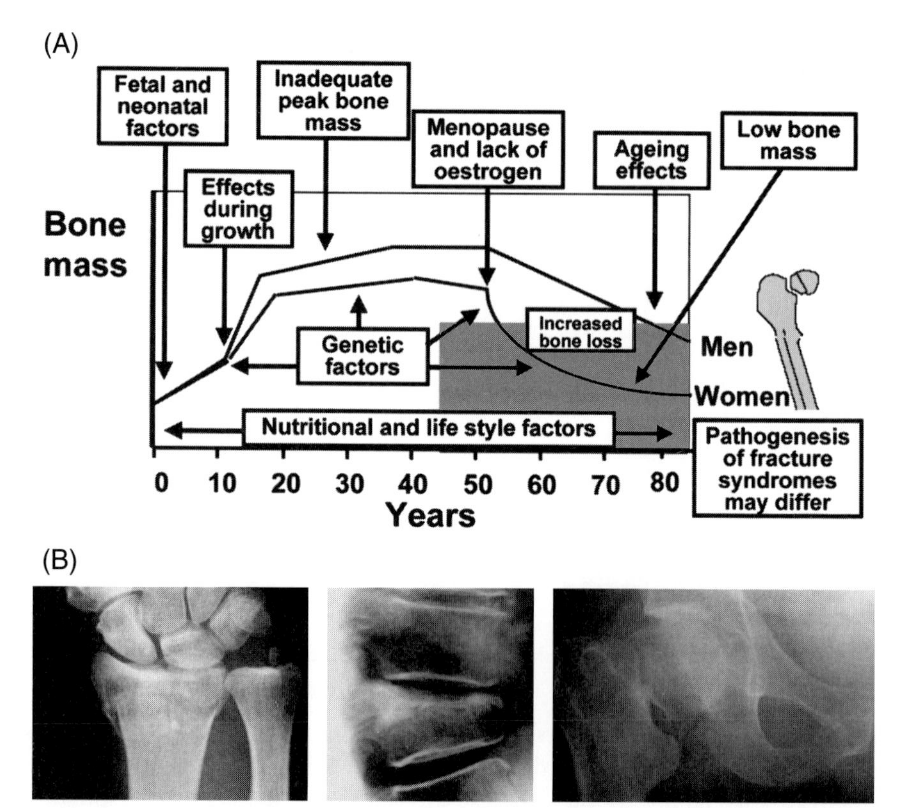

FIG. 4. Osteoporosis affects 1 in 3 women over the age of 50 and leads to fractures, which cause significant suffering and disability. The cost to the NHS amounts to about £1 billion per annum. (A) Multiple mechanisms involved in the pathogenesis of osteoporosis. (B) X-rays of the commonest sites of fracture: the wrist, spine and hip (left to right).

Bone metastases. These commonly occur from dissemination from breast cancer and other tumours, including lung and kidney. In breast cancers, bone resorption may be stimulated by factors like parathyroid hormone-related peptide (PTHrP), which also acts as a hormonal mediator of hypercalcaemia of malignancy.

Inflammatory bone loss. Local bone loss is a feature of several inflammatory diseases. Thus, erosive conditions occur in bone in osteomyelitis, rheumatoid arthritis and periodontal disease. The pathogenic mechanisms are gradually being elucidated. Inflammatory cytokines such as IL-1 and TNF are probably key, and recent studies show that synovial cells may express RANKL. Reactive oxygen species (ROS), such as hydrogen peroxide, and oxygen radicals have been implicated as resorption mediators in the anoxic environment of rheumatoid joints.

TABLE 1 Current and future drugs for osteoporosis

Current treatments	*New treatments*
• Oestrogens (hormone-replacement therapy) — several types, with or without progestogens • Selective oestrogen receptor modulators (SERMs) — Raloxifene • Bisphosphonates — Etidronate (Didronel) as cyclical therapy — Alendronate or risedronate as continuous therapy • Ca^{2+}, often plus Vitamin D • Calcitriol, 1,25 dihydroxy vitamin D • Calcitonins — not used much in UK. Nasal rather than injectable formulations may be preferred • Others drugs not specifically registered for osteoporosis (fluoride, anabolic steroids, testosterone in men, etc.)	• New oestrogen formulations — e.g. pulmonary • New SERMs • 'New' bisphosphonates — Zoledronate, ibandronate • Parathyroid hormone and analogues (PTHrP peptides) • Calcitonins — new formulations and analogues, e.g. oral or nasal formulations • Others, e.g. strontium salts, ipriflavone, growth factors • New discoveries

Osteoporosis. The past decade has witnessed a remarkably greater awareness of osteoporosis as a major health problem that is associated with profound socioeconomic consequences (Fig 4). There have been impressive advances in understanding the epidemiology and pathogenesis of osteoporosis and its associated fractures, in the application of physical and biochemical methods to its diagnosis and evaluation, and in the therapeutic approaches to prevention and treatment of postmenopausal and other forms of osteoporosis (Table 1). There are several recent good reviews (Compston & Fogelman 1999, Ralston 1997, Royal College of Physicians 1999). Despite these advances much remains to be done, and the development of better and more cost-effective methods of treatment must remain a high priority. The essential features are bone loss, particularly due to lack of oestrogen after the menopause in women. Although osteoporosis is usually thought of as a woman's disease, it also occurs in men at about one quarter of the rate in women.

Current therapy and drugs in development

Table 2 provides a list of current therapies and those likely to become available in the near future. Current therapies (Eastell 1998) are based on drugs that are inhibitors of bone resorption and remodelling rather than stimulators of bone

TABLE 2 Some of the genes that may contribute to osteoporosis by influencing bone mass and rates of bone loss (Ralston 1997)

- Vitamin D receptor
- Oestrogen receptor
- Sp1 site in $\alpha 1$ chain of Type I collagen
- TGFβ
- Parathyroid hormone receptor
- Interleukin 1 receptor antagonist
- Other loci identified in mice

formation. The major current therapies include vitamin D and Ca^{2+} supplements, oestrogens and related compounds, and the bisphosphonates (Fig. 5).

Oestrogens and SERMs

Recent advances in the uses of oestrogens include the development of novel delivery systems, particularly by the transdermal route. Future developments may include additional novel routes of administration, e.g. pulmonary

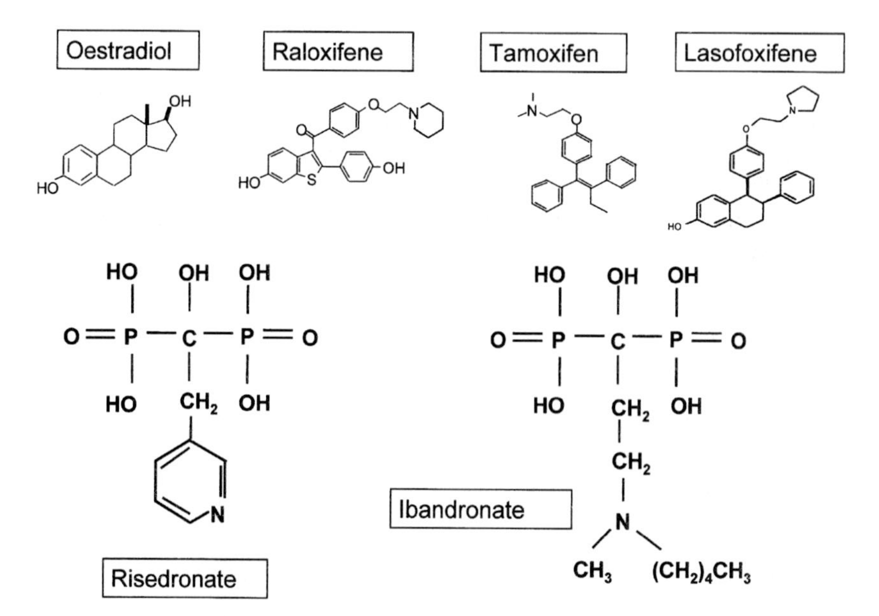

FIG. 5. Chemical structures of drugs relevant to osteoporosis. The SERMs raloxifene, tamoxifen, and lasofoxiphene are shown alongside the natural oestrogen, oestradiol. The two bisphosphonates shown here are risedronate and ibandronate.

inhalation. One of the more interesting recent achievements has been the development of more tissue-selective oestrogens or SERMs (selective oestrogen receptor modulators). The ideal compound of this type would possess all the good properties of oestrogens but none of the bad (MacGregor & Jordan 1998). Such an agent would therefore be effective in osteoporosis, ischaemic heart disease and Alzheimer's disease, without adverse effects on the breast or uterus in terms of increasing cancer risk, and with no risk of inducing venous thromboembolism. The first compound in this class is raloxifene. Although raloxifene does not yet fulfil all these criteria, it is effective in reducing bone loss and vertebral fracture, and also the incidence of newly diagnosed breast cancers. Other SERMs may follow relatively soon.

It is encouraging that the opportunities for chemical innovation in this area are enormous. The biological basis for tissue selectivity is beginning to be understood. Although there are now known to be at least two functional oestrogen receptors, α and β isoforms, the differential tissue distribution of these receptors does not account for the differences in action among the known SERMs. It is likely that the differential tissue effects are related to altered conformations of the oestrogen receptors to which they bind, followed by different associations with other transcriptional regulatory proteins in individual cell types. In theory it may be possible to develop drugs with the ideal properties if the basic biology allows it.

Bisphosphonates

Bisphosphonates (BPs) are now well established as successful anti-resorptive agents for the prevention and treatment of osteoporosis (Bijvoet et al 1995, Russell 1999). In particular, etidronate, alendronate and, more recently, risedronate, are approved therapies in many countries and both can increase bone mass and reduce fracture rates at the spine, hip and other sites by up to 50% in post menopausal women.

The use of BPs in osteoporosis is relatively recent compared with the many years of experience in other diseases such as Paget's disease of bone, and bone metastases, for which compounds such as pamidronate and clodronate have been used extensively. The clinical pharmacology of BPs is characterized by low intestinal absorption, but highly selective localization and retention in bone.

In osteoporosis, the effects of BPs on bone mass may account for some of their action to reduce fractures, but it is likely that their ability to reduce activation frequency, and possibly to enhance osteon mineralization, may also be related to the reduction in fractures.

Current issues with BPs include the use of intermittent rather than continuous dosing, intravenous versus oral therapy, the optimal duration of therapy, the combination with other drugs such as oestrogens, and their extended use in

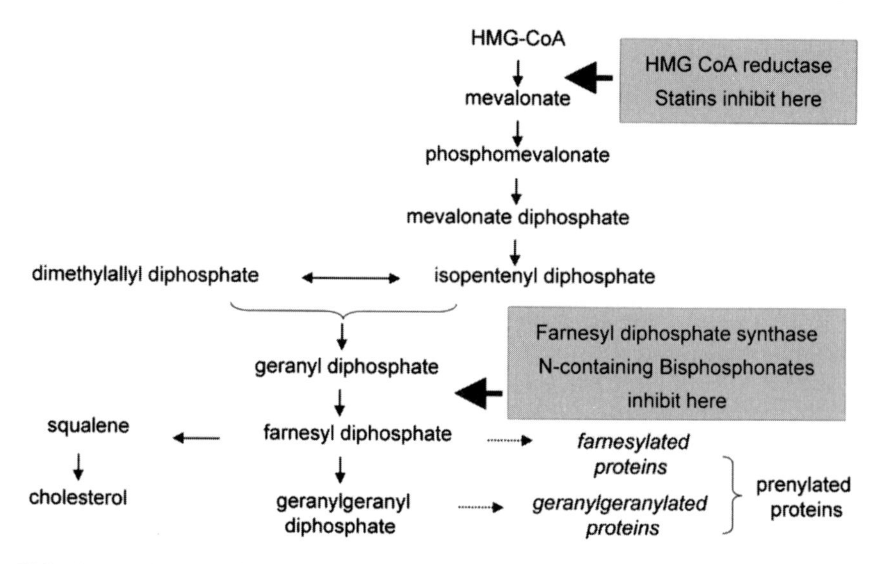

FIG. 6. Mechanism of action of nitrogen-containing bisphosphonates as inhibitors of the mevalonate pathway.

related indications, e.g. glucocorticosteroid-associated osteoporosis, male osteoporosis, childhood osteopenic disorders and arthritis. There is therefore much that needs to be done to improve the way in which existing drugs can be used as well as introducing new ones.

BPs inhibit bone resorption by being selectively taken up and adsorbed to mineral surfaces in bone, where they interfere with the action of osteoclasts. It is likely that bisphosphonates are internalised by osteoclasts and interfere with specific biochemical processes. Recent mechanistic studies show that BPs can be classified into at least two groups with different modes of action. Those that most closely resemble inorganic pyrophosphate (PPi), e.g. clodronate and probably etidronate, can be incorporated into toxic ATP analogues. In contrast, more potent nitrogen-containing BPs, such as alendronate and risedronate, interfere with other reactions, e.g. those in the mevalonate pathway (Fig. 6). These may affect cellular activity such as apoptosis by interfering with protein prenylation (Fig. 7), and therefore disrupt the intracellular trafficking of key regulatory proteins (Fisher et al 1999, van Beek et al 1999). There may therefore be subtle differences between compounds in terms of their clinical effects.

Calcitonins

Despite many years of use in some countries, there is still controversy about whether calcitonin reduces fractures in osteoporosis, although recent studies

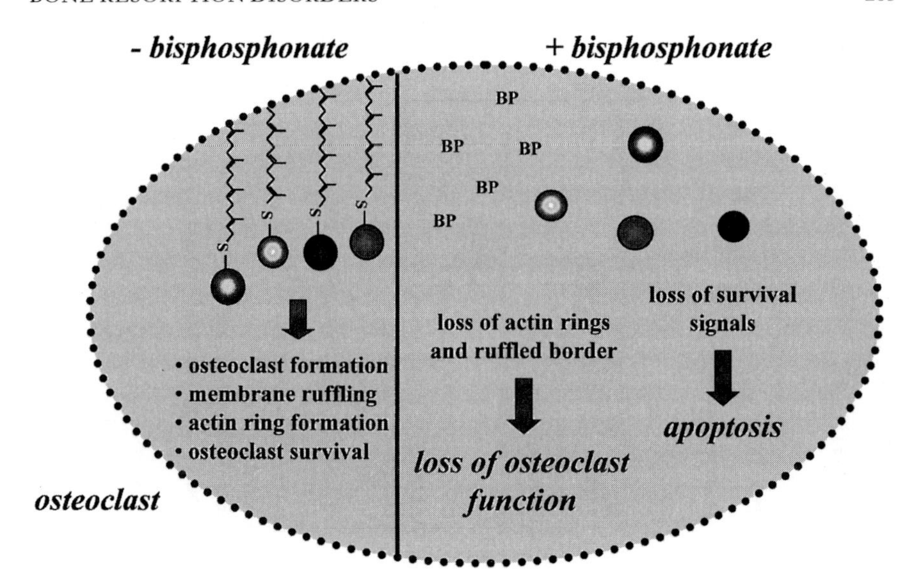

FIG. 7. Mechanism of action of bisphosphonates within cells as inhibitors of protein prenylation (Rogers 2000).

suggest that this may be the case for selected doses of salmon calcitonin. With more powerful and less expensive options available for inhibiting bone resorption, the impetus behind developing new therapies based on calcitonins or its peptide analogues has diminished, although there is still some interest in improving drug delivery, e.g. by the oral route.

Other anti-resorptive compounds

Other compounds currently under study include ipriflavone, a synthetic isoflavenoid derived from soya beans that has some oestrogen-like properties. The potential place for other phyto-oestrogens in the management of osteoporosis remains unclear, although differences in the dietary intake of such compounds may contribute to the geographical variations observed in the prevalence of osteoporosis.

The need for anabolic agents: parathyroid hormone, high-dose oestrogens, strontium salts and other options

It would be a significant advance if anabolic agents could be developed that would enhance the formation of new bone and therefore produce bigger changes in bone mass and strength than can be achieved with current drugs, such as the

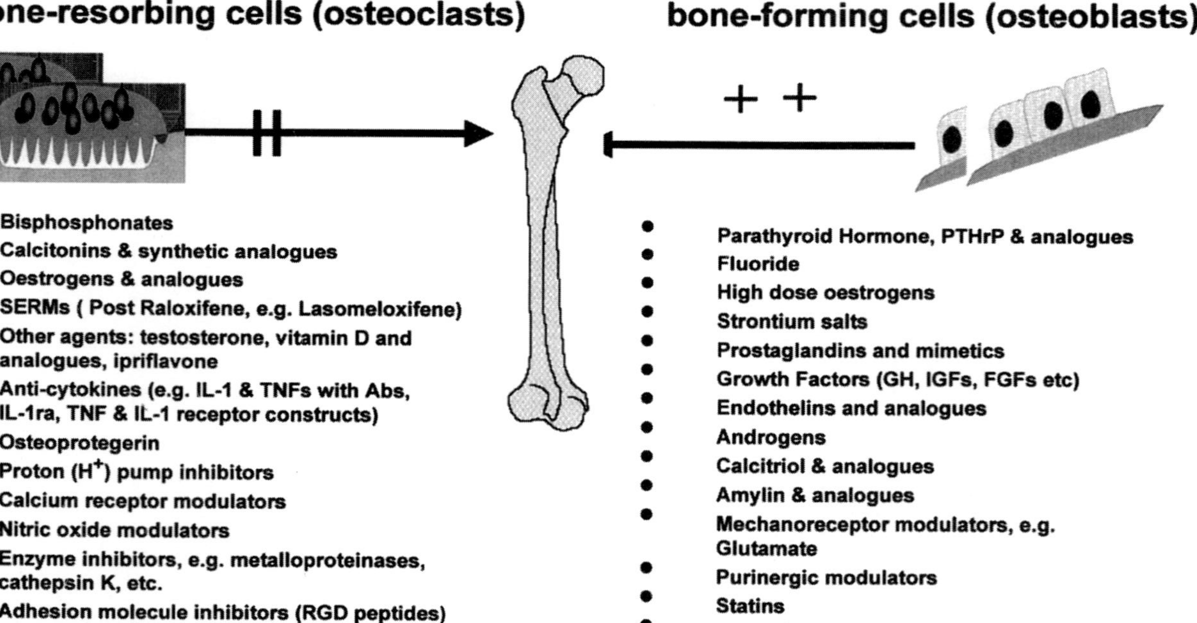

FIG. 8. This illustrates some of the many possibilities for drugs to be developed to increase bone mass either by suppressing bone resorption, or by augmenting bone formation.

bisphosphonates and oestrogens, that seem to only inhibit bone resorption. There is some evidence both from experimental studies in animals, and from clinical studies in women, that oestrogens may have a significant anabolic effect on bone, particularly at higher doses.

The only anabolic agent that has been used over many years in some countries is fluoride, which at appropriate doses produces an increase in bone mass. However convincing effects on reducing fractures have not been seen in trials performed so far, and there is little prospect of substantial backing from industry for an agent that cannot be protected from generic use by patents. With regard to the future, the remarkable effect of fluoride may provide clues to biochemical mechanisms that might be exploited to produce anabolic effects on bone. At present the mode of action of fluoride is only partially understood. Its effects on bone mineral to produce fluoroapatite may be significant, but there are probably additional effects on intracellular signalling pathways in bone cells.

The best known of the newer potentially anabolic agents for bone is parathyroid hormone (PTH), which has been studied on and off for about 30 years. There have been several recent initiatives by different companies to see through trials of a sufficient size not only to verify the large effects on bone mass but also to determine whether there is a significant effect on fractures. It now seems likely that PTH will become available as an anabolic agent in the near future.

A strontium (Sr) salt (strontium ranelate) has been shown to be anabolic in animal and clinical studies in that it produces increases in bone mass without corresponding increases in bone resorption. The gains in bone mass appear to be greater than can be accounted for by substitution of Ca^{2+} by Sr in the hydroxyapatite crystal lattice. The mode of action of Sr remains unclear.

Among other known agents that show anabolic effects in experimental models there are several peptide hormones and growth factors. These include IGFs, BMPs, FGFs and amylin. Non-peptide factors such as prostaglandins (e.g. PGE2), and statins that induce BMP2, also have anabolic effects (Mundy et al 1999, Rogers 2000, Sugiyama et al 2000). It is perhaps unlikely that any of these can be directly utilized clinically, but identifying low molecular weight secretagogues for growth factors that might be selective for bone and could be given by mouth is an experimental approach worthy of exploration.

The challenges of developing new treatments

Within the next few years, several more new drugs are likely to be licensed for use in osteoporosis and other disorders of bone resorption. Unfortunately, the process of drug discovery and development is slow. It usually takes at least 10 years from the discovery of a new compound for it to be studied experimentally, for its safety to be established, and for clinical trials to be completed.

The clinical trial stage for drugs in osteoporosis requires five years or more from start to finish, in order to meet the current regulatory requirements for demonstrating a reduction in fractures.

Prospects for novel drugs in the future

Looking further ahead, there are several other ways in which novel agents might be developed based on the rapidly increasing knowledge of bone biology and the pathogenesis of osteoporosis. Greater knowledge of the agonists, antagonists and receptors for osteoblasts and osteoclasts and their cellular precursors, as well as defining the functional machinery within these cells, should offer many opportunities for rational drug design. Figure 8 illustrates some of the practical and theoretical options available. Work towards defining the human genome will also contribute to this process of discovery, as will the use of informative transgenic animal models.

Other developments in biotechnology should make the process faster. For example, high-throughput screening methods and combinatorial chemistry have vastly increased the rate at which new pharmacological candidates may be identified.

In any case we are moving rapidly from an era when only few drugs were available to one in which more and better methods of treatment should soon exist.

References

Bijvoet O, Fleisch H, Canfield RE, Russell RGG 1995 Bisphosphonate on Bones. Elsevier Science Publishers, Amsterdam

Compston JE, Fogelman I 1999 Key advances in the effective management of osteoporosis. Royal Soc Med

Croucher PI, Russell RGG 1999 Regulatory Components: Growth Factors. In: Seibel MJ, Robins SP, Bilezekian JP (eds) Dynamics of bone and cartilage metabolism, chapter 6. Academic Press, New York p 83–95

Damien E, Price JS, Lanyon LE 1998 The estrogen receptor's involvement in osteoblasts' adaptive response to mechanical strain. J Bone Miner Res 13:1275–1282

Eastell R 1998 Treatment of postmenopausal osteoporosis. N Engl J Med 338:736–746

Favus MJ, Christakos S, Goldring SR 1999 Primer on the metabolic bone diseases and disorders of mineral metabolism, 4th edn. Lippincott, Williams & Wilkins, Philadelphia

Felix R, Cecchini MG, Fleisch H 1990 Macrophage colony stimulating factor restores in vivo bone resorption in the op/op osteopetrotic mouse. Endocrinology 127:2592–2594

Fisher JE, Rogers MJ, Halasy JM et al 1999 Alendronate mechanism of action: geranylgeraniol, an intermediate in the mevalonate pathway, prevents inhibition of osteoclast formation, bone resorption, and kinase activation in vitro. Proc Natl Acad Sci 96:133–138

Hofbauer LC, Khosla S, Dunstan CR, Lacey DL, Boyle WJ, Riggs BL 2000 The roles of osteoprotegerin and osteoprotegerin ligand in the paracrine regulation of bone resorption. J Bone Miner Res 15:2–12

Istova V, Caamano J, Loy J, Yang Y, Lewin A, Bravo R 1997 Osteopetrosis in mice lacking NF-kappaB1 and NF-kappaB2. Nat Med 3:1285–1289
Kong YY, Yoshida H, Sarosi I, et al 1999 OPGL is a key regulator of osteoclastogenesis, lymphocyte development and lymph-node organogenesis. Nature 397:315–323
Lanyon LE 1998 Amplification of the osteogenic stimulus of load-bearing as a logical therapy for the treatment and prevention of osteoporosis, In: Russell RGG, Skerry TM, Kollenkirchen U (eds) Novel approaches to treatment of osteoporosis. Springer-Verlag, Berlin, p 199–209
Macgregor JI, Jordan VC 1998 Basic guide to the mechanisms of antiestrogen action. Pharmacol Rev 50:151–196
Mason DJ, Suva LJ, Genever PG et al 1997 Mechanically regulated expression of a neural glutamate transporter in bone: a role for excitatory amino acids as osteotropic agents? Bone 20:199–205
Mundy G, Garrett R, Harris S et al 1999 Stimulation of bone formation in vitro and in rodents by statins. Science 286:1946–1949
Plotkin LI, Weinstein RS, Parfitt AM, Roberson PK, Manolagas SC, Bellido T 1999 Prevention of osteocyte and osteoblast apoptosis by bisphosphonates and calcitonin. J Clin Invest 104:1363–1374
Ralston SH 1997 The genetics of osteoporosis. QJM 90:247–251
Rodan GA, Harada S 1997 The Missing Bone. Cell 89:677–680
Rogers MJ 2000 Statins: lower lipids and better bones? Nat Med 6:21–23
Royal College of Physicians 1999 Osteoporosis. Royal College of Physicians, London.
Russell RGG 1999 The bisphosphonate odyssey. A journey from chemistry to the clinic. (In: Ebetino FH, McKenna CE (eds) Proc. XIV International Conference Phosphorus Chemistry. Ohio, July 1998). Phosphorus Sulfur Relat Elem 144–146:793–820
Schwartzberg PL, Xing L, Hoffmann O et al 1997 Rescue of osteoclast function by transgenic expression of kinase-deficient Src in *src-/-* mutant mice. Genes Dev 11:2835–2844
Suda T, Kobayashi K, Jimi E, Udagawa N, Takahashi N 2001 The molecular basis of osteoclast differentiation and activation. In: The molecular basis of skeletogenesis. Wiley, Chichester (Novartis Found Symp 232) p 235–250
Sugiyama M, Kodama T, Konishi K, Abe K, Asami S, Oikawa S 2000 Compactin and simvastatin, but not pravastatin, induce bone morphogenetic protein-2 in human osteosarcoma cells. Biochem Biophys Res Commun 271:688–692
van Beek E, Pieterman E, Cohen L, Lowik C, Papapoulos S 1999 Farnesyl pyrophosphate synthase is the molecular target of nitrogen-containing bisphosphonates. Biochem Biophys Res Commun 264:108–111
Wozney JM, Rosen V 1998 Bone morphogenetic protein and bone morphogenetic protein gene family in bone formation and repair. Clin Orthop 346:26–37
Yamashita H, Ten Dijke P, Helden CH, Miyazono K 1996 Bone morphogenetic protein receptors. Bone 19:569–574

DISCUSSION

Kingsley: One surprise from recent developmental biology studies has been the role of cholesterol modification in signalling pathways such as the Hedgehog pathway. In the various tests of the effects of some of the nitrogen-containing bisphosphonates on prenylation of proteins, you might expect that you would also see affects on cholesterol-modified proteins. Has this been looked at?

Russell: To my knowledge it has not been looked at. The key to the effects of the bisphosphonates is that they are highly selective for bone, if not exclusive to bone. When you give them to humans there is usually no effect on cholesterol levels in the blood: this is almost certainly because these drugs penetrate most cells very poorly and they localize so effectively to bone that it's really only osteoclasts that see them in any significant intracellular quantities. On the other hand, the statins were developed as being liver-selective. On the basis of that biochemistry, the statins can be extremely toxic if they get into other cell types. It is just possible there might be effects, because these drugs will localize in the growing limb. If the rapidly growing young animal is injected with radiolabelled bisphosphonate, it will be hot in the metaphyseal region. Diffusion of compounds into the growth plate is certainly possible.

Mundlos: When we see patients with resorptive defects such as osteopetrosis, the striking thing for me is that it gets worse with time, and if these patients lack functioning osteoclasts I don't understand how they would get bone marrow in the first place. How do they actually get a growth plate in the first place? Osteopetrotic patients are often normal and then after a year or so we pick them up because of their short stature and it gets worse and worse. If you want to invade the calcified cartilage you need osteoclasts, and if there are none present how does a growth plate form in the first place?

Kronenberg: Humans with osteopetrosis are a heterogeneous group. If there is a marrow space in osteopetrosis there must have been some osteoclast differentiation.

Mundlos: What about in mice?

Kronenberg: In those, there is no marrow: they are like the RANK knockouts.

Mundlos: How does the bone get inside?

Burger: If the stacks of hypertrophic cartilage cells disappear because of apoptosis, they leave a small tunnel that is wide enough for a blood vessel to come in. You don't need osteoclasts to allow that. The osteoclasts remodel the metaphysis so that it also disappears once it isn't needed anymore.

Mundlos: But there is lots of extracellular cartilagenous matrix around these hypertrophic chondrocytes that has to be removed if you want to put bone in there.

Burger: Chondrocytes can remove the horizontal layers in between cells; it's only the vertical layers where the mineral is deposited.

Poole: No, chondrocytes cannot remove the transverse septa. This is a process that's entirely dependent upon chondroclasts and probably also septoclasts and capillaries.

Burger: But at least you don't need osteoclasts for this.

Poole: There is a similar situation in hypophosphatasia for example, where there is an inability to resorb the cartilage calcified matrix.

Kingsley: Can you comment on what you think is the likely future clinical use of the bisphosphonates and some of the oestrogen derivatives?

Russell: Bisphosphonates have become widely used — they are probably already the major drugs used in osteoporosis, and in tumour-related bone disease they are the only really effective drugs. The oestrogen analogues are creating a lot of excitement. The main issue with raloxifene at the moment is that it only gives quite a small increase in bone mass, which is not as much as you get with oestrogen itself. None the less, there is the same level of reduction of vertebral fractures, but apparently little anti-fracture effect at any other site. Most importantly, there is no effect on hip fracture reduction. Whether you would get that if the trials were bigger and went on for longer is under debate, but no company can sell a drug for osteoporosis as well as they would if they had a hip fracture benefit. The bisphosphonates do have an effect on hip fracture reduction, and there will be a large study reported next year with risedronate that will probably confirm that nicely. The development of the so-called SERMs has been a battlefield for the pharmaceutical industry, because there have been at least three drugs that have tried to follow raloxifene, all of which have perished in development, mainly because of uterine side effects. There are lots of new compounds being made which will presumably go through clinical development, but an osteoporosis drug costs about US$500 million dollars to develop, so you want to be sure you have got something which is at least as good if not better than what's out there now. I think that in the biphosphonate arena there will be two more drugs and that will be it, leaving the total at seven or eight. With the SERMs there's only one at this point, unless people are tempted to use tamoxifen, which actually does good things to bone. But tamoxifen has the commercial problem of being an old and cheap drug.

OPG (osteoprotegerin) is fascinating, because fusing it with the Fc region of immunoglobulin, as the Amgen group have done, results in a very prolonged biological half-life. This is similar to what has been done for the TNF receptor constructs used in rheumatoid arthritis. It is likely that injections of the OPG constructs every one to two months will have a good clinical effect. Already, the studies in post-menopausal women show suppression of bone resorption markers as effectively as is seen with any of the other drugs. The real attraction of OPG will be in the tumour field and perhaps in children, where people are reluctant to use bisphosphonates. There are some dramatic results with osteogenesis imperfecta, in which there are surprisingly large increases in bone mass and apparent reductions in fractures if patients are treated with bisphosphonates. But the worry here is that these drugs go into bone and stay for very long periods. The use of OPG, which would have an on/off effect if given every few months, would be a highly attractive option.

Karsenty: Can you comment on the effects of bisphosphonates on bone formation?

Russell: The usual conclusion is that bisphosphonates don't have any effects on bone formation that are not an indirect result of their inhibition of resorption. Every now and again there are studies that suggest that they may have small effects on bone formation. The enhanced bone formation with statins might be reproducible with bisphosphonates if they got inside osteoblasts. Their great virtue is that they don't get into cells unless cells have a very special way of internalizing them. My own belief is that their effects on bone formation are, if present at all, very minor and not a major contribution to their clinical effect.

Karsenty: Do they affect the number of osteoclasts?

Russell: No, not as an immediate effect, but there are long term reductions in osteoclast numbers in patients treated for Paget's disease or osteoporosis. In addition, there are some interesting new studies appearing which have taken the field off in a slightly different direction. Bellido and Manolagas have been studying an osteocyte cell culture system, in which they can induce apoptosis with glucocorticoids (Plotkin et al 1999). In experimental steroid-induced osteoporosis, osteocytes are killed prematurely by this means, and apparently bisphosphonates protect against this apoptosis at very low concentrations. They are postulating that these are ion channel effects, affecting Ca^{2+} entering to the cell and then the kinase activation that follows. It is curious: they must be cell surface effects.

These effects are therefore different from the pro-apoptotic effects seen with bisphosphonates on osteoclasts.

Poole: As I understand it, you said that in Paget's disease you get an increase in the size of bones. This is intriguing, when you have such a driven resorptive process. Is this actually stimulating bone formation much more than would normally happen?

Russell: Trying to understand what the abnormality in bone remodelling is in Paget's is an interesting challenge. The osteoclasts are definitely larger, they have more nuclei, and one postulate is that they are like this because they don't undergo apoptosis at the usual rate, and they just accumulate nuclei and survive a lot longer. Their resorptive lifespan may therefore be greatly prolonged. This is why bisphosphonates might work so effectively, putting all that into reverse. But in remodelling of Pagetic bone you don't get normal lamellar bone, you get excessive amounts of what's called woven bone, which is disordered collagen deposition. This is probably the reason that the amount of bone that gets put back is more than was resorbed. When you treat, the remodelling rate goes back towards normal, so the bone can actually go back to its original shape.

Poole: Is the reason for the formation of excessive woven bone understood?

Russell: No, other than at a very descriptive level. There is an excessive number of osteoblasts and they make matrix at a faster than normal daily apposition rate. The fascinating thing talking about the need for osteoclasts to initiate remodelling, is that if you supprcss osteoclast activity, the bone will remodel back to normal lamellar bone and it will have a lower volume.

Reference

Plotkin LI, Weinstein RS, Parfitt AM, Roberson PK, Manolagas SC, Bellido T 1999 Prevention of osteocyte and osteoblast apoptosis by bisphosphonates and calcitonin. J Clin Invest 104:1363–1374

Final discussion

Hall: The first question I would like us to address in our final discussion concerns how patterning is related to specification of where limbs will develop. We think we know, but in reality I don't think we really know this at all. What we do know is that if we look along the neural axis of a vertebrate, there are specific patterns of *Hox* gene expression. And we know that if you look at the boundaries of those patterns of *Hox* gene expression, you can specify that the forelimb bud will develop at a particular position in relation to a particular pattern of *Hox* gene expression and the hindlimb bud will develop in another position. We think that explains how the forelimb and the hindlimb limb buds arise in those positions, but of course it doesn't: it describes a pattern, a correlation between *Hox* gene expression boundaries and where forelimb and hindlimb will develop. That this is not explaining a mechanism is illustrated by the sorts of experiments that people like Cheryll Tickle have done, when one takes a bead soaked in FGF and places it into the flank between where the wing and the leg will develop in the chick embryo. This results in a limb developing with absolutely no relation to these patterns. Clearly this mechanism of initiating the limb bud is totally independent of the mechanism which is determining position. If you think a little further about the experiments that have been done, whereby the closer you put the bud to the wing, the structure that develops is in fact wing, and if you put the bud closer to where the leg would normally be, the structure that develops is a leg — clearly that decision to be a wing or a leg is not being made by the signal which has been imposed, but is made by some properties of cells here, properties which we call 'wingness' and 'legness'. These properties are obviously absolutely fundamental to determining where a wing develops and where a leg develops, and we have no understanding of them at all. What I would like to know is what is the connection between the *Hox* genes and this territory that is set up for wing and for leg? And what is the signal that specifies that somewhere in the middle of that territory a limb bud will be initiated?

M. Cohn: It is not strictly true that the induction of limbs from different positions does not relate to *Hox* expression along the main body axis. We have shown that reprogramming flank cells to form limb involves re-setting *Hox* boundaries to reproduce the wing or leg pattern of *Hox* gene expression in the flank. In other words, there is a correlation between the identity of these cells with respect to forelimb, flank or hindlimb, and patterns of *Hox* gene expression. We still don't know how these *Hox* gene expression patterns relate to the activation of signalling molecules like FGFs.

Tickle: There are FGFs which are expressed specifically in limb-forming regions, such as FGF10, in the mesenchyme in both wing and leg. One would assume that somehow the patterns of *Hox* gene expression lead to initiation of expression of FGF10 in those particular regions. Other genes have been identified in mice, initially by Virginia Papaioannou. These are the *Tbx* genes, which are expressed in a striking pattern specifically in the wing or in the leg. Elegant recent work from a number of labs, including that of Cliff Tabin, of Juan-Carlos Izpisúa Belmonte and of Ogura, has shown that misexpressing the leg-specific *Tbx* gene in the wing will give leg-like structures. Thus, at the moment the picture may be hazy, but we are in quite a good position to begin to understand how the head-to-tail axis of the embryo is set up and determines where the limbs will form. There are a number of very testable hypotheses. One thing we need to know is what establishes the *Hox* gene expression, and what are the consequences of this. This would involve experiments altering patterns of *Hox* gene expression, and then seeing if you could alter limb position. There is some work in mice which suggests that changes in expression of certain *Hox* genes can shift limb position slightly. Now we need to unravel relationships between some of the key genes that we have now identified.

There is some old work on the potential for chondrogenesis: cells in limb-forming regions appear to acquire the potential to become chondrogenic, but flank cells do not. The molecular basis of this potential is interesting, and I don't know of any clues to that at all.

Hall: What Cheryll is referring to is that during one or two stages of chick development before you get an AER, the epithelium is signalling to the underlying mesenchyme and that signalling is required for those cells to have the potential to become chondrogenic later on. There is some sort of pre-patterning that is being imposed there. This isn't being imposed by what will be flank ectoderm, but by ectoderm that is subsequently going to become AER ectoderm. You are right: we know nothing about the molecular basis of that interaction.

Tickle: An allied question which I think is very interesting, and again I don't think that we know much about, is the cell lineage within the limb lateral plate mesoderm: whether all cells have the potential to be cartilage, and later on what the cell lineage is in terms of the skeleton. Is there a multipotential cell that will give rise to cartilage, tendon and various other connective tissues, or are there individual precursor cells for each of these different kinds of connective tissue? This is pretty fundamental, and as far as I know we don't have the answer.

Hall: The muscle that forms in the limb actually comes from cells that come in from the somites, and so it has a completely separate origin.

Kingsley: I would amplify the same question for the detailed cell types that are found in joints.

Hall: You start off with a nice condensation, and somewhere in the middle of that you are going to get a joint from cells that look exactly the same as the cells which are going to make cartilage, which in turn look exactly the same as the cells that are going to make perichondrium around the edge.

Bard: What happens if you take a needle and stir up the mesenchyme?

Morriss-Kay: You can disaggregate the mesenchyme and then reaggregate it again and put it back in an ectodermal jacket. Cartilage elements, muscle and connective tissue form.

Bard: In the right pattern?

Tickle: Not in the right pattern without adding a polarizing region: it becomes more like a *talpid3* limb.

Bard: If you put in as much of the signalling as you can, presumably you get a pretty much normal limb. In other words, the pattern comes from the signalling not from the mesenchyme.

Tickle: That's what it looks like, but even without a signal, you can still get patterns: you can still get digits without the polarizing region.

Kingsley: It is important to realize that the type of limb that grows out depends not on the signal but on the type of mesenchyme.

Bard: I'm thinking of the patterning of what you get in there: is it in any sense dependent on the pattern that the cells take up under normal circumstances, which might reflect, for example, patterns of *Hox* genes? If you stir those up, do you still get normal development?

Newman: You get a kind of nondescript limb, if you have disaggregated and reaggregated cells. It looks like a limb but you can't identify what kind of limb and you can't even identify what kind of digits. As Cheryll said, you need to add a polarizing zone, which entrains production of Hox gradients and so on, and then you start getting a recognizable limb.

M. Cohn: Do tendons develop in these reaggregated limbs?

Tickle: I think they do. One of the other things I've always been interested in is the question of different cartilage identities. The myogenic cells all seem to be equivalent, and muscle patterning really determined by the surrounding connective tissue. I've often wondered whether cartilage is not the same. Are the cartilage cells in the different skeletal elements intrinsically different? The cartilage cells could be all the same and it could be the connective tissue (the perichondrium) around each element that determines what they're going to do, in a very similar way to muscle patterning.

Bard: Are they different populations originally?

Tickle: We don't know.

Ornitz: One of the gaps that we still have in determining the outgrowth of the limb is the connection between the Hox code (which is well established) and the pattern of FGF10 expression in lateral plate mesoderm. One idea is that there is a second FGF signal that comes from the mesonephros that may signal lateral plate mesoderms to turn on FGF10. Alternatively, FGF10 may be turned on directly by *Hox* genes. When you consider experimental approaches to figure this out, one approach would be to look at FGF10 gene regulation and ask whether that is under direct Hox control or whether some other signalling molecule regulates patterning.

Hall: So you might expect FGF10 to be coming out preferentially from axial tissues in areas where the limbs are going to develop.

Ornitz: Possibly FGF8 could signal to lateral plate mesoderm and activate FGF10. The other alternative is that the anterior–posterior Hox code determines where FGF10 is turned on.

Tickle: This raises a fundamental issue about lining up the limb-forming regions in lateral plate mesoderm with the appropriate regions of the neural tube. The idea that you're talking about is that there's some kind of medial to lateral way of connecting these things. The neural tube becomes regionally different very early on, and yet this is only manifested much later. At that time, it seems to be the somites in the adjacent tissue that seem to be controlling neural tube pattern (Ensini et al 1998). The somites are in a good position to control both neural tube on the one hand and lateral plate mesoderm on the other hand. This will ensure that the limb arises opposite the right bit of the neural tube, otherwise the whole thing will be totally useless.

Hall: Certainly, what seems to be happening in the head is that you have a bit of both: there are neural crest cells coming out from the developing brain which are carrying information with them about a Hox code that was in the neural tube, and they're going down to the visceral arches and coming into contact with pharyngeal endoderm which it appears also has a coded expression of genes. Thus there is both a central code and a peripheral code, with the two meeting in the visceral arches.

Morriss-Kay: The head is fundamentally different from the trunk in terms of patterning of the neural tube and crest.

I have the impression that very little is known about target genes of Hox transcription factors as a whole. Is this true? Everybody talks about the results of a particular *Hox* code in terms of downstream events such as FGF expression or morphogenesis, but what is known about the direct transcriptional targets of *Hox* genes?

Newman: A little, but not much. There's a paper by Yokouchi et al (1995) from Kuroiwa's lab in Japan, which talks about *HoxA13*. They mis-expressed it and found that the adhesive properties of the cells changed. They argued that the association of this particular *Hox* gene expression with the kinds of small elements that occur in the distal region of the limb is associated with this adhesive difference. This is in the context of previous work from Edelman's group, that also suggested that both extracellular matrix and adhesive proteins are modulated in their expression by various Hox proteins (Jones et al 1992).

Morriss-Kay: In a sense this is looking at the phenomenological output. Is there knowledge about specific target genes involved?

Newman: NCAM was the target gene in this case.

Wilkins: Edelman's group found that some of the *Hox* genes actually have opposite effects on NCAM expression, which may be interesting in terms of combinatorial models about why certain things happen at Hox gene expression boundaries.

Tickle: Susan Darling has studied the *Hypodactyly* mutant in which there is lack of function of *HoxA13*. The outcome of this is massive cell death, so it looks like another role of *Hox* genes is controlling cell survival.

Newman: We know that we can initiate pre-cartilage condensations by members of the TGF family, which induce extracellular matrix and adhesive proteins, but I don't think we know anything about what restricts the expansion of those condensations. There must be some kind of inhibitory signal that's induced around the same time, and acts laterally to this initiating signal. If we understood what that inhibitory molecule is, then we would get an idea of why we get the kind of repeating patterns that we see in the limb. But we don't know anything about that inhibitor.

Kingsley: One of the surprises in the BMP field over the last several years has been how many different molecules studied by developmental biologists actually turn out to impinge upon the BMP pathway by acting as an inhibitor.

Newman: As far as I know nothing of that sort has appeared for the TGFs per se, as opposed to the BMPs.

Hall: What about the role of the dorsal and the ventral ectoderm in the limb? Do we have any idea of molecules there that might be confining condensations to particular positions?

Tickle: Lmx1 is not expressed in cartilage. It is expressed dorsally. It could be that.

Perrin-Schmitt: There are also genes of the BHLH family, such as *dermo1*. This is not very well named because it is more predominantly expressed in future sclerotome. There would also be genes that are acting between mesoderm formation, commitment and differentiation of the skeleton.

Hall: One of the other aspects of cell lineage that struck me from Henry Kronenberg's chimeric studies, was that it looked as if you had independent columns of cells running down through the growth plate. One could conceive these columns as being independent of the cells that were on either side of them. These clonal columns of cells could be differentiating and undergoing hypertrophy irrespective of what was going on in the column beside them. Is that a sensible way to think about the growth plate?

Kronenberg: Once you've got condensation and start to form a cartilage, that's just the beginning of the mystery. We don't know why certain chondrocytes are chosen to become the first hypertrophic chondrocytes. It is usually the ones furthest from the ends of bones. What are the signals that determine that, and what are the signals that make those columns so orderly and polarised? And of course, where precisely the osteoblasts come from is a complete mystery.

Hall: Do we know anything about the precursors for those osteoblasts?

Kronenberg: I don't think we have rigorous data. We have to keep an open mind about where osteoblasts initially come from. Stromal cells can be injected in the bloodstream and they then go places and survive, so there's a potential for that. We have very few markers of the early differentiation states of osteoblasts. It will be important to do lineage studies of osteoblasts.

Chen: There is a theory from Cancedda's group, who claim that hypertrophic chondrocytes can differentiate into osteoblasts. Do you see any labelling in the osteoblasts?

Kronenberg: We're looking at that. There are other more straightforward ways of doing lineage analysis. This is a lineage question, and it hasn't really been addressed using lineage-type tools. In the chick it looks as though there are some settings in which cells of the chondrocyte lineage can become osteoblasts. It is an unsettled question.

Beresford: It is the chick that provides the best evidence. It is very difficult to find evidence of transdifferentiation being an important mechanism for generating bone-forming cells in other species.

Meikle: There's no fundamental reason why chondrocytes and osteoblasts couldn't switch: it is just a matter of changing the amount of proteoglycans or collagen synthesized.

Kronenberg: There's nothing that we would know that would make that terribly surprising or unlikely, it is just that it has only been seen in a couple of special circumstances in the chick.

Kingsley: This is also what I thought it was going to be when the *Cbfa* expression patterns were first described. Rather than being a bone-specific marker, the early expression pattern of that gene is in both cartilage and bone precursors.

Kronenberg: This leads to a hypothesis about the further differentiation of early chondrocytes. One of the key steps is that the early progenitor of both the chondrocytes and the osteoblasts makes *Cbfa1*. Karsenty might say that the key to the further differentiation of the chondrocyte could be the down-regulation of *Cbfa1*. Perhaps if you maintain *Cbfa1* expression at a high level, you would get osteoblasts. It is conceivable that *Cbfa1* is part of a switch for chondrocyte lineages and osteoblast lineages.

Bard: I would like to know what is going on in the condensation when it forms. What are the influences that spatially organize it? If you look at the early condensations, they are not like the final bones.

Hall: I will quote from H. Fleisch's comments in one of the discussions at the Ciba Foundation Symposium on the Cell and molecular biology of vertebrate hard tissues (Ciba Foundation 1988), held 12 years ago, because I think it raises an interesting issue:

'We should take advantage of the knowledge developed in the haemopoietic field on the development from stem cells to mature blood cells and we should investigate whether a cascade of factors is necessary' (p288). It is clear from the osteoclast studies that this has been an incredibly powerful approach for studying osteoclast lineage, but is it a useful approach for getting at osteoblasts or chondroblasts? It seems that perhaps it is not.

Kronenberg: It hasn't been done, but there is every reason to think that it will be very useful.

Beresford: I don't know. I think one of the reasons that we have made such little progress in the last 12 years, is precisely because we have been too influenced by the model of the haematopoietic stem cell compartment, and the way it branches to give rise to lines of committed progenitors. This has enslaved our thinking about the lineage hierarchy of the marrow stromal cell system. The truth is that the harder we look, the less and less good is the evidence that the experience of the haematopoietic field is actually relevant. We're dealing with something else, possibly a single pool of incredibly plastic progenitors that, depending on the environment that they find themselves in and the cues that they are receiving from adjacent cells and tissues, do different things. There may actually be quite a significant potential for modulation of what is apparently a differentiated phenotype.

Kronenberg: So you are postulating more flexibility in pathways of differentiation, rather than strict lineages.

Beresford: We use the term commitment very loosely. As far stromal cells are concerned, there is no real evidence for irreversible commitment as a process.

Hall: Certainly, the fact that cartilage and bone are the most common ectopic tissues to arise in the body is in agreement with that.

Beresford: What we may be looking at is default pathways. The trick is actually repressing or stopping them unravelling, rather than actually having to positively initiate a specific pathway.

Morriss-Kay: I'd like to ask about the definition of stem cells. I used the term 'stem cell' in relation to the proliferating cells in the suture, but when you talk about stem cells in the haematopoietic system, you're talking about cells that are truly multipotent. If you're talking about cells that only have the ability to go to become osteoblasts, and there's no evidence that they would become anything else, is it correct to call them stem cells?

Beresford: Yes, I think so. The skin is an example of a stem cell compartment with only a single functional end cell type, the keratinocyte.

Kingsley: Your diagram illustrated multiple possibilities, with a cell producing one daughter cell that differentiates and another that keeps generating the proliferating cell. This would fit a stem cell definition.

Beresford: Some pretty good lateral thinking is going to have to be done here to solve the lineage. We mustn't lose sight of the fact that the reason the haematopoietic people know so much is that it is possible to ablate the entire immune system of a recipient animal, and then reconstitute it with donor cells. This is not possible with the stromal system. There is no evidence that it is possible to produce a negative background on which you could then assay the developmental potential of a donor cell population, as has been done for the haematopoietic precursors. This is a real obstacle.

Wilkins: More comparative information might be useful if it yielded an evolutionary perspective. There might be some value in exploring bone cell and cartilage formation in some of the primitive bony fishes, using the molecular markers that we have for these cell types.

Hall: I wondered about this in relation to David Ornitz's paper, where he showed that different Gdfs were involved in patterning. If you think of what's happening in how tetrapod limbs evolved from fish fins, there are cartilagenous elements at the base of the fin which we think are homologues of the elements in the limb, and the digits have been added on as novel elements. We might therefore expect the digits in the tetrapod limb to be patterned and organized differently at the genetic level from the more proximal cartilagenous elements which are the ones that were inherent to fish ancestors. This might be a useful comparison.

Meikle: Most of the discussion over the last couple of days has been related to the limbs, but those of us who work with heads are dealing with fundamentally different mechanisms. The bones are different and Hox genes aren't expressed there. Those interested in limbs and condensations are in a much more difficult position than people who study the head. It seems reasonably clear that bones and teeth and other structures in the head result from epithelial–mesenchymal interactions. Can one evoke similar kinds of mechanisms in terms of limb development? It seems to me that one probably can't.

Hall: I was struck by the difference between the molecules that are involved in joint formation, and molecules that are involved in sutures, which are clearly not the same. They are both boundary conditions between skeletal elements but specified by different mechanisms.

Perrin-Schmitt: While there are parallels between the haemopoeitic system and the development of the skeleton from stromal stem cells, the differentiation of the haemopoietic cells is a simpler system, because they are much more independent of their environment. In the differentiation of the bones, the initial stage of condensation is a general process that is similar to haemopoiesis, but then the differentiation of bones is no longer just a general system, because the cells receive positional information. Even for the joints, some are going to receive positional information from one of the members of the TGF family, and the other joints are going to be receiving other positional information, and then they differentiate into some other structures. The initial condensation is probably a property of the mesenchymal cells, but the later differentiation depends on positional information from the ectoderm.

Burger: Regarding the idea about positional information via gradients of factors, has it ever been shown that such a gradient would bring about a form? If you want to invoke positional information, a gradient that is in principle continuous, must somehow be transformed into something that has very steep boundaries. Cartilage or non-cartilage is a steep transition, but gradients are continuous. Is there a threshold effect at a certain concentration of factor?

Perrin-Schmitt: This occurs during the segmentation process in *Drosophila* development. Initially there are just two or three grossly demarcated regions, but then there are segments, and afterwards anterio-posterior segments delimited by waves of expression of genes. Perhaps it could be the same problem at least partially for the differentiation of the condensation of the bone cells.

Burger: But has it been pinpointed that at certain concentrations suddenly cellular differentiation changes?

Perrin: Yes.

Kronenberg: Different genes are turned on by different levels of factors. It's a classical morphogen test.

Newman: This kind of mechanism may occur, but it's not universal. There are other kinds of mechanisms as well where there are underlying pre-patterns: periodic arrangements that eventually correspond to periodic structures.

Perrin-Schmitt: Yes, but you can give periodic expression and then afterwards define the pattern by other types of molecules along a second axis as a gradient.

Reddi: We haven't yet discussed whether there are fundamental differences between membrane bone formation of the type involved in head, and endochondral bone formation. Are there any fundamental biochemical or mechanistic differences when the first osteoblasts do their thing? Second, vascular invasion is a pre-requisite for most cases of bone formation. Is this vascularization just bringing in nutrients, or is it bringing in a haematopoietic-associated cell type. It seems to me that we need to label this angioblast-type cell before it comes in if we are going to do lineage studies. When this first lineage of vascular elements approaches the growth plate or membrane bone formation area, do we see any direct transformation of pericytes to osteoblast-like phenotypes? There are an increasing number of common molecular markers between smooth muscle cells and osteoblasts, and it is entirely possible that the smooth muscle cell may have a contribution to the bone formation.

Meikle: It is particularly interesting that there is interest in the calvaria by those researching craniosynostosis. The problem is if you read craniosynostosis papers, they're all focused on the sutures, and there's very little beyond that. In fact, as far as I'm aware there is nothing to do with mammals which is involved in the molecular induction of calvarial bone formation. Since the calvarial bones are the phylogenetic descendants of the dermal shield, then epithelial signals of the same kind that invoke the outgrowth of the limb or even the initiation of tooth formation will be similar, but we don't have any data on that. There are obviously differences between membrane bone formation and endochrondrial bone formation.

Wilkie: Something that has only been touched on perhaps a couple of times at this meeting, is the role of mechanical factors in feeding back to fine-tune morphology. Two questions particularly come to mind in my own field. How does the brain signal to control the growth of the overlying skull, and how does this position the cranial sutures? We know very little about that. A related question in the limbs is how you get from a crudely formed structure to a finely tuned bone? It was mentioned that if you paralyse an animal *in utero*, the joints don't form properly.

Hall: We have virtually no information on the molecular links between the mechanical signals and the transduction of differentiation.

Kronenberg: Another broad group of questions that we haven't talked about concerns how the molecules and pathways that are involved in the initial modelling of bone are important in controlling the remodelling of bone in the adult. Is a whole new group of regulatory molecules and loops involved, or is it partly new and partly the same?

Kingsley: One of the old studies I liked best in short ear mice, was the study of fracture repair in adult animals. This was a question that Margaret Green asked way back in the 1950s: whether the same genes that were disrupting embryonic cartilage and bone formation would also disrupt the repair process in the adult. The studies were very clear that the same mutant that disrupted early skeletal structures also delayed the ability of the adult mice to repair bone.

M. Cohn: Another area we haven't really touched on is the control of size and shape of skeletal elements during development. It seems likely, especially from David Kingsley's paper, that a key player in the development of bone size and shape has to be the joint regions. I use the term joint regions' in a broad sense to refer not only to the spaces between the bones, but also to the growth plates on either side of the joint. There are so many interesting signalling events happening at these interfaces.

Kingsley: I think that's a very interesting area. This also relates to some of what you covered in your paper, which is the loops that may be set up to determine the rates at which cells go through growth plates. There have been interesting studies on brachypodism animals looking at the postnatal effects on proliferation and growth plates. It is clear that the same genes that turn on pretty stripes along inner zones in developing embryos can also control the proliferation of cells in the growth plate of bones after birth. This is a good example of growth control from that region. I've been to interesting bone meetings, where Cornelia Farnum from Cornell has gone through some of her data on the detailed bookkeeping of cell transition rates through different growth plates at different bones. Interestingly, it looks like there are characteristic differences in the rate at which cells progress through growth plates at different anatomical locations in the body. This corresponds to interesting differences in the growth of skeletal elements and is probably related to the determination of overall bone shape. I don't think we know anything about the way that the kind of basic signals that we have been talking about may be modulated at different anatomical locations to set specific growth rates that determine regional bone morphology.

Kronenberg: And that are modulated through time, as well. In fetal life the rate of these things is enormously faster than a few weeks later.

Hall: What was very surprising to me, when I first learned it, was that each end of a bone can be growing at a different rate.

Kingsley: I got interested one time in trying to see what was known about why the fingers and digits of bats are so elongated. I found one paper on the embryology of digit formation in bats. In the early embryonic stages they don't look that different from other mammals. Then the enormous elongation in the digits to make a wing comes from dramatic changes in the proliferation rates of cartilage at later stages of development.

Hall: One sees the same thing in horses toes, which start off with condensations of equal size, and then the middle one grows rapidly, leaving the lateral toes behind.

References

Ciba Foundation 1988 Cell and molecular biology of vertebrate hard tissues. Wiley, Chichester (Ciba Found Symp 136)

Ensini M, Tsuchida TN, Belting H-G, Jessell TM 1998 The control of rostrocaudal pattern in the developing spinal cord: specification of motor neuron subtype identity is initiated by signals from paraxial mesoderm. Development 125:969–982

Jones FS, Prediger EA, Bittner DA, De Robertis EM, Edelman GM 1992 Cell adhesion molecules as targets for Hox genes: neural cell adhesion molecule promoter activity is modulated by cotransfection with Hox-2.5 and -2.4. Proc Natl Acad Sci USA 89:2086–2090

Yokouchi Y, Nakazato S, Yamamoto M et al 1995 Misexpression of Hoxa-13 induces cartilage homeotic transformation and changes cell adhesiveness in chick limb buds. Genes Dev 9:2509–2522

Index of contributors

Non-participating co-authors are indicated by asterisks. Entries in bold indicate papers; other entries refer to discussion contributions.

B

Bard, J. 19, 20, 21, 45, 46, 92, 96, 97, 116, 119, 139, 189, 191, 192, 193, 211, 228, 233, 274, 278
Beresford, J. 59, 116, 134, 153, 187, 247, 248, 249, 277, 278, 279
Blair, H. 37, 59, 95, 153, 168, 224, 248, 249, 250
Burger, E. H. 40, 95, 153, 154, 156, 222, 229, 231, 247, 248, 249, 250, 268, 280

C

Chen, Q. 40, 156, 168, 169, 211, 228, 231, 277
*Chung, U-I. **144**
Cohn, D. H. 77, 78, 92, 168, **195**, 210, 211, 212
Cohn, M. J. 45, **47**, 57, 58, 59, 60, 61, 98, 99, 157, 232, 233, 272, 274, 281
*Croucher, P. **251**

H

Hall, B. K. 17, 20, 36, 38, 40, 41, 61, 62, 79, 80, 93, 95, 97, 98, 99, 100, 119, 154, 168, 170, 185, 187, 191, 192, 212, 222, 223, 228, 233, 250, 272, 273, 274, 275, 276, 277, 278, 279, 281, 282

I

*Iseki, S. **102**

J

*Jimi, E. **235**
*Johnson, D. **102**

K

Karsenty, G. **6**, 17, 18, 38, 91, 92, 154, 166, 191, 223, 226, 231, 249, 250, 270
Kingsley, D. M. 17, 18, 37, 42, 44, 60, 76, 78, 79, 80, 93, 94, 95, 96, 97, 117, 118, 120, 135, 139, 140, 152, 154, 155, 191, 192, 193, 210, **213**, 222, 223, 224, 225, 226, 227, 228, 229, 230, 231, 232, 267, 269, 273, 274, 276, 277, 279, 281, 282
*Kobayashi, K. **235**
*Kojima, T. **158**
Kronenberg, H. M. 21, 38, 45, 57, 58, 76, 79, 93, 98, 117, 118, 136, 139, 140, **144**, 152, 153, 154, 155, 156, 157, 166, 169, 185, 187, 223, 230, 231, 233, 249, 268, 277, 278, 280, 281, 282

M

*Maxson, R. E. Jr. **122**
Meikle, M. C. 21, 44, 60, 77, 92, 95, 119, 167, 231, 277, 279, 281
Morriss-Kay, G. M. 40, 41, 44, 45, 58, 80, 92, 93, 97, **102**, 116, 117, 118, 119, 120, 141, 156, 274, 275, 276, 278
*Mueller, G. **251**
Mundlos, S. 37, 38, 40, 60, **81**, 91, 92, 93, 94, 95, 96, 97, 99, 119, 137, 140, 156, 169, 170, 226, 227, 231, 268
*Mwale, F. **158**

N

Newman, S. A. 17, 19, 20, 21, 38, 39, 59, 95, 96, 98, 134, 136, 137, 138, 139, 140, 141, 187, 192, 193, 212, 224, 225, 233, 274, 275, 276, 280
*Niswander, L. **23**

O

*Oldridge, M. **122**
Ornitz, D. M. 18, 20, 21, 44, 45, 58, 60, 61, **63**, 76, 77, 78, 79, 80, 92, 97, 119, 133, 135, 136, 137, 139, 140, 156, 186, 229, 274, 275

P

Perrin-Schmitt, F. 99, 100, 276, 279, 280
Pizette, S. **23**, 36, 37, 38, 39, 40, 41, 42, 61, 98, 119, 228
Poole, A. R. 38, 119, 152, **158**, 166, 167, 168, 169, 170, 230, 231, 250, 268, 270

R

Reddi, A. H. 18, 20, 39, 44, 225, 229, 232, 280
Russell, G. 77, 117, 153, 166, 167, 211, 247, 248, 250, 251, 268, 269, 270, 271

S

*Sampaio, A. V. **171**
*Shipman, C. **251**
Suda, T. **235**, 247, 248, 249, 250

T

*Takahashi, N. 235
*Tang, Z. **122**
*Tchetina, E. **158**
Tickle, C. 18, 20, 42, 44, 45, 62, 97, 99, 118, 154, 186, 273, 274, 275, 276

U

*Udagawa, N. **235**
Underhill, T. M. 40, **171**, 185, 186, 187

W

*Weston, A. D. **171**
Wilkie, A. O. M. 60, 77, 78, 79, 93, 94, 95, 118, 120, **122**, 134, 135, 136, 137, 138, 139, 140, 226, 281
Wilkins, A. 45, 58, 60, 92, 118, 119, 136, 192, 276, 279
*Wu, W. **158**

Y

*Yasuda, T. **158**

Subject index

A

Acanthostega gunnari 48
achondrogenesis
  type IB 205, 210
  type II 197
achondroplasia (ACH) 64–66, 77, 78, 80, 137
  mouse models 70
acid box 128
acromesomelic chondrodysplasia
  Grebe-type 216
  Hunter–Thompson-type 216
activating protein 1 (AP1) 186
activin 32
aggrecanases 166
alanine 83, 85
amino acid substitution 65
anabolic agents 263–265
anterior–posterior (A–P) axis 6–7, 17
anti-resorptive compounds 263
Apert phenotype 128
Apert syndrome 68, 78, 120, 123, 137, 138, 140
  clinical features 125
  mutations 128–129
apical ectodermal ridge (AER) 7, 44, 51–52, 59, 60, 98, 172, 273
apoptosis
  in bone turnover 255
  interdigital 181–182
APS (adenosine-5′-phosphosulfate) 206
Arg172His 131
Arg248Cys 129, 135
arthritis 223
  rheumatoid arthritis 167, 231, 250
articular cartilage 229–231
ATP 206
autophosphorylation 66
autopod 32–34, 36

B

BaF3 cells 67
basic helix-loop-helix (BHLH) proteins 118, 276
Basilosaurus 58
bisphosphonates 153, 260–262, 269, 270
BMP (bone morphogenic protein) 11, 17–20, 24, 30, 32–34, 37, 39–41, 44, 72, 76, 118, 140, 154, 173, 191, 192, 214, 218–219, 222, 224, 276
  and chondrogenesis 173
  signalling 181–182
  BMP1B 37
  BMP2 7, 8, 24, 173, 175, 181, 219
  BMP4 8, 24, 72, 79, 173, 181, 219
  BMP5 214–218
  BMP6 219
  BMP7 8, 24, 163, 182, 219
  BMP8 219
BMPR (bone morphogenic protein receptor) 30
  BMPR1A 8
  BMPRIB 25–27, 32, 33
bone, types 1
bone cells 253–256
bone chrondrogenesis 72
bone dysplasia 197
bone formation 82, 255, 264, 270
  genetic control 213–234
  modes 2
  osteoblasts in 253
  see also endochondral bone formation
bone formation rate 1–2
bone fusions 140
bone loss 258, 260
bone marrow stromal cells *see* osteoblasts
bone mass 256, 260, 264, 269
bone metabolism, monitoring 256
bone metastases 258
bone morphogenetic protein *see* BMP
bone morphogenetic protein receptor *see* BMPR
bone remodelling 1, 196, 250, 255–256, 270
bone repair 281

bone resorption 82, 255, 264
clinical disorders 251–271
diminished 256
increased 256–259
major diseases 256–259
osteoclasts in 253
bone status, clinical assessment 256
bone turnover 256
apoptosis in 255
Booidea 55
brachydactyly 85, 86, 94, 200
type C 226
brachypodism 154, 190, 215–218, 225, 226, 228
branching 232, 233
BrdU 25
bromodeoxyuridine 110

C

Ca^{2+} 251–252, 270
CAG trinucleotide 84
calcitonin 251, 262–263
calcitonin receptors (CTR) 237
calvaria 119–120
cartilage 210, 213, 277, 282
defects 199
degradation 166
differentiation 227, 228
formation 27, 31
functions during development 171
osteoarthritic 231
types 1
cartilage–bone interface 250
cartilage collagens and osteochondrodysplasias 200–201
cartilage element 228
cartilage extracellular matrix 199, 203
architecture 207
structural proteins 204
cartilage matrix protein (CMP) 231
cartilage matrix resorption 158–170
current studies 161–162
cartilage matrix structure 160
cartilage oligomeric matrix protein (COMP) 201–205, 211, 212
cartilaginous condensations 189–190
CBFA1 10, 18, 83, 87, 93, 277
CBFA2 87
CBFA3 87
CBMP1 226
CDMP1 216, 226
cell–cell interactions 145, 148
cell differentiation 8–13
cell proliferation–differentiation balance 102–121
c-*fos* 12
chimeric FGFRs 76
chimeric mice 148–150
cholesterol 153, 267
chondroblasts 278
differentiation 180, 188
chondrocytes 8–10, 24, 93, 144, 146, 160, 161, 172, 204, 210–212, 223, 231, 232, 268
differentiation 9–10, 30, 63–80, 90
and Noggin 27–30
growth 63–80
see hypertrophic chondrocytes, periarticular chondrocytes, prehypertrophic chondrocytes
hypertrophy 154, 163
proliferation 199
chondrodysplasia, mixed cases 65
chondrogenesis 8, 23–43, 72, 173, 174–176, 178, 181, 183, 187
and BMPs 173
Noggin misexpression in 25, 30, 32–34
chondrogenic condensation 190
chondrogenic progenitors 27
chondroprogenitors 23–24, 28
differentiation 182–183
c-Jun N-terminal kinase (JNK) 244
clavicle 92–93
cleft palate 127
cleidocranial dysplasia (CCD) 10, 87–92
clodronate 262
COL2A1 196–198, 200
COL9A1 200
COL9A2 200
COL9A3 200
COL10A1 200
COL11A1 200
COL11A2 200
colII 25, 29, 30
collagen
type I 93, 134, 170, 198
type II 37, 38, 63, 76, 93, 159, 161, 162, 166, 168, 170, 196–201, 203, 211, 228
type IIa 38
type IIb 8
type IX 159, 169, 170, 199, 201, 204

type X 8, 153, 159, 162, 168, 169, 230, 231
type XI 199, 200
collagen fibrils 159, 160, 168, 201, 205
type II 211
collagenases 161
collagenase 1 (MMP1) 161, 168, 170
collagenase 2 (MMP8) 161
collagenase 3 (MMP13) 161, 167, 168, 169
collagenopathies, type II 196–200
colony-stimulating factors (CSFs) 253
Colubroidea 55
COMP
see Cartilage oligomeric matrix protein
condensations 189–190, 213–214, 218, 228, 233, 280
cartilaginous 189–190
de novo 214–218
formation and development 191
genetic control 215
precartilage 17, 29–30
prechondrogenic 29, 36, 40, 42
congential parietal foramina 130
coronal suture 103–105, 107
cell proliferation 111
experimental application of FGF2 110–113
expression of *Fgfr* and *Twist* genes 109
coronal synostosis 103, 104
cortical tunnelling 248
coupling 3
cranial sutures 281
FGF ligands critical for signalling 129–130
cranial vault 122
craniosynostosis 118, 120, 122–143, 281
FGFR signalling 127–129
mixed cases 65
mutations 123–127
Crouzon-like syndrome 137
crystal structure 136
CYP26 187, 188
cytokines 167
gene polymorphisms 255
cytoplasmic retinoic acid binding proteins (CRABPs) I and II 174–175

D

degenerative joint disease 204
de novo condensations 214–218
DEXA (dual energy X-ray absorptiometry) 256
diastrophic dysplasia 205, 210
differentiation 1, 223, 228
diffusion 44
digoxygenin 116
1α,25-dihydroxyvitamin D3 (calcitriol) 236, 251
DNA binding 87
Dollo's Law 60
dorsal ectoderm 7
dorsal–ventral (D–V) axis 6–7, 23
doxycycline 168
Drosophila 10, 84, 92, 103, 117, 192, 193, 280
DTD (diastrophic dysplasia) 206
DTDST (diastrophic dysplasia sulphate-transporter) 205, 206
dura mater 119
dwarfing chondrodysplasias 64
dwarfism 82
dysplasia
see bone dysplasia, chondrodysplasia, thanatophoric dysplasia (TD), Cleidocranial dysplasia (CCD), diastrophic dysplasia, kniest dysplasia, spondyloepimetaphyseal, dysplasia, spondyloepiphyseal dysplasia

E

Ehlers–Danlos syndrome (EDS)
type VI 205
type VIIA, VIIB and VIIC 205
elbow fusion 141
embryonic cartilage 172
embryonic stem cells 148
endochondral bone formation 144–145
IHH as master coordinator 150–151
IHH role in 146–148
PTHrP regulation 145–146
endochondral bones 189
endochondral ossification 2, 63, 195–196
collagenases in 161
rate-limiting step 71–72
Engrailed 1 (En1) 7
etidronate 153, 262
exon 1 84
exoskeleton 2
extracellular cartilagenous matrix 268
extracellular growth factor (EGF) 211

extracellular matrix 159, 162, 212, 213
molecules 20
structural proteins 195–212
primary defects 196–205
secondary defects 205–207

F

familial expansile osteolysis (FEO) 257
FGF (fibroblast growth factor) 9, 18, 20, 21, 53, 59, 69–70, 99, 100, 113, 154, 167
FGF1 45, 129, 130
FGF2 44, 45, 69, 70, 105, 107, 116, 117, 129, 130, 135, 136, 163
FGF4 7, 44, 45, 98, 130, 135, 139, 175
FGF6 130
FGF7 61, 129
FGF8 7, 44, 45, 53, 70, 275
FGF9 135, 136
FGF10 45, 58, 61, 129, 139, 273–275
FGF17 70
signalling inhibition of chondrocyte growth 71
signalling pathways interaction 72
synthesis 98
FGFR (fibroblast growth factor receptor) 69–70, 76, 100, 109, 123, 223
extracellular and cell membrane components 105–109
mutations 65–69, 72, 79, 126
cellular consequences 127
signalling pathways 102–121
FGFR1 76, 77, 103, 109, 110, 113, 116–118, 120, 123, 126, 129, 134, 136, 155, 167
FGFR2 58, 77, 78, 103, 105, 109, 110, 113, 114, 116–118, 120, 123, 126, 128, 129, 133, 134, 138–140
FGFR3 9, 63–80, 92, 103, 105, 109, 110, 113, 114, 116, 123, 129, 135, 137, 138, 156
loss of function mutations in mice 69
mutations activating signalling 65
FGFSHH 60
fibroblast growth factor *see* FGF
fibroblast growth factor receptor *see* FGFR
fibrodysplasia ossificans progressiva (FOP) 34
fibula 178, 185–186

G

G380R 64, 66, 72, 79
β-galactosidase 148
GDF (growth differentiation factor) 39, 40, 232
GDF5 8, 30, 32, 33, 36, 37, 38, 40, 154, 173, 215–218, 223, 224, 226–229
GDF6 216, 218, 219, 223, 225, 227–229
GDF7 215, 216, 218, 219, 225, 229
gelatinase B (MMP9) 164
gene expression patterns, functional correlates 109–110
genetic control of bone and joint formation 213–234
genetic studies, skeletal maintenance and disease 219–220
glial cells 80
glycine 198
Grebe-type acromesomelic chondrodysplasia 216
growth 81
growth/differentiation factor *see* GDF
growth factors 265
growth hormone 72
growth plate 69–70, 76, 144–157, 199, 204, 211, 276, 282
cartilage 229
function defects 81–82
growth rates 81–82

H

hallux 99, 100
haploinsufficiency 92, 204
heart 222
heartless (*htl*) 103
Hedgehog interacting protein (HIP) 11
Hedgehog pathway 267
heparan sulfate 20
heparan sulfate 2-sulfotransferase 207
heparan sulfate proteoglycan (HSPG) 107, 116
heparinases 119
homeostasis 81
homozygosity 206, 210
HOX136 185
HOX 7–8, 17, 18, 48, 49, 55, 56–58, 60, 61, 97, 192, 272–275
HOXA9–HOXA13 100
HOXA13 275
HOXB5 49, 50

HOXB6 176
HOXC6 49, 57
HOXC8 49, 57, 58, 59
HOXD9 99, 175
HOXD11 84, 85, 157
HOXD11–13 99
HOXD12 84, 85
HOXD13 82–85, 157, 175
Humero-radial synostosis 141
humero-ulnar synostosis 141
Hunter–Thompson-type acromesomelic chondrodysplasia 216
huntingtin 84
Huntington disease 84
hyaline cartilage 87
hyaluronic acid 159
hydrocephalus 119
hydrogen peroxide 258
hypercalcaemia of malignancy 258
hyperparathyroidism 248
hypersegmentation 226
hypertrophic cartilage 268
hypertrophic cells 79
hypertrophic chondrocytes 63–64, 144–145, 152, 159, 162, 231, 268, 277
hypertrophic phenotype 152, 230
hypochondrogenesis 197, 203
hypochondroplasia (HCH) 64, 65
hypodactyly 85, 86, 276

I

Ig-like domains 127, 128
IGFs (insulin-like growth factors) 145
IGF-1 (insulin growth factor 1) 72
IgII 68, 139
IgIII 68
IgIIIa 139
IgIIIb 129, 133, 138, 139, 141
IgIIIc 138, 139, 141
IL-1 13, 167, 236, 244, 245, 247, 249, 258
  in osteoclast activation 239–240
IL-1α 241
IL-1R 240, 241, 244
IL-1RA 167
IL-6 13
IL-11 13, 236
Indian Hedgehog (IHH) 9–11, 72, 76, 79, 144–157, 163, 228
  and endochondral bone formation 146–148
  feedback loop 147
  master coordinator of endochondral bone formation 150–151
Indian Hedgehog (IHH)–PTHrP interactions 146–150
inflammatory bone loss 258
inorganic pyrophosphate (PPi) 262
in situ hybridization 25, 149, 187, 215, 228
insulin-like growth factor I *see* IGF-I
insulin-like growth factors *see* IGFs
interdigital apoptosis 181–182
intermediate mesoderm 61
interstitial collagenase 161
interzone 216
ipriflavone 263

J

joint formation 37
  genetic control 213–234
joint morphology 229
joint specification and condensation 36

K

K650E 64–66, 71
Kniest dysplasia 197, 198

L

lacZ 37, 219
limb
  cartilage elements 24
  cells 82
  condensations 190–193
  defects 176–181
  development 82, 232
    embryonic cartilage in 172–173
    mechanisms 47–62
  development overview 172
  evolution 47–62
  mesenchyme 30
  morphogenesis 24, 232
  outgrowth 7
  patterning 23–43, 82
  reduction 49
  skeleton 48
LMX1 7

M

macrophage colony-stimulating factor (M-CSF) 12, 236, 238, 241–242, 248

matrix metalloproteinase *see* MMP
MED (multiple epiphyseal dysplasia) 201, 203–205
mesenchymal cells 23–24, 144, 150, 172, 213
mesenchymal condensation 8, 136
mesenchyme 98, 102, 109, 133, 274
mesoderm 2
mesodermal cells 172–173
mesonephros 61
mi (microphthalmia) 12
microcephalus 119
microdontia 95
micromass cultures 178
mineralization 152, 153, 162, 166, 196, 230
MMP (matrix metalloproteinase) 161
MMP1 161, 168, 170
MMP8 161
MMP9 164
MMP13 161, 162, 164, 167–169
molecular genetics 1
mononuclear osteoclasts 250
morphogen 21
morphogenesis 3
MSX1 175
Msx1/Msx2 double knockout 41
MSX2 123, 130, 137
M-twist 99, 100
mutation analysis 87
myeloma 257
myogenic cells 274

N

N540K 65
NC4 domain 170
NCAM 276
nephron formation 189–190
neural crest 2
neutrophil collagenase 161
NF-κB 12, 240, 244
nitric oxide 253
nitric oxide synthase (NOS) 253
Noggin 24–34, 38–40, 46
  and chondrocyte differentiation 27–30
  and formation of prechondrogenic condensations 25–26
  misexpression in chondrogenesis 25, 30, 32–34
Noggin-mediated antagonism 32
non-syndromic craniosynostosis (NSC) 65, 68

O

ODF (osteoclast differentiation factor) 13, 236, 239, 247–250
  in osteoclast differentiation and activation 239
  molecular cloning 237–238
  signalling receptor 238–239
  soluble form (sODF) 239
ODF-RANK interaction 241–242, 244
odontoblasts 95
oestrogens 260–261
oligodactyl 86
oncostatin-M 13
op/op 12
OPG (osteoprotegrin) 13, 236–238, 269
OPGL 166, 237–238
organogenesis 81, 87–90
ossification *see* endochondral ossification
osteoarthritic cartilage 231
osteoarthritis 167
osteoblasts 10–11, 82, 89, 235, 277, 278
  in bone formation 253
  differentiation 10
  formation 90
osteochondrodysplasia 195–212
  and cartilage collagens 200–201
  molecular analysis 196
  molecular basis 207–208
  phenotypes 196
  primary structural protein defects 197
  sulfation pathway defects in 205–207
osteoclast differentiation factor *see* ODF
osteoclastogenesis inhibitory factor (OCIF) 13, 236–237
osteoclasts 11–13, 82, 156–157, 268
  activation 235–250
    regulation by TNF ligand family members 244
  in bone resorption 253
  differentiation 235–250
    hypothetical concept 237
    regulation by TNF ligand family members 244
    schematic representation 240
osteocytes, response to mechanical effects on bone 253–254
osteogenesis imperfecta 95
Osteonectin 105, 110
osteopetrosis 268
Osteopontin 105, 110

osteoporosis 223, 259, 270
challenges of developing new treatments 265–266
current and future drugs 259–266
osteoprotegerin ligand 166, 237–238
osteoprotegerin (OPG) 13, 236–237, 269
otospondylomegaepiphyseal dysplasia (OSMED) 200

P

P450RA 187
Pachyrhachis problematicus 55
Paget's disease 256–257, 270
PAPS (3′-phosphoadenosine-5′-phosphosulfate) 206
PAPS synthase genes
PAPSS1 206, 207
PAPSS2 206
paracrine factors 145
parathyroid hormone *see* PTH
parathyroid hormone-related peptide/protein *see* PTHrP
Patched (*Ptc*) 150, 154
patterning 1, 6–8, 17, 23–44, 81, 82, 102–103, 190–193, 218, 223, 272, 274
peanut agglutinin (PNA) 25–28
peptide hormones 265
periarticular chondrocytes 145
perichondrial cells 144–145
perichondrium 2, 24, 76, 79
periodontal ligament 168–169
periosteum 2, 93
perlecan 107, 116
Pfeiffer syndrome (PS) 68
phenotypes 85, 93, 94, 96, 128, 130–131, 134, 135, 137, 140, 145, 190, 225, 226
physeal cartilage 159
plasminogen activator (PA) 255
polyalanine 84
polydactyly 82, 84, 86, 140
polyglutamine 84
polysyndactyly 187
positional information 280
post-translational maturation defects 205–207
precartilage condensation 17, 29–30
prechondrogenic condensation markers 29
prechondrogenic condensations 36, 40, 42
formation of, and Noggin 25–26
prechondrogenic mesenchyme 41
prehypertrophic chondrocytes 162
Pro253Arg 127, 129, 139
procollagen, type II 198
progress zone (PZ) 172
progressive ankylosis (*Ank*) 220
proliferating cell nuclear antigen (PCNA) 110
prostaglandin E2 (PGE2) 236
protease inhibitors 167
proteases 166
proteinase 248
proteoglycan 119
proteolytic processing 205
proximal–distal (P–D) axis 6, 17, 33–34
pseudoachondroplasia (PSACH) 203, 204
PTH 9, 251, 265
PTH receptors 145, 146, 148–152, 155, 156, 230
PTHrP 9, 72, 76, 144–157, 163, 223, 229, 230, 258
feedback loop 147
regulation of endochondral bone formation 145–146
PTHrP receptors 145, 146, 148–152, 155, 156, 230
PU1 11–12
Puffer fish 61
python 49
axial skeleton 50
cells 99
hindlimbs of 51
mesenchyme 57, 59

R

R248C 64–67
RA response elements (RAREs) 174, 181
Radical fringe (*Rfng*) 7, 52
raloxifene 269
RANK 13, 238–239, 241, 244, 247–250, 252, 253, 257, 268
RANKL 238, 253, 258
RAR 174–175, 178, 185, 187
signalling 180–182
RARα 174–180, 182, 187
RARβ 40, 97, 174–176, 182
RAREs 174, 181
RARγ 40, 174–176, 179, 187
RCAS-Noggin infected autopods 33
RCAS-Noggin infected limbs 27, 29
RCAS-Noggin infected zeugopod 33

RCAS-Noggin virus 28, 30
reaction–diffusion models 96, 233
reactive oxygen species (ROS) 258
retinoic acid (RA) 41, 97, 100, 173–174, 187
  signal transduction 174
retinoic acid receptor *see* RAR
retinoid signalling 171–188
retinoid-X-receptors *see* RXR
retinoids 98, 191
retroviral infection 25
reverse transcriptase (RT)-PCR 243
rheumatoid arthritis 167, 231, 250
RNA *in situ* hybridization 25
RXR 174–175, 185, 187
RXRβ 175

S

Saethre–Chotzen syndrome 127
Sasquatch 94
segmentation 214, 232
  defects 96
  genetic control 215
SEMD (spondyloepimetaphyseal dysplasia)
    Pakistani type 206
Ser249Cys 129
Ser252Trp 127, 139
Ser-Pro dipeptide 128
SERMs (selective oestogen receptor
    modulators) 260–261, 269
short-digit mouse 94
short-ear mouse 214–218, 281
skeletal development, genetic control 6–22
skeletal disease, genetic studies 219–220
skeletal growth 63
skeletal maintenance, genetic studies 219–
    220
skeletal morphogenesis, developmental
    processes 81
skeletal morphology, histories 47
skeletal patterning *see* patterning
skeletal tissues 3, 208
  embryological origins 2
skeletogenesis
  changing views 3–5
  defects 81–101
  major stages 3
  overview 1–5
skeletogenic cells, origins 2
skeletogenic mesenchyme 82
skeletogenic tissues 1–2
skeleton, overview 1–5
skull 80, 281
  development 123
skull vault 102–121
  pattern of growth 102–103
snakes, evolution 55
SOFA (stromal osteoclast forming activity)
    236
somite differentiation 96
Sonic hedgehog (SHH) 7, 24, 52, 59, 60, 94,
    95, 97–99, 175, 186–187
SOX9 9–10, 25
SPD *see* syndactyly type II
spinocerebellar ataxia 84
splicing 139
spondyloepimetaphyseal dysplasia 211–212
spondyloepiphyseal dysplasia 211–212
spondyloepiphyseal dysplasia congenita 197
Sry-related *Sox9* 25
STAT1 70, 126
stem cells 148, 278–279
Stickler syndrome 197, 198, 200
strontium ranelate 265
stylopod 24
subcutaneous mesenchyme 113
sulfation pathway
  defects in osteochondrodysplasias
      205–207
supernumerary teeth 93, 95
syndactyly 86, 120, 136, 139
  FGFR signalling 127–129
  type II (SPD) 82–86
  type III/IV 85, 86
synostosis 117
  humero-radial 141
  humero-ulnar 141
synovial joint 216
synpolydactyly 82–83, 86
synpolydactyly homologue (spdh) 83

T

Talpid3 154
Tbx4 58, 59
Tbx5 58
teeth formation 95
tendons 274
teratogenesis 175
TGF (transforming growth factor) 173
TGFβ 11, 19, 32, 118, 134, 136, 138–140,
    145, 154, 167, 190, 214, 226

TGFβ receptor, type II 11
TGFβ1 163, 173
TGFβ2 11, 19, 32, 33, 163, 173
TGFβ3 173
thanatophoric dysplasia (TD) 64–66, 68, 114, 137
tibia 178
TIMP (tissue inhibitors of metalloproteinases) 167
TNF (tumour necrosis factor) 13, 258
  signal transduction 240–241
TNF receptor *see* TNFR
TNF receptor-associated factors (TRAFs) 13, 240–241, 244
TNFα 240, 245, 247, 249
  in osteoclast differentiation 241–244
TNFR 236–238, 269
TNFR1 240, 242–243
TNFR2 240, 242–243
TRAIL receptors 236
TRANCE 13, 238
transforming growth factor *see* TGF
transgenic mesenchyme 185
transmembrane segment (TM) 128
TRAP 237, 242, 244
triiodothyromine (T_3) 163
tumour necrosis factor *see* TNF
tumour necrosis factor receptors *see* TNFR
Turing mechanism 19, 20–21, 192, 193
TWIST 103, 104, 109, 110, 114, 117, 118, 123, 127, 130
tyrosine kinase domains 64, 128

U

ubiquitin 84

V

vascular invasion 152
ventral ectoderm 7
vertebrate limbs *see* limb
v-*fos* 12
vitamin A metabolites 173
vitamin D deficiency 166
vitamin D receptor 153

W

Wnt3a 52
Wnt7a 7

Y

Y-shaped bifurcation 213
Y-shaped pattern 232

Z

zebrafish 48
zeugopod 33, 34
zone of polarizing activity (ZPA) 7, 23, 24, 52, 172